METAL MATRIX COMPOSITES: *MECHANISMS AND PROPERTIES*

A Volume of the *Treatise on Materials Science and Technology*

Contents

PART I. STRENGTHENING MECHANISMS

1   Plasticity Theories for Fibrous Composite Materials

George J. Dvorak

2   Strengthening of Metal Matrix Composites Due to Dislocation Generation Through CTE Mismatch

R. J. Arsenault

3   Strengthening Behavior of In Situ Composites

T. H. Courtney

PART II. MECHANICAL PROPERTIES

4   **Tensile and Compressive Properties of Metal Matrix Composites**

R. J. Arsenault

5   **Mechanical Behavior of Metal Matrix Composites under High Strain Rates and Impact Loadings**

M. Taya

6   **Creep Behavior of Metal Matrix Composites**

M. Taya

PART III. FRACTURE AND FATIGUE

7   **Fracture Toughness of Particulate Metal Matrix Composites**

D. L. Davidson

METAL MATRIX COMPOSITES:
MECHANISMS AND PROPERTIES

Edited by

R. K. EVERETT

Naval Research Laboratory
Washington, D. C.

R. J. ARSENAULT

The University of Maryland
College Park, Maryland

ACADEMIC PRESS
Harcourt Brace Jovanovich, Publishers

Boston San Diego New York
London Sydney Tokyo Toronto

669- 419 MET

This book is printed on acid-free paper.

ACADEMIC PRESS, INC.
1250 Sixth Avenue, San Diego, CA 92101

United Kingdom Edition published by
ACADEMIC PRESS LIMITED
24–28 Oval Road, London NW1 7DX

LIBRARY OF CONGRESS CATALOG CARD DATA

Metal matrix composites. Mechanisms and properties/edited by
R. K. Everett and R. J. Arsenault.
     p.   cm.—(Treatise on materials science and technology)
    Includes bibliographical references.
    ISBN 0-12-341833-X (alk. paper)
    1. Metallic composites. I. Everett, R. K. II. Arsenault, R. J.
III. Series.
TA403.T74
[TA481]
620.1'1 s—dc20   90-399
  [620.1'6]   CIP

PRINTED IN THE UNITED STATES OF AMERICA
91 92 93   9 8 7 6 5 4 3 2 1

PART IV. PHYSICAL AND CHEMICAL PROPERTIES

12   Electrical Conductivity in Continuous-Fiber Composites

Jacques E. Schoutens

13   Corrosion

Patricia P. Trzaskoma

Contributors

Numbers in parentheses indicate the pages on which the authors' contributions begin.

R. J. ARSENAULT (79, 133), *Engineering Materials Program, Metallurgical Materials Laboratories, University of Maryland, College Park, Maryland, 20742*

K. K. CHAWLA (235), *Department of Materials and Metallurgy Engineering, New Mexico Institute of Mining and Technology, Socorro, New Mexico, 87801*

T. H. COURTNEY (101), *Department of Materials Science, University of Virginia, Charlottesville, Virginia 22901*

D. L. DAVIDSON (217), *Southwest Research Institute, P.O. Drawer 28510, San Antonio, Texas 78284*

GEORGE J. DVORAK (1), *Department of Civil Engineering and Institute Center for Composite Materials and Structures, Rensselaer Polytechnic Institute, Troy, New York 12180*

THOMAS A. HAHN (329), *Naval Research Laboratory, Code 6373, Washington D.C. 20375-5000*

R. O. RITCHIE (255), *Department of Materials Science and Mineral Engineering, University of California, Berkley, California 94720*

JACQUES E. SCHOUTENS (357), *Kaman Sciences Corporation, 816 State Street, P.O. Drawer QQ, Santa Barbara, California 93102*

JIAN KU SHANG (255), *Department of Materials Science and Engineering, University of Illinois at Urbana-Champaign, Urbana, Illinois 61801*

M. TAYA (169, 189), *Department of Mechanical Engineering, University of Washington, Mail Stop Fu-10, Seattle, Washington 98195*

PATRICIA P. TRZASKOMA (383), *Naval Research Laboratory, Code 6322, Washington, D.C. 20375-5000*

A. WOLFENDEN (287), *Mechanical Engineering Department and Amorphous Materials Research Group, Texas A&M University, College Station, Texas 77843*

J. M. WOLLA (287), *Composites and Ceramics Branch, Naval Research Laboratory, Code 6372, Washington D.C. 20375-5000*

Preface

Interest in metal matrix composites (MMC) as engineering materials is increasing. As current functional materials reach their performance limits, designers are looking to composites to provide the extra strength, stiffness, and higher-temperature capabilities required for advanced applications. Whereas few commercial products are currently manufactured from MMCs, that trend is bound to change as technologies mature. Many people working in the materials sector are intrigued with MMCs and want to learn more about this class of composite materials.

Basic and applied research into MMCs has increased along with the general increase in interest. Many journal articles and conference proceedings are published every year with composites-oriented articles. However, this wealth of information is available only to a few researchers. Keeping abreast of this literature can be a herculean task. A need exists to condense and summarize the current state of knowledge concerning MMCs. By distilling this information, trends in research and development can be discerned and, we hope, wasteful duplication avoided.

It is to address this need for a more unified source of information that we have undertaken the production of this book. Our goal was fourfold: to present the state-of-the-art of MMC knowledge in terms of processing, properties, and mechanisms; to indicate current issues and trends in MMCs; to provide a coherent, historical background for today's issues; and to provide bibliographic data on selected topics. Although the wide scope of activities in the MMC field is difficult to summarize in just two volumes, a representative selection of topics has been prepared so that the reader may gain a good understanding of the field. Subjects of particular interest may be pursued by studying the reference materials cited throughout. Some material has been deliberately left out. One example is the topic of reinforcements. A two-volume work could be written on that subject alone. That being the case, it was believed better to concentrate on the composite materials and the general principles which could be learned from them, rather than on the constituents and their properties. It is hoped that by choosing this approach, the book will not become outdated as quickly as some reviews can.

To present as much material on composites as possible and to keep explanations brief, we have assumed a degree of familiarity with the concepts of materials science. An intimate working knowledge of MMCs is not necessary, but a firm grasp of the fundamentals is essential. However, there should be sufficient diversity of information that both composites researchers, who may have in-depth knowledge of specific topics, and students will find this book useful.

This coverage of metal matrix composites has been divided into two volumes. The first volume, subtitled *Processing and Interfaces*, contains three sections. First, an introductory overview by W. Harrigan provides some historical perspective on the current state of the composites field. Second, an extensive section is devoted to synthesis and processing, which details, as much as possible, the standard techniques used to fabricate composites and some novel techniques which may become increasingly important in the years to come. The new emphasis placed on composite interfaces is recognized by the final section. A review of techniques to probe, model, and modify composite interfaces is included. This volume, subtitled *Mechanisms and Properties*, contains four sections. Strengthening mechanisms and mechanical properties are covered in the first two. Fracture and fatigue phenomena are analyzed in the third. Physical properties and environmental effects are discussed in the final section.

The purpose of this somewhat unusual organization is to present the information by a "bottom-up" approach. Thus, composite fabrication takes on increased emphasis, since it is the first step taken in studying any material, and processing details irrevocably affect the material and subsequent properties. The interface, being dependent on the fabrication, is next followed by the active strengthening mechanisms and the mechanical properties, which depend on the interface, and so on. If read sequentially, the sections should build on each other. However, individual chapters and sections are reasonably independent and can be useful by themselves.

This project could not have been completed without the help of many people. The time and talents of the authors as well as additional inputs from colleagues, too numerous to name individually, helped to relieve the burden. To all of these associates, we owe a debt of gratitude.

R. J. Arsenault, would like to acknowledge several specific individuals who have supported his investigations of metal matrix composites. The first is Dr. Bruce MacDonald who is presently at the National Science Foundation, but was previously at the Office of Naval Research where he was the first to support Arsenault's fundamental studies. This support was continued by Dr. Steven Fishman of the Office of Naval Research. Also, Arsenault would like to acknowledge the support of several other individuals, Drs. Y. Flom and J. R. Feng, Mr. N. Shi, L. Wang, and J. Romero; the assistance and patience shown by Mrs. Judy Anderson in typing up various chapters in this two-volume endeavor.

Dr. Samuel Johnson (1709–1784) wrote that, "Knowledge is of two kinds. We know a subject ourselves, or we know where we can find information upon it." It is our hope that this book will help promote both kinds of knowledge and become a frequently utilized reference for the community.

R. K. Everett
R. J. Arsenault

Plasticity Theories for Fibrous Composite Materials

GEORGE J. DVORAK

Department of Civil Engineering and
Institute Center for Composite Materials and Structures
Rensselaer Polytechnic Institute
Troy, New York

I. Introduction

Fabrication, processing, and effective use of metal matrix composites often cause inelastic deformations in the material. In many actual systems, the elastic-strain range of the elastic–plastic matrix is much smaller than the failure strain of the elastic-brittle fiber. Similarly, the temperature changes that may cause yielding in a stress–free composite are often smaller than those encountered in service. However, the total strains seen in fibrous systems are also small, i.e., they seldom exceed the failure strain of the fiber, which is usually found in the range of 0.01–0.02. Therefore, in contrast to metals, plasticity of fibrous composites is usually limited to small strains, but it may affect much of the useful load-bearing capacity of structural parts designed to utilize the high strength of the fibers.

The purpose of this chapter is to give a brief review of the elastic–plastic response of fiber composites and of its implications for the mechanical behavior of these materials. First, we shall discuss some general features of thermoelastic behavior, such as evaluation of overall thermomechanical properties, phase concentration factors, thermal concentration factors, and transformation stress and strain concentration factors. Next, the elastic–plastic behavior of macroscopically homogeneous metals and composites is outlined. This provides a basis for a discussion of the plasticity of fiber composites, which includes a number of new, exact results. Then, some specific micromechanical models for plasticity analysis are described, together with their experimental verification and implementation in finite element programs for structural analysis.

The approach is based on micromechanics of heterogeneous media. The objective is to evaluate overall instantaneous properties of the medium from information about local properties and microstructural geometry and to establish various general connections between local and overall response. The consequences of plasticity in such phenomena as dimensional stability, fatigue, and fracture of fibrous metal matrix laminates are briefly described.

The notation used is similar to that introduced by Hill [1, 2]. Vectors are denoted by lower-case boldface letters, e.g., $\boldsymbol{\sigma}$, $\boldsymbol{\varepsilon}$, $\mathbf{a}$, $\mathbf{b}$; matrices are denoted by upper-case boldface Roman letters, e.g., $\mathbf{L}$, $\mathbf{M}$. In the contracted notation used, those will typically be (6×1) vectors and (6×6) matrices. $\mathbf{L}^{-1}$ denotes the inverse of $\mathbf{L}$, defined if it exists to satisfy $\mathbf{L}\mathbf{L}^{-1} = \mathbf{I} = \mathbf{L}^{-1}\mathbf{L}$, where $\mathbf{I}$ is the unit matrix. Further details of the notation are explained in the Appendix.

II. Elastic Response

A. Overall Properties and Local Fields

We start the exposition with an outline of a general procedure for the evaluation of overall thermomechanical properties of two-phase composite media in terms of thermoelastic constants and volume fractions of the phases. The elastic response contributes to the total strain during plastic loading, and it is the sole source of this strain in any elastic unloading step.

Consider a fibrous composite material consisting of two distinct continuous phases of cylindrical shape, which are aligned parallel to the x_1-axis of a Cartesian coordinate system. The phases remain bonded and are free of voids and cracks during deformation. A representative volume V with surface S is chosen so that under certain boundary conditions, it represents the macroscopic response of the composite. Within V, each phase $r = \alpha, \beta$ occupies a volume V_r, and the volume fractions $c_r = V_r/V$ satisfy $c_\alpha + c_\beta = 1$.

The volume V is subjected to certain uniform overall stresses $\bar{\sigma}$ or strains $\bar{\varepsilon}$, and to a uniform temperature change θ. Suppose that the constitutive relations of the phases are known, e.g., from experiments on neat matrix samples and on the fibers, and are written in the form

$$\sigma(\mathbf{x}) = \mathbf{L}_r \varepsilon(\mathbf{x}) + \mathbf{l}_r \theta, \qquad \varepsilon(\mathbf{x}) = \mathbf{M}_r \sigma(\mathbf{x}) + \mathbf{m}_r \theta, \tag{1}$$

where $\mathbf{L}_r$, $\mathbf{M}_r$ are instantaneous stiffness and compliance matrices, which have full diagonal symmetry; $\mathbf{L}_r = \mathbf{L}_r^T$, $\mathbf{M}_r = \mathbf{M}_r^T$; $\mathbf{LM} = \mathbf{I}$; $\mathbf{l}_r$ and $\mathbf{m}_r$ are the thermal stress and strain vectors, such that $\mathbf{l}_r = -\mathbf{L}_r \mathbf{m}_r$. As long as the phase remains elastic, the coefficients of these matrices are constant in V_r. Note that the components of $\mathbf{m}_r$ are the linear coefficients of thermal expansion of the phase, which are not affected by deformation and are assumed to be independent of temperature.

The local fields in (1) are generally not uniform. Therefore, it is often convenient to work with volume averages of the nonuniform fields defined by the integral

$$\{\cdot\}_{V'} = \frac{1}{V'} \int_{V'} (\cdot)\, dV. \tag{2}$$

Phase volume averages of local fields follow from (2) if one takes $V' = V_r$ and integrates both sides in (1),

$$\sigma_r = \mathbf{L}_r \varepsilon_r + \mathbf{l}_r \theta, \qquad \varepsilon_r = \mathbf{M}_r \sigma_r + \mathbf{m}_r \theta, \tag{3}$$

where

$$\sigma_r = \{\sigma(\mathbf{x})\}_{V_r}, \tag{4}$$

$$\varepsilon_r = \{\varepsilon(\mathbf{x})\}_{V_r}. \tag{4'}$$

Since $\mathbf{L}_r$, $\mathbf{M}_r$ are constant in V_r, (3) are exact analogs of (1) for phase volume averages. One can also obtain the overall stresses and strains as averages of the respective local quantities over the representative volume V and write the overall constitutive relations as

$$\bar{\sigma} = \mathbf{L}\bar{\varepsilon} + \mathbf{l}\theta, \qquad \bar{\varepsilon} = \mathbf{M}\bar{\sigma} + \mathbf{m}\theta, \tag{5}$$

where

$$\bar{\sigma} = \{\sigma(\mathbf{x})\}_V, \tag{6}$$

$$\bar{\varepsilon} = \{\varepsilon(\mathbf{x})\}_V. \tag{6'}$$

The implication is that the representative volume V of the composite aggregate is regarded as a macroscopically homogeneous medium and that under uniform overall stress or strain, the $\mathbf{L}, \mathbf{M}$ are the elastic overall stiffness and compliance matrices, and $\mathbf{m}$, $\mathbf{l} = -\mathbf{Lm}$ are the overall thermal strain and stress vectors of this composite medium.

To determine $\mathbf{L}$, $\mathbf{M}$, $\mathbf{l}$, and $\mathbf{m}$ of an elastic composite, one can proceed as follows. Suppose that for the composite system considered, one could evaluate the actual local fields (1) and write them in the form

$$\boldsymbol{\sigma}(\mathbf{x}) = \mathbf{B}_r(\mathbf{x})\bar{\boldsymbol{\sigma}} + \mathbf{b}_r(\mathbf{x})\theta, \qquad \boldsymbol{\varepsilon}(\mathbf{x}) = \mathbf{A}_r(\mathbf{x})\bar{\boldsymbol{\varepsilon}} + \mathbf{a}_r(\mathbf{x})\theta. \tag{7}$$

Of course, in many practical situations, one cannot find the actual fields, but it is usually possible to evaluate an estimate of average local fields in the two phases under the prescribed load increment. The result is the integral (2) of (7), taken over V_r,

$$\boldsymbol{\sigma}_r = \mathbf{B}_r\bar{\boldsymbol{\sigma}} + \mathbf{b}_r\theta, \qquad \boldsymbol{\varepsilon}_r = \mathbf{A}_r\bar{\boldsymbol{\varepsilon}} + \mathbf{a}_r\theta, \tag{8}$$

where $\mathbf{A}_r$, $\mathbf{B}_r$, $\mathbf{a}_r$, $\mathbf{b}_r$ are certain mechanical and thermal strain and stress concentration factors. If those are known, one can utilize (2) to write for $r = \alpha, \beta$,

$$\bar{\boldsymbol{\sigma}} = c_\alpha \boldsymbol{\sigma}_\alpha + c_\beta \boldsymbol{\sigma}_\beta, \tag{9}$$

$$\bar{\boldsymbol{\varepsilon}} = c_\alpha \boldsymbol{\varepsilon}_\alpha + c_\beta \boldsymbol{\varepsilon}_\beta. \tag{9'}$$

Then, for $\theta = 0$, equations (3), (8), and (9) give the overall mechanical properties

$$\mathbf{L} = \mathbf{L}_\alpha + c_\beta(\mathbf{L}_\beta - \mathbf{L}_\alpha)\mathbf{A}_\beta, \qquad \mathbf{M} = \mathbf{M}_\alpha + c_\beta(\mathbf{M}_\beta - \mathbf{M}_\alpha)\mathbf{B}_\beta,$$
$$\mathbf{L} = c_\alpha \mathbf{L}_\alpha \mathbf{A}_\alpha + c_\beta \mathbf{L}_\beta \mathbf{A}_\beta, \qquad \mathbf{M} = c_\alpha \mathbf{M}_\alpha \mathbf{B}_\alpha + c_\beta \mathbf{M}_\beta \mathbf{B}_\beta, \tag{10}$$

together with the results

$$c_\alpha \mathbf{A}_\alpha + c_\beta \mathbf{A}_\beta = \mathbf{I}, \qquad c_\alpha \mathbf{B}_\alpha + c_\beta \mathbf{B}_\beta = \mathbf{I},$$
$$\mathbf{A}_r \mathbf{M} = \mathbf{M}_r \mathbf{B}_r, \qquad \mathbf{B}_r \mathbf{L} = \mathbf{L}_r \mathbf{A}_r.$$

This sequence, first outlined by Hill [1, 2], enables evaluation of the overall instantaneous $\mathbf{L}$ and $\mathbf{M}$ in terms of one mechanical concentration factor. The overall thermal strain and stress vectors $\mathbf{m}$ and $\mathbf{l}$ can be evaluated from known overall mechanical moduli and compliances or from local properties and concentration factors [3–7]. Convenient forms are

$$\mathbf{m} = c_\alpha \mathbf{B}_\alpha^T \mathbf{m}_\alpha + c_\beta \mathbf{B}_\beta^T \mathbf{m}_\beta \tag{11}$$

$$= (\mathbf{M} - \mathbf{M}_\beta)(\mathbf{M}_\alpha - \mathbf{M}_\beta)^{-1}\mathbf{m}_\alpha + (\mathbf{M} - \mathbf{M}_\alpha)(\mathbf{M}_\beta - \mathbf{M}_\alpha)^{-1}\mathbf{m}_\beta \tag{11^{I}}$$

$$= c_\alpha(\mathbf{M}_\alpha \mathbf{b}_\alpha + \mathbf{m}_\alpha) + c_\beta(\mathbf{M}_\beta \mathbf{b}_\beta + \mathbf{m}_\beta), \tag{11^{II}}$$

$$\mathbf{l} = c_\alpha \mathbf{A}_\alpha^T \mathbf{l}_\alpha + c_\beta \mathbf{A}_\beta^T \mathbf{l}_\beta \tag{11^{III}}$$

$$= (\mathbf{L} - \mathbf{L}_\beta)(\mathbf{L}_\alpha - \mathbf{L}_\beta)^{-1}\mathbf{l}_\alpha + (\mathbf{L} - \mathbf{L}_\alpha)(\mathbf{L}_\beta - \mathbf{L}_\alpha)^{-1}\mathbf{l}_\beta \tag{11^{IV}}$$

$$= c_\alpha(\mathbf{L}_\alpha \mathbf{a}_\alpha + \mathbf{l}_\alpha) + c_\beta(\mathbf{L}_\beta \mathbf{a}_\beta + \mathbf{l}_\beta). \tag{11^{V}}$$

Using (10), one can establish that the first two forms are equivalent.

In a recent paper, Benveniste and Dvorak [8] had shown that the local thermal fields in the phases can be derived from the mechanical fields. Their result is

$$\mathbf{a}_r(\mathbf{x}) = [\mathbf{I} - \mathbf{A}_r(\mathbf{x})](\mathbf{L}_\alpha - \mathbf{L}_\beta)^{-1}(\mathbf{l}_\beta - \mathbf{l}_\alpha),$$

$$\mathbf{b}_r(\mathbf{x}) = [\mathbf{I} - \mathbf{B}_r(\mathbf{x})](\mathbf{M}_\alpha - \mathbf{M}_\beta)^{-1}(\mathbf{m}_\beta - \mathbf{m}_\alpha).$$

$$(12)$$

Of course, since all local and overall properties are assumed to be constant, (12) can be integrated as in (2) to find the thermal concentration factors in terms of the mechanical concentration factors. In each case, these factors are phase volume averages of the fields $\mathbf{A}_r(\mathbf{x})$, $\mathbf{B}_r(\mathbf{x})$, $\mathbf{a}_r(\mathbf{x})$, or $\mathbf{b}_r(\mathbf{x})$ in (7).

B. Micromechanical Models

1. INCLUSION PROBLEMS

In Section II.A, the averages of local fields and the overall elastic properties have been found in terms of stress or strain concentration factors. These factors can be evaluated in several different ways. First we describe an approach based on the solution of an inclusion problem, which will be useful in some of the micromechanical models discussed in the following.

A single inclusion of ellipsoidal shape is embedded in a large volume of a different homogeneous elastic material. There is a perfect bond between the phases, hence the tractions and displacements must be continuous across the interface. A circular cylindrical fiber is a particular example of such an inclusion. Let $\mathbf{L}'$ and $\mathbf{L}''$ denote the stiffness matrices of the inclusion and the surrounding medium, neither of which need to be isotropic. These matrices have full diagonal symmetry, and their inverses are denoted by $\mathbf{M}'$ and $\mathbf{M}''$, respectively. The surrounding medium is subjected to a uniform stress $\bar{\boldsymbol{\sigma}}$ or strain $\bar{\boldsymbol{\varepsilon}}$ at infinity. The objective is to find the stresses and strains in the inclusion.

Eshelby [9, 10] pointed out that in problems of this kind, the stress and strain fields in the inclusion are also uniform. Therefore, one can write the result in a form analogous to (8) as

$$\boldsymbol{\sigma}' = \mathbf{B}'\bar{\boldsymbol{\sigma}}, \qquad \boldsymbol{\varepsilon}' = \mathbf{A}'\bar{\boldsymbol{\varepsilon}},$$

$$\boldsymbol{\sigma}'' = \mathbf{B}''\bar{\boldsymbol{\sigma}}, \qquad \boldsymbol{\varepsilon}'' = \mathbf{A}''\bar{\boldsymbol{\varepsilon}},$$

$$(13)$$

where $\boldsymbol{\sigma}'$ and $\boldsymbol{\varepsilon}'$ denote the uniform stresses and strains in the inclusion, and $\boldsymbol{\sigma}''$ and $\boldsymbol{\varepsilon}''$ the averages of the fields in the surrounding medium. Since the inclusion causes only a local perturbation of the overall field, its contribution to the overall averages $\bar{\boldsymbol{\sigma}}$ and $\bar{\boldsymbol{\varepsilon}}$ is vanishingly small, and it follows that $\mathbf{A}'' = \mathbf{B}'' = \mathbf{I}$.

The evaluation of $\mathbf{B}'$ was outlined by Eshelby, and in a more general case by Kinoshita and Mura [11]; a recent survey of pertinent results appears in Mura's monograph [12]. Here we limit our attention to a particularly simple approach to the solution proposed by Hill [2]. The fact that the inclusion fields are uniform suggests that the medium surrounding the inclusion is loaded at the interface by certain surface tractions derived from a uniform stress field. Therefore, it is useful to formulate an auxiliary problem in which the inclusion is removed from the medium, the wall of the ellipsoidal cavity is loaded by certain surface tractions derived from a unit uniform stress field σ^*, and the remotely applied stresses vanish. Suppose that the unit stress components that generate the surface tractions are applied in a sequential manner and that six such solutions are obtained. For each solution, one can find the displacements of the cavity wall and convert them into strain components ε^*, which represent a uniform strain of the ellipsoidal cavity. The result can thus be written as

$$\sigma^* = -\mathbf{L}^*\varepsilon^*, \qquad \varepsilon^* = -\mathbf{M}^*\sigma^*, \tag{14}$$

where each column of $\mathbf{L}^*$ was generated by one of the six solutions. Both $\mathbf{L}^*$ and $\mathbf{M}^*$ have full diagonal symmetry, and their coefficients depend on the aspect ratios of the ellipsoid and on the moduli $\mathbf{L}''$ of the surrounding medium; they may be regarded as stiffness and compliance matrices of the cavity. In Hill's terminology, $\mathbf{L}^*$ is the overall constraint tensor.

The solution of the inclusion problem then follows from that of the auxiliary problem if one replaces the inclusion in the surrounding medium and writes

$$\sigma' - \bar{\sigma} = -\mathbf{L}^*(\varepsilon' - \bar{\varepsilon}), \qquad \varepsilon' - \bar{\varepsilon} = -\mathbf{M}^*(\sigma' - \bar{\sigma}). \tag{15}$$

Then, from the local constitutive relations and (13),

$$\mathbf{A}' = (\mathbf{L}^* + \mathbf{L}')^{-1}(\mathbf{L}^* + \mathbf{L}''), \qquad \mathbf{B}' = (\mathbf{M}^* + \mathbf{M}')^{-1}(\mathbf{M}^* + \mathbf{M}''). \tag{16}$$

Note that if the local properties $\mathbf{L}'$ and $\mathbf{L}''$ are known, $\mathbf{L}^*$ alone yields the solution. It is advantageous to separate that part of the solution that depends only on $\mathbf{L}''$, or $\mathbf{M}''$ and to write

$$\mathbf{P} = (\mathbf{L}^* + \mathbf{L}'')^{-1}, \qquad \mathbf{Q} = (\mathbf{M}^* + \mathbf{M}'')^{-1}. \tag{17}$$

This suggests that $\mathbf{P} = \mathbf{P}^T$, $\mathbf{Q} = \mathbf{Q}^T$. One can also establish that $\mathbf{P}\mathbf{L}'' + \mathbf{M}''\mathbf{Q} = \mathbf{I}$.

The forms of $\mathbf{P}$ that correspond to different ellipsoidal shapes in various anisotropic materials can be found in the literature [11–15]. Of particular interest here is the result for an inclusion in the shape of a circular cylinder and for the surrounding medium, which is transversely isotropic about x_1, the cylinder axis. This result can be recorded as follows:

$$P_{22} = P_{33} = \frac{k'' + 4m''}{8m''(k'' + m'')}, \qquad P_{23} = P_{32} = \frac{-k''}{8m''(k'' + m'')},$$

$$P_{44} = P_{55} = \frac{1}{2p''}, \qquad P_{66} = \frac{k'' + 2m''}{2m''(k'' + m'')}, \tag{18}$$

where k'', m'', and p'' are Hill's [16] elastic moduli of the surrounding transversely isotropic medium, which belong to L'' and are defined in the Appendix. The remaining P_{ij} vanish.

2. THE SELF-CONSISTENT METHOD

So far, we have considered only the problem of a single inclusion, embedded in an elastic medium, as a stepping stone to the more important problem of finding the stresses in the constituents of a composite medium. The latter problem can be solved in several different ways. One approach is based on the self-consistent approximation [2, 17–19], which assumes that the stress and strain field averages in the fiber are equal to those found in the inclusion problem above, provided that the fiber is embedded in a homogeneous medium that has the properties of the composite. In the notation of Section II.B.1, the L'', M'' are identified with the overall properties L and M, and the L' and M' with the fiber properties, which we denote here by L_α, and M_α. The matrix properties are denoted by L_β, M_β.

Rewrite (9) in the form

$$c_\alpha(\sigma_\alpha - \bar{\sigma}) + c_\beta(\sigma_\beta - \bar{\sigma}) = 0, \qquad c_\alpha(\varepsilon_\alpha - \bar{\varepsilon}) + c_\beta(\varepsilon_\beta - \bar{\varepsilon}) = 0. \tag{19}$$

The above postulate of the self-consistent method and (15), (19) suggest that

$$(\sigma_\alpha - \bar{\sigma}) = -L^*(\varepsilon_\alpha - \bar{\varepsilon}), \qquad (\sigma_\beta - \bar{\sigma}) = -L^*(\varepsilon_\beta - \bar{\varepsilon}). \tag{20}$$

It is now apparent that both phases are regarded on the same footing, i.e., the concentration factor for the matrix phase is derived from the same L^* as the concentration factor of the fiber. The expressions follow from (16) and (17),

$$A_r^{-1} = I + P(L_r - L), \qquad B_r^{-1} = I + Q(M_r - M), \tag{21}$$

where P is given by (18) provided that the moduli k'', m'', p'' are replaced by the unknown overall moduli k, m, and p. Since the coefficients in (18) were derived for a transversely isotropic medium, the substitution of this particular form into (21) is permissible only if the fibrous composite itself conforms to the usual assumption of overall transverse isotropy.

Equations (21) and (10) yield estimates of overall moduli of the composite,

$$\mathbf{L} = \mathbf{L}_\alpha + c_\beta(\mathbf{L}_\beta - \mathbf{L}_\alpha)[\mathbf{I} + \mathbf{P}(\mathbf{L}_\beta - \mathbf{L})]^{-1},$$
$$\mathbf{M} = \mathbf{M}_\alpha + c_\beta(\mathbf{M}_\beta - \mathbf{M}_\alpha)[\mathbf{I} + \mathbf{Q}(\mathbf{M}_\beta - \mathbf{M})]^{-1}. \tag{22}$$

Self-consistency of the result can be established by showing that $\mathbf{L}^{-1} = \mathbf{M}$. Note, however, that (22) is a system of six implicit algebraic equations for the overall moduli. For fibrous composites that are transversely isotropic and are made of two transversely isotropic phases, Hill [19] derived a different set of equations for evaluations of the self-consistent estimates of the overall moduli, which yield the same results as (22):

$$\frac{c_\alpha k_\alpha}{k_\alpha + m} + \frac{c_\beta k_\beta}{k_\beta + m} = 2\left[\frac{c_\alpha m_\beta}{m_\beta - m} + \frac{c_\beta m_\alpha}{m_\alpha - m}\right],$$
$$\frac{1}{2p} = \frac{c_\alpha}{p - p_\beta} + \frac{c_\beta}{p - p_\alpha}, \tag{23}$$
$$\frac{1}{k + m} = \frac{c_\alpha}{k_\alpha + m} + \frac{c_\beta}{k_\beta + m}.$$

In addition, Hill [16] shows that regardless of the method used to obtain their estimates, only three of the five overall moduli of such composites are actually independent. The moduli, k, l, and n are related by so-called universal connections,

$$\frac{k - k_\alpha}{l - l_\alpha} = \frac{k - k_\beta}{l - l_\beta} = \frac{l - c_\alpha l_\alpha - c_\beta l_\beta}{n - c_\alpha n_\alpha - c_\beta n_\beta} = \frac{k_\alpha - k_\beta}{l_\alpha - l_\beta}, \tag{24}$$

between overall and phase moduli and volume fractions. Therefore, only one of the three moduli is independent. Relations (23) reflect this; they provide a cubic equation for m, and quadratic equations for p and k, the latter in terms of m. Once k is known, l and n follow from (24). If one or both phases are isotropic, then there are only two independent moduli in each such phase, and k_r, l_r, n_r, m_r, and p_r can be written in terms of the engineering phase moduli, as shown in the Appendix. Hill also gives a proof that the estimates (23) lie between rigorous bounds on the moduli, discussed below.

3. The Mori–Tanaka Method

Another and somewhat simpler approach to the evaluation of phase concentration factors of composite media was proposed by Mori and Tanaka [20] and restated recently in a more tractable form by Benveniste [21]. In a binary, matrix-based composite system, the method asumes that the stress or strain in the inclusion can be evaluated from a solution of the problem in Section II.B.1, provided that $\mathbf{L}_\alpha$ is identified with the inclusion stiffness

$\mathbf{L'}$, $\mathbf{L''}$ with the matrix stiffness $\mathbf{L}_\beta$, and the matrix average stress $\boldsymbol{\sigma}_\beta$ is applied in place of $\bar{\boldsymbol{\sigma}}$ at infinity.

This suggests that one must first find the partial concentration factors

$$\boldsymbol{\varepsilon}_\alpha = \mathbf{T}\boldsymbol{\varepsilon}_\beta, \qquad \boldsymbol{\sigma}_\alpha = \mathbf{W}\boldsymbol{\sigma}_\beta, \tag{25}$$

which follow from (16) as

$$\mathbf{T} = (\mathbf{L^*} + \mathbf{L}_\alpha)^{-1}(\mathbf{L^*} + \mathbf{L}_\beta), \qquad \mathbf{W} = (\mathbf{M^*} + \mathbf{M}_\alpha)^{-1}(\mathbf{M^*} + \mathbf{M}_\beta), \tag{26}$$

where the $\mathbf{L^*}$ and $\mathbf{M^*}$ are now functions of the coefficients $\mathbf{L}_\beta$ and $\mathbf{M}_\beta$ and are derived from (17). $\mathbf{P}$ is again found from (18) if the moduli k'', m'', p'' are replaced by k_β, m_β, p_β, respectively.

Once $\mathbf{T}$ and $\mathbf{W}$ are known, one can utilize (9) and (25) to establish that

$$\boldsymbol{\sigma}_\beta = [c_\alpha \mathbf{W} + c_\beta \mathbf{I}]^{-1}\bar{\boldsymbol{\sigma}}, \qquad \boldsymbol{\varepsilon}_\beta = [c_\alpha \mathbf{T} + c_\beta \mathbf{I}]^{-1}\bar{\boldsymbol{\varepsilon}}, \tag{27}$$

Finally, using (10), the overall properties can be obtained in the form

$$\begin{aligned}
\mathbf{L} &= \mathbf{L}_{\hat{\alpha}} + c_{\hat{\beta}}(\mathbf{L}_{\hat{\beta}} - \mathbf{L}_{\hat{\alpha}})[c_\alpha \mathbf{T} + c_\beta \mathbf{I}]^{-1}, \\
\mathbf{M} &= \mathbf{M}_{\hat{\alpha}} + c_{\hat{\beta}}(\mathbf{M}_{\hat{\beta}} - \mathbf{M}_{\hat{\alpha}})[c_\alpha \mathbf{W} + c_\beta \mathbf{I}]^{-1}.
\end{aligned} \tag{28}$$

Note that in contrast to (22), these are explicit algebraic equations for the overall properties. Proofs of self-consistency $\mathbf{LM} = \mathbf{I}$ and full diagonal symmetry of $\mathbf{L}$, $\mathbf{M}$ for binary fibrous media are available; also, the estimates (28) lie within rigorous bounds on overall moduli [21–23].[1]

As yet unpublished results of Chen and Dvorak [24] give explicit expressions of the Mori–Tanaka estimates of the overall moduli for a transversely isotropic fibrous medium made of transversely isotropic phases. With $r = \alpha$ representing the fiber, and $r = \beta$ the matrix, these expressions are

$$\begin{aligned}
k &= \frac{k_\alpha k_\beta + m_\beta(c_\alpha k_\alpha + c_\beta k_\beta)}{c_\alpha k_\beta + c_\beta k_\alpha + m_\beta}, \\[2mm]
l &= \frac{c_\alpha l_\alpha(k_\beta + m_\beta) + c_\beta l_\beta(k_\alpha + m_\beta)}{c_\alpha(k_\beta + m_\beta) + c_\beta(k_\alpha + m_\beta)}, \\[2mm]
n &= c_\alpha n_\alpha + c_\beta n_\beta + (l - c_\alpha l_\alpha - c_\beta l_\beta)\frac{l_\alpha - l_\beta}{k_\alpha - k_\beta}, \\[2mm]
m &= \frac{m_\alpha m_\beta(k_\beta + 2m_\beta) + k_\beta m_\beta(c_\alpha m_\alpha + c_\beta m_\beta)}{k_\beta m_\beta + (k_\beta + 2m_\beta)(c_\alpha m_\beta + c_\beta m_\alpha)}, \\[2mm]
p &= \frac{2c_\alpha p_\alpha p_\beta + c_\beta(p_\alpha p_\beta + p_\beta^2)}{2c_\alpha p_\beta + c_\beta(p_\alpha + p_\beta)}.
\end{aligned} \tag{29}$$

---

[1] See also a more recent paper by Benveniste et al. [103].

These are analogous to (23) and also satisfy the universal connections (24). Again, if needed, the expressions in the Appendix may be used to evaluate the phase moduli of isotropic phases in terms of engineering elastic constants.

4. BOUNDS ON OVERALL MODULI

The methods discussed in the previous two sections lead to simple estimates of local fields and overall properties, but they are heuristic in nature and thus do not provide an assurance of accuracy of the results. The legitimacy of the self-consistent, Mori–Tanaka, and similar estimates is derived from proofs that show that, in certain cases of practical interest, the estimates are bracketed by rigorous bounds on overall moduli. A simple illustration of the bounding procedure is the derivation of the Voigt and Reuss bounds [1]. The Voigt assumption is that the strain field in the entire representative volume V is uniform, i.e., $\mathbf{A}_r = \mathbf{I}$ in (8). If this field is derived from continuous displacements that coincide with the prescribed surface displacements on S, it is a kinematically admissible field, and the principle of minimum potential energy then gives the inequality $\bar{\varepsilon}\mathbf{L}\bar{\varepsilon} < \bar{\varepsilon}(c_\alpha \mathbf{L}_\alpha + c_\beta \mathbf{L}_\beta)\bar{\varepsilon}$, where the first term is the potential energy of the actual state, and the second term is the energy of the admissible state. There follows an upper bound on $\mathbf{L}$,

$$\mathbf{L} < \mathbf{L}_\mathrm{V} = c_\alpha \mathbf{L}_\alpha + c_\beta \mathbf{L}_\beta, \tag{30}$$

where $\mathbf{L}_\mathrm{V}$ is the Voigt estimate of the overall stiffness.

The dual assumption, introduced by Reuss, is that the stress field in V is uniform, i.e., $\mathbf{B}_r = \mathbf{I}$ in (8). The overall field that is in equilibrium with the surface tractions is a statically admissible field, and the principle of minimum complementary energy leads to the inequality that gives an upper bound on $\mathbf{M}$,

$$\mathbf{M} < \mathbf{M}_\mathrm{R} = c_\alpha \mathbf{M}_\alpha + c_\beta \mathbf{M}_\beta. \tag{31}$$

Each of the principles actually gives dual bounds that can be summarized as

$$\mathbf{L}_\mathrm{R} < \mathbf{L} < \mathbf{L}_\mathrm{V}, \qquad \mathbf{M}_\mathrm{R} > \mathbf{M} > \mathbf{M}_\mathrm{V}. \tag{32}$$

Of course, the Voigt and Reuss bounds are insensitive to microstructural geometry and thus not particularly sharp. More restrictive bounds, which reflect essential features of the geometry, were originally derived by Hashin and Shtrikman [25–28], and for fibrous composites by Hashin and Rosen [29]. Alternative derivations were given by Hill [16, 30], Walpole [31, 32], Willis [33, 34], and Milton and Kohn [35]; they also appeared in Christensen's monograph [36] and in several reviews [34, 37–39]. The procedure again relies on the elastic extremum principles and on certain polarization stress and strain fields that are estimated from solutions of inclusion problems in a homogeneous comparison material of stiffness L_0, compliance $M_0 = L_0^{-1}$, from which follow the constraint tensors L_0^*, M_0^* in (14–17). Walpole [*31, 32, 37*] gives a particularly simple form of the bounds:

$$[\sum c_r(L_r + L_0^*)^{-1}]^{-1} - L_0^* > L \qquad \text{if} \quad L_0 > L_r,$$
$$[\sum c_r(M_r + M_0^*)^{-1}]^{-1} - M_0^* > M \qquad \text{if} \quad M_0 > M_r, \tag{33}$$

where L_0 and M_0 can be assembled from the appropriate terms of either L_r or M_r.

C. *Transformation Strain*

1. UNIFORM STRAIN FIELDS IN FIBROUS MEDIA

Local and overall deformation in elastic composite media can be caused either by mechanical loads or by transformation strains in the phases. Thermal changes and phase transformations are the most common sources of such strains; however, both local and overall plastic strains also can be regarded as transformation strains.

Problems of this kind are best analyzed with the help of uniform fields in fibrous composite media, which were recently discovered by Dvorak [*7, 40*]. To introduce the subject, we rewrite (3) in the more general form

$$\varepsilon_r(x) = M_r\sigma_r(x) + \mu_r, \tag{34}$$

$$\sigma_r(x) = L_r\varepsilon_r(x) + \lambda_r, \tag{34'}$$

where μ_r denotes uniform transformation strains in the phases, which would be the only strains present if the phases were free to deform without mutual constraints, and $\lambda_r = -L_r\mu_r$. The corresponding form of the overall relation (5) is

$$\bar{\varepsilon} = M\bar{\sigma} + \mu, \tag{35}$$

$$\bar{\sigma} = L\bar{\varepsilon} + \lambda, \tag{35'}$$

where $\bar{\sigma}, \bar{\varepsilon}, \lambda$, and μ are the overall volume averages of stress and strain, L and M are the tensors of elastic moduli and compliances, and $\lambda = -L\mu$. Note that if the medium were unloaded in an elastic manner, then μ would be the remaining overall strain caused by the eigenstrains μ_r; λ is the overall stress caused by μ_r in a fully constrained medium.

In the presence of both mechanical overall stress or strain, and uniform phase eigenstrains, the local fields in the phases can be written as

$$\varepsilon_r(\mathbf{x}) = \mathbf{A}_r(\mathbf{x})\bar{\varepsilon} + \mathbf{D}_{r\alpha}(\mathbf{x})\boldsymbol{\mu}_\alpha + \mathbf{D}_{r\beta}(\mathbf{x})\boldsymbol{\mu}_\beta,$$

$$\sigma_r(\mathbf{x}) = \mathbf{B}_r(\mathbf{x})\bar{\sigma} + \mathbf{F}_{r\alpha}(\mathbf{x})\boldsymbol{\lambda}_\alpha + \mathbf{F}_{r\beta}(\mathbf{x})\boldsymbol{\lambda}_\beta, \tag{36}$$

where $\mathbf{D}_{r\alpha}(\mathbf{x})$ and $\mathbf{D}_{r\beta}(\mathbf{x})$ are the strain fields caused in the phases by unit phase transformation strains while the representative volume V of the aggregate is under zero overall strain. Similarly, $\mathbf{F}_{r\alpha}(\mathbf{x})$ and $\mathbf{F}_{r\beta}(\mathbf{x})$ are stress fields in the respective phases due to unit phase transformation stresses $\boldsymbol{\lambda}_r$ in V when the overall stress is zero.

With these definitions, we pose the following problem: For a fibrous two-phase composite medium that is subjected to uniform transformation strains $\boldsymbol{\mu}_r$ in the phases, find an overall stress $\hat{\sigma}$ such that the superposed local strains are uniform in the entire volume V and equal to the overall strain $\hat{\varepsilon}$,

$$\hat{\varepsilon}_\alpha(\mathbf{x}) = \hat{\varepsilon}_\beta(\mathbf{x}) = \hat{\varepsilon}, \tag{37}$$

The problem can be solved with the following decomposition procedure. The phases are separated and subjected to the respective transformation strains and to as yet unknown surface tractions that cause only uniform stresses $\hat{\sigma}_r$ in the separated phases. Before reassembly, the tractions must be in equilibrium at the cylindrical interfaces such that the auxiliary phase and overall stresses satisfy the conditions

$$\hat{\sigma}_1^\alpha \neq \hat{\sigma}_1^\beta \qquad\qquad \hat{\sigma}_j^\alpha = \hat{\sigma}_j^\beta \quad \text{for} \quad j = 2, 3, \ldots 6,$$

$$\hat{\sigma}_1 = c_\alpha \hat{\sigma}_1^\alpha + c_\beta \hat{\sigma}_1^\beta \qquad \hat{\sigma}_j = \hat{\sigma}_j^\alpha = \hat{\sigma}_j^\beta \quad \text{for} \quad j = 2, 3, \ldots 6 \tag{38}$$

in the usual contracted notation. This means that the axial normal stress may be piecewise uniform in V, whereas all other stress components are uniform in V.

The components of the unknown overall stress then follow from (34_1), (37), and the above equilibrium conditions,

$$M_{i1}^\alpha \hat{\sigma}_1^\alpha - M_{i1}^\beta \hat{\sigma}_1^\beta + \sum_{j=2}^{6}(M_{ij}^\alpha - M_{ij}^\beta)\hat{\sigma}_j + \mu_i^\alpha - \mu_i^\beta = 0; \quad i = 1, 2, \ldots 6. \tag{39}$$

There are seven unknown stresses in this system of six equations; if a solution is found, it guarantees the existence of a uniform strain in the aggregate under $\boldsymbol{\mu}_r$ and $\hat{\sigma}$. In general, even the homogeneous set associated with (39) has a solution that gives a proportional overall stress path that creates a uniform field in the medium when $\boldsymbol{\mu}_r = \mathbf{0}$.

Whereas the solution of (39) may be obtained for any material symmetry of the phases, we restrict our attention to systems in which both phases are transversely isotropic about x_1. Of course, one or both phases may be isotropic if the elastic moduli of the phases are defined as indicated in the Appendix. In any case, the phases and the aggregate share the same transverse plane of symmetry, the x_2x_3-plane. This is the only symmetry condition that must be satisfied on the macroscale; in single-crystal elasticity, it is associated with monoclinic crystals. For this system, the solution of (39) was obtained by Dvorak [40]; here, we only summarize the results in the following form,

$$\hat{\sigma} = \hat{\sigma}_0 + \hat{\sigma}_\mu, \qquad \hat{\sigma}_r = \hat{\sigma}_r^0 + \hat{\sigma}_r^\mu, \qquad \hat{\epsilon} = \hat{\epsilon}_0 + \hat{\epsilon}_\mu, \tag{40}$$

such that for $\mu_r = 0$,

$$\hat{\sigma}_\mu = \hat{\sigma}_r^\mu = 0, \qquad \hat{\epsilon}_\mu = 0. \tag{41}$$

The homogeneous solution is

$$\hat{\sigma}_0 = \mathbf{f}_0 S_T, \qquad \hat{\sigma}_r^0 = \mathbf{f}_r^0 S_T, \qquad \hat{\epsilon}_0 = \mathbf{d}_0 S_T, \tag{42}$$

where $S_T = (\hat{\sigma}_2 + \hat{\sigma}_3)/2$ is selected as a free parameter in the solution. The coefficients of the above vectors are obtained in terms of Hill's moduli of transversely isotropic solids defined in the Appendix. Also, the following additional symbols are used:

$$\Delta k = k_\alpha - k_\beta, \qquad m^* = m_\alpha m_\beta/(m_\beta - m_\alpha); \quad m_\alpha \neq m_\beta,$$

$$\Delta l = l_\alpha - l_\beta, \qquad p^* = p_\alpha p_\beta/(p_\beta - p_\alpha); \quad p_\alpha \neq p_\beta,$$

$$q^{-1} = (l_\alpha k_\beta - k_\alpha l_\beta) = 2k_\alpha k_\beta(v_L^\alpha - v_L^\beta) \neq 0.$$

Then, the nonvanishing coefficients of $\mathbf{f}_0$, $\mathbf{f}_r^0$, and $\mathbf{d}_0$ are

$$f_1^0 = q[c_\alpha(l_\alpha \Delta l - n_\alpha \Delta k) + c_\beta(l_\beta \Delta l - n_\beta \Delta k)], \qquad f_2^0 = f_3^0 = 1,$$

$$(f_r^0)_1 = q(l_r \Delta l - n_r \Delta k), \qquad (f_r^0)_2 = (f_r^0)_3 = 1, \tag{43}$$

$$d_1^0 = -q\Delta k, \qquad d_2^0 = d_3^0 = q\Delta l/2.$$

The terms $\hat{\sigma}_\mu$, $\hat{\sigma}_r^\mu$, and $\hat{\epsilon}_\mu$ in (40) can be expressed in the form

$$\hat{\epsilon}_\mu = \mathbf{K}_\alpha \mu_\alpha + \mathbf{K}_\beta \mu_\beta = -\mathbf{K}_\alpha \mathbf{M}_\alpha \lambda_\alpha - \mathbf{K}_\beta \mathbf{M}_\beta \lambda_\beta, \tag{44}$$

$$\hat{\sigma}_\mu = \mathbf{N}_\alpha \mu_\alpha + \mathbf{N}_\beta \mu_\beta = -\mathbf{N}_\alpha \mathbf{M}_\alpha \lambda_\alpha - \mathbf{N}_\beta \mathbf{M}_\beta \lambda_\beta, \tag{45}$$

$$\hat{\sigma}_r^\mu = \mathbf{N}_{r\alpha} \mu_\alpha + \mathbf{N}_{r\beta} \mu_\beta = -\mathbf{N}_{r\alpha} \mathbf{M}_\alpha \lambda_\alpha - \mathbf{N}_{r\beta} \mathbf{M}_\beta \lambda_\beta, \tag{46}$$

where the nonvanishing coefficients of $\mathbf{K}_r$, $\mathbf{N}_r$, $\mathbf{N}_{r\alpha}$, $\mathbf{N}_{r\beta}$ are

$$K_{11}^\alpha = ql_\alpha k_\beta, \qquad\qquad K_{12}^\alpha = K_{13}^\alpha = qk_\alpha k_\beta, \quad K_{21}^\alpha = -ql_\alpha l_\beta/2,$$

$$K_{22}^\alpha = -m^*/2m_\beta - qk_\alpha l_\beta/2, \qquad\qquad K_{23}^\alpha = m^*/2m_\beta - qk_\alpha l_\beta/2,$$

$$K_{31}^\alpha = K_{21}^\alpha, \qquad\qquad K_{32}^\alpha = K_{23}^\alpha, \qquad K_{33}^\alpha = K_{22}^\alpha,$$

$$K_{44}^\alpha = -m^*/m_\beta, \qquad\qquad K_{55}^\alpha = -p^*/p_\beta, \qquad K_{66}^\alpha = K_{55}^\alpha,$$

and

$$N^\alpha_{11} = q(c_\alpha k_\alpha l_\beta E^\alpha_L + c_\beta k_\beta l_\alpha E^\beta_L),$$
$$N^\alpha_{12} = N^\alpha_{13} = qk_\alpha k_\beta(c_\alpha E^\alpha_L + c_\beta E^\beta_L),$$
$$N^\alpha_{ij} = N^{r\alpha}_{ij} \quad \text{for} \quad i \neq 1,$$

together with

$$N^{r\alpha}_{11} = qk_r E^r_L l_\rho, \quad (\rho = \alpha \;\; \text{if} \;\; r = \beta; \quad \rho = \beta \;\; \text{if} \;\; r = \alpha),$$
$$N^{r\alpha}_{12} = N^{r\alpha}_{13} = qk_r E^r_L k_\rho, \qquad N^{r\alpha}_{22} = -N^{r\alpha}_{23} = -m^*,$$
$$N^{r\alpha}_{32} = -N^{r\alpha}_{33} = -N^{r\alpha}_{44} = m^*, \qquad N^{r\alpha}_{55} = N^{r\alpha}_{66} = -p^*.$$

In addition, one can establish that

$$\mathbf{K}_\alpha + \mathbf{K}_\beta = \mathbf{I}, \tag{47}$$

$$\mathbf{N}_\alpha + \mathbf{N}_\beta = \mathbf{0}, \qquad \mathbf{N}_{r\alpha} + \mathbf{N}_{r\beta} = \mathbf{0}, \tag{48}$$

$$\mathbf{N}_r = c_\alpha \mathbf{N}_{\alpha r} + c_\beta \mathbf{N}_{\beta r}. \tag{49}$$

A similar solution could be obtained for other material symmetries of the phases; of course, the coefficients in the matrices in (43) to (49) would change. This is also true for the case of $q^{-1} = 0$.

2. LOCAL AND OVERALL EFFECTS

The existence of the uniform fields opens the way for the derivation of local fields caused by the uniform phase eigenstrains, and of the overall stress λ and strain μ in (35). Consider a two-phase fibrous system that is initially stress free and introduce uniform eigenstrains into the phases. Apply overall auxiliary stress or strain that makes the strain field uniform in V. With reference to (40), disregard the homogeneous solution and use only the overall stress $\hat{\sigma}_\mu$ or strain $\hat{\varepsilon}_\mu$ in this loading step. Finally, reduce the overall strain or stress to zero.

In terms of the overall quantities, this sequence can be recorded as

$$\hat{\varepsilon}_\mu - \mathbf{M}\hat{\sigma}_\mu = \mu,$$
$$\hat{\sigma}_\mu - \mathbf{L}\hat{\varepsilon}_\mu = \lambda, \tag{50}$$

and in local terms as

$$\varepsilon_r(\mathbf{x}) = \hat{\varepsilon}^\mu_r - \mathbf{A}_r(\mathbf{x})\hat{\varepsilon}_\mu = (\mathbf{I} - \mathbf{A}_r(\mathbf{x}))\hat{\varepsilon}_\mu,$$
$$\sigma_r(\mathbf{x}) = \hat{\sigma}^\mu_r - \mathbf{B}_r(\mathbf{x})\hat{\sigma}_\mu, \tag{51}$$

with reference to (37).

Next, substitute into (50) for the auxiliary quantities from (44) to (46) and recover the following explicit forms of (35), with the overall μ and λ being expressed in terms of local and overall moduli and the phase eigenstrains,

$$\varepsilon = \mathbf{M}\sigma + (\mathbf{K}_\alpha - \mathbf{MN}_\alpha)\mu_\alpha + (\mathbf{K}_\beta - \mathbf{MN}_\beta)\mu_\beta, \tag{52}$$

$$\sigma = \mathbf{L}\varepsilon - (\mathbf{LK}_\alpha - \mathbf{N}_\alpha)\mu_\alpha - (\mathbf{LK}_\beta - \mathbf{N}_\beta)\mu_\beta. \tag{52'}$$

Recall that (11) presented an analogous result, albeit for thermal phase strains. This can be generalized in a self-evident way to the case of general eigenstrains if $\mathbf{m}\theta$ and $\mathbf{m}_r\theta$ in (11) are replaced by μ and μ_r, and $\mathbf{l}\theta$ and $\mathbf{l}_r\theta$ by λ and λ_r. Then, compare the result with (52) and extract the following expressions for the mechanical phase concentration factors:

$$\begin{aligned}
c_\alpha \mathbf{B}_\alpha^T &= \mathbf{K}_\alpha - \mathbf{MN}_\alpha = \mathbf{I} - \mathbf{K}_\beta + \mathbf{MN}_\beta, \\
c_\beta \mathbf{B}_\beta^T &= \mathbf{K}_\beta - \mathbf{MN}_\beta = \mathbf{I} - \mathbf{K}_\alpha + \mathbf{MN}_\alpha, \\
c_\alpha \mathbf{A}_\alpha^T &= (\mathbf{LK}_\alpha - \mathbf{N}_\alpha)\mathbf{M}_\alpha = \mathbf{I} - (\mathbf{LK}_\beta - \mathbf{N}_\beta)\mathbf{M}_\beta, \\
c_\beta \mathbf{A}_\beta^T &= (\mathbf{LK}_\beta - \mathbf{N}_\beta)\mathbf{M}_\beta = \mathbf{I} - (\mathbf{LK}_\alpha - \mathbf{N}_\alpha)\mathbf{M}_\alpha.
\end{aligned} \tag{53}$$

These equations give the same magnitudes of $\mathbf{A}_r$ and $\mathbf{B}_r$ as one would obtain from (10) or (11), but they can be easily expanded so that one can see the dependence of the individual coefficients A_{ij}, B_{ij} on L_{ij} and M_{ij}, respectively.

Proceed now to the evaluation of local fields and their phase averages. Substitute from (36) into (51) and use (44) to (46) for the evaluation of the auxiliary strain and stress. After some algebra, derive the following relations between transformation and mechanical fields in the phases

$$\begin{aligned}
\mathbf{D}_{r\alpha}(\mathbf{x}) &= (\mathbf{I} - \mathbf{A}_r(\mathbf{x}))\mathbf{K}_\alpha, & \mathbf{F}_{r\alpha}(\mathbf{x}) &= (\mathbf{B}_r(\mathbf{x})\mathbf{N}_\alpha - \mathbf{N}_{r\alpha})\mathbf{M}_\alpha, \\
\mathbf{D}_{r\beta}(\mathbf{x}) &= (\mathbf{I} - \mathbf{A}_r(\mathbf{x}))\mathbf{K}_\beta, & \mathbf{F}_{r\beta}(\mathbf{x}) &= (\mathbf{B}_r(\mathbf{x})\mathbf{N}_\beta - \mathbf{N}_{r\beta})\mathbf{M}_\beta,
\end{aligned} \tag{54}$$

together with the connections

$$\mathbf{D}_{r\alpha}(\mathbf{x}) + \mathbf{D}_{r\beta}(\mathbf{x}) = (\mathbf{I} - \mathbf{A}_r(\mathbf{x})), \qquad \mathbf{F}_{r\alpha}(\mathbf{x})\mathbf{L}_\alpha + \mathbf{F}_{r\beta}(\mathbf{x})\mathbf{L}_\beta = 0. \tag{55}$$

Find phase volume averages (2), (4) of the above fields and employ (53) to replace the mechanical concentration factors. This connects the eigenstrain and eigenstress concentration factors directly to the overall $\mathbf{L}$ and $\mathbf{M}$,

$$\begin{aligned}
c_r\mathbf{D}_{r\alpha}^T &= \mathbf{K}_\alpha^T[c_r\mathbf{I} - (\mathbf{LK}_r - \mathbf{N}_r)\mathbf{M}_r], & c_r\mathbf{F}_{r\alpha}^T &= \mathbf{M}_\alpha[\mathbf{N}_\alpha^T(\mathbf{K}_r - \mathbf{MN}_r) - c_r\mathbf{N}_{r\alpha}^T], \\
c_r\mathbf{D}_{r\beta}^T &= \mathbf{K}_\beta^T[c_r\mathbf{I} - (\mathbf{LK}_r - \mathbf{N}_r)\mathbf{M}_r], & c_r\mathbf{F}_{r\beta}^T &= \mathbf{M}_\beta[\mathbf{N}_\beta^T(\mathbf{K}_r - \mathbf{MN}_r) - c_r\mathbf{N}_{r\beta}^T].
\end{aligned} \tag{56}$$

One also finds that these factors satisfy

$$c_\alpha \mathbf{D}_{\alpha\alpha} + c_\beta \mathbf{D}_{\beta\alpha} = \mathbf{0}, \qquad c_\alpha \mathbf{F}_{\alpha\alpha} + c_\beta \mathbf{F}_{\beta\alpha} = \mathbf{0},$$
$$c_\alpha \mathbf{D}_{\alpha\beta} + c_\beta \mathbf{D}_{\beta\beta} = \mathbf{0}, \qquad c_\alpha \mathbf{F}_{\alpha\beta} + c_\beta \mathbf{F}_{\beta\beta} = \mathbf{0}. \tag{57}$$

These results, and particularly Eqns. (54), can be utilized in the evaluation of the local fields. Equations (36) are the key; they can be restated as follows:

The local strain fields in a composite subjected to an overall strain $\bar{\varepsilon}$ and to the phase eigenstrains $\boldsymbol{\mu}_r$ are given by

$$\boldsymbol{\varepsilon}_r(\mathbf{x}) = \mathbf{A}_r(\mathbf{x})\bar{\varepsilon} + \mathbf{D}_{r\alpha}(\mathbf{x})\boldsymbol{\mu}_\alpha + \mathbf{D}_{r\beta}(\mathbf{x})\boldsymbol{\mu}_\beta. \tag{58}$$

The local stress fields in a composite loaded by an overall stress $\bar{\sigma}$ and by the phase eigenstresses $\boldsymbol{\lambda}_r = -\mathbf{L}_r\boldsymbol{\mu}_r$ are given by

$$\boldsymbol{\sigma}_r(\mathbf{x}) = \mathbf{B}_r(\mathbf{x})\bar{\sigma} + \mathbf{F}_{r\alpha}(\mathbf{x})\boldsymbol{\lambda}_\alpha + \mathbf{F}_{r\beta}(\mathbf{x})\boldsymbol{\lambda}_\beta. \tag{59}$$

Phase averages of the local fields follow if $\mathbf{A}_r(\mathbf{x})$, $\mathbf{B}_r(\mathbf{x})$, $\mathbf{D}_{r\alpha}(\mathbf{x})$, etc., are replaced by $\mathbf{A}_r$, $\mathbf{B}_r$, $\mathbf{D}_{r\alpha}$, and so on.

3. EVALUATION PROCEDURES

Applications of the various relations that describe the elastic behavior of fibrous composites must be preceded by a specification of phase thermo-elastic properties and volume fractions. The first step is the evaluation of mechanical stress and strain concentration factors and overall moduli. This can be done in two different ways that, for a given approximation, lead to identical results. The factors can be evaluated directly by either the self-consistent method from (21) or by the Mori–Tanaka procedure from (25) to (27). The overall moduli and compliances then follow from (22) or (28), respectively. A more direct way to this result is indicated by (23) and (29), which offer the magnitudes of the moduli. Using the equations in the Appendix, one can readily assemble the overall $\mathbf{L}$ and $\mathbf{M}$. Alternatively, (33) can be used to find bounds on these moduli. In any event, each set of moduli obtained from one of these approaches must conform with the universal connections (24). Also, each set may be substituted into (10) or (53), which yield the concentration factor tensors. Of course, if (53) is used, one must first find the coefficients of $\mathbf{K}_r$, $\mathbf{N}_r$, and so on, in (44) to (46). The relations following (10) and (47)–(49) may be used to verify the results.

The response to a uniform change in temperature can be described in terms of local thermoelastic properties and overall $\mathbf{L}$ or $\mathbf{M}$; equations (11) give the overall properties, and (12) give the local fields. These results represent exact connections between the various thermal and mechanical terms. However, the latter are not known exactly; they are known only in terms of bounds or estimates. Hence the overall thermal properties and phase field averages are also approximations produced by the various procedures.

The effect of uniform transformation strains on overall stress or strain and on phase field averages can also be described by local and overall moduli and by mechanical concentration factors. Equations (52) indicate the overall stress and strain in terms of the concentration factors (53); and (58), (59) give the local fields. Of course, these results reduce to those caused by a temperature change if the transformation strains are replaced by phase thermal strains.

The overall thermoelastic properties and averages of local fields can be obtained with minimal information about the microstructural geometry of the fibrous medium, essentially in terms of local properties and phase volume fractions. A much more difficult problem is the evaluation of local fields in the phases. Obviously, the outcome depends very much on the details of the geometry of the phases, and as such it calls for solution of specific boundary-value problems for prescribed geometries. The actual local geometry usually varies in any given system, hence exact solutions are beyond reach. However, the parts of the local fields that are of interest in applications are usually located in the fiber and in the proximity of the fiber–matrix interfaces. Estimates of such fields cannot be found by the self-consistent method unless it is modified by introduction of a matrix interlayer to surround the fiber, nor can they be deduced from the bounds. Only the Mori–Tanaka method offers a direct estimate of the interface stress and of those in the surrounding matrix. Recall that the method is based on the solution of an inclusion problem in which the fiber is embedded in an infinite matrix subjected to the average matrix stress or strain (27) at infinity; this field can be used as an estimate of the actual field. The fields are not described herein, but can be found in available texts [*12, 36*].

Once the estimate of the mechanical fields (27) is found, it is an easy matter to use it in the evaluation of the thermal stress fields via (12) or of the transformation fields (58) and (59), with the connections (54).

III. Elastic–Plastic Response

A. *Homogeneous Materials*

1. YIELD AND RELAXATION SURFACES

When a metal or a metal composite part is loaded beyond a certain stress magnitude, the total strain at each material point may consist of both elastic and inelastic contributions. If the loading rates are such that the response can be regarded as independent of time, the inelastic part is represented by a plastic strain. At a given applied stress, the plastic part of the total strain, if any, is defined as the strain that would remain after complete local unloading, in which the material would be constrained to deform only elastically. In other words, the plastic strain in a homogeneous volume of material under uniform stress is the difference between the total strain and the elastic strain that would be generated by the current applied stress in a completely elastic solid.

In polycrystalline solids that are assumed to be elastically homogeneous on the macroscale, the plastic strains caused by prescribed stress histories can be evaluated, in principle, within the framework of classical plasticity [41–44]. The results are useful in the formulation of constitutive equations for the plastically deforming phases in composite materials. Although the theory is largely phenomenological and thus not necessarily sensitive to the effects that the heterogeneous microstructure may have on overall behavior, its general aspects are relevant to the plasticity of composite materials. We are concerned only with situations where the total strains are small and where the thermoelastic properties are independent of temperature. The yield stress, however, may change with temperature. Only uniform temperature changes will be applied.

Consider a representative volume of an elastic–plastic solid such that the selected sample can be regarded as elastically homogeneous on the macroscale. Starting from an undeformed, stress-free state, apply uniform overall stresses along a prescribed path that terminates at the current state $\bar{\sigma}$. In most metals and composites, there is an elastic domain in stress space where no plastic strains are generated by load cycles from the current stress. In principle, its boundary can be found as a locus of points that can be reached by purely elastic loading excursions from $\bar{\sigma}$. The outcome is described by a scalar-valued yield or loading function $f(\bar{\sigma}, H)$, which depends on the current stress and also on a functional of past history of inelastic deformation H. The function represents a closed surface in the six-dimensional stress space. If the sample is heterogeneous on the microscale, the surface may consist of a finite number of smooth branches.

Experimental evaluation of yield surfaces has been performed most extensively on polycrystalline metals, particularly on commercially pure aluminum, on some aluminum alloys, and on other metals [45–49]. Only recently have similar results become available for metal matrix composites [50]. The results suggest that the surface translates and distorts during plastic loading. The distortions are often much less pronounced than the translation; hence in the first approximation, they may be neglected, and

$$f(\bar{\sigma}, H) = f(\bar{\sigma} - \bar{\alpha}), \tag{60}$$

where $\bar{\alpha}$ denotes the position of the center of the surface. In the absence of prior inelastic deformation, $\bar{\alpha} = 0$, and (60) defines the initial yield surface.

Alternatively, the material volume considered can be viewed as subjected to a macroscopically uniform state of strain. After a deformation sequence leading to the current state $\bar{\varepsilon}$, elastic strain excursions from the current state may be employed to determine the outer boundary of the elastic strain domain. For materials that obey (60), the outcome is represented by the relaxation surface in strain space

$$\phi(\bar{\varepsilon}, H) = \phi(\mathbf{M}\bar{\sigma} + \bar{\varepsilon}^p, H) = f(\bar{\sigma}, H), \tag{61}$$

where H, in an appropriate form, is again a functional of past plastic strain history.

The specific analytic forms of ϕ or f, which approximate experimental observations, are closely related to elastic symmetry of the material volume. In the undeformed state, and after deformation histories that do not cause internal rearrangements affecting material symmetry, the functions (60) and (61) must be form-invariant under the group of symmetry transformations associated with the particular solid. For example, most metals are usually regarded as macroscopically isotropic, and the functions are expressed in terms of the familiar isotropic stress or strain invariants. The assumption of plastic incompressibility eliminates the dependence on the first invariant, and thus on the stress $\bar{\sigma}$ in favor of the deviatoric stress $\bar{s}$. The third invariant of $\bar{s}$ is convenient in special situations, but its role is neglected in typical applications. This leaves only the second invariant $J_2 = s_{ij}s_{ij}/2$, and f is then represented by the Mises form of the yield function

$$f(\bar{\sigma} - \bar{\alpha}) \equiv (\bar{\sigma} - \bar{\alpha})^T \mathbf{C}(\bar{\sigma} - \bar{\alpha}) - Y^2 = 0, \tag{62}$$

where the nonvanishing coefficients of the (6×6) matrix $\mathbf{C}$ are

$$C_{11} = C_{22} = C_{33} = 1, \ C_{23} = C_{13} = C_{12} = -\tfrac{1}{2}, \ C_{44} = C_{55} = C_{66} = 3,$$

and Y is the yield stress in simple tension; its magnitude may depend on temperature and/or the plastic strain history.

In contrast to the isotropic metals, fibrous composite materials are usually transversely isotropic, i.e., their properties remain invariant under rigid body rotations about the fiber axis x_1 and also under the transformation $x_1 = -x_1$. Hill [41] and Mulhern et al. [51] point out that the functions (60) and (61) must then depend on the corresponding invariants. Green and Atkins [52] and Spencer [53] give the appropriate forms; for transverse isotropy the stress invariants are

$$I_1 = (\bar{\sigma}_{22} + \bar{\sigma}_{33})/2,$$

$$I_2 = \bar{\sigma}_{11},$$

$$I_3 = (\bar{\sigma}_{31}^2 + \bar{\sigma}_{21}^2), \tag{63}$$

$$I_4 = [\tfrac{1}{4}(\bar{\sigma}_{33} + \bar{\sigma}_{22})^2 + \bar{\sigma}_{32}^2]/2,$$

$$I_5 = [(\bar{\sigma}_{33} - \bar{\sigma}_{22})(\bar{\sigma}_{31}^2 - \bar{\sigma}_{21}^2) + 4\bar{\sigma}_{21}\bar{\sigma}_{31}\bar{\sigma}_{32}]/2.$$

Other appropriate sets must be selected for other than transversely isotropic composites. Such considerations impose specific restrictions on the admissible form of (60) and (61) that must be respected in modeling. In particular, $f(\bar{\sigma}, H) = f(I_1, I_2, I_3, I_4, I_5, H)$. In properly constructed micro-mechanical models, this restriction should be automatically satisfied when the overall yield or relaxation surfaces reflect the onset of inelastic deformation, e.g., in the matrix phase.

2. PLASTIC STRAINS AND HARDENING

When the overall stress is taken as the independent variable, the plastic strain is defined as the difference between total strain and the elastic strain recovered during purely elastic unloading. Conversely, when the overall strain is independent, plastic deformation is associated with the existence of a relaxation stress $\bar{\sigma}^R$, defined as the difference between the elastic and current stress at the prescribed overall strain. When this is applied to strain and stress increments, it suggests the additive decompositions

$$d\bar{\varepsilon} = \mathbf{M}\,d\bar{\sigma} + d\bar{\varepsilon}^p, \tag{64}$$

$$d\bar{\sigma} = \mathbf{L}\,d\bar{\varepsilon} - d\bar{\sigma}^R. \tag{64'}$$

Unlike $\mathbf{M}\,d\bar{\sigma}$ or $\mathbf{L}\,d\bar{\varepsilon}$, the $d\bar{\varepsilon}^p$ or $d\bar{\sigma}^R$ are not linear functions of stress or strain. Therefore, the total quantities generated under a prescribed load or deformation history must be found by integration along the actual path.

The transition from elastic to plastic deformation can be defined with reference to the loading or relaxation surfaces. In what follows, we focus on the stress space formulation; an analogous strain space formulation can be found in [54–56]. By definition, the surface (60) must always contain the loading point during plastic deformation; this is assured by the consistency condition

$$df = \left(\frac{\partial f}{\partial \bar{\sigma}}\right)^T (d\bar{\sigma} - d\bar{\alpha}) + \frac{\partial f}{\partial H}\,dH + \frac{\partial f}{\partial \theta}\,d\theta = 0. \tag{65}$$

The possible states reached by loading excursions from the current elastic state are:

$$d\bar{\varepsilon}^p = 0 \text{ for } f \le 0, \quad \frac{\partial f}{\partial \bar{\sigma}} \cdot d\bar{\sigma} + \frac{\partial f}{\partial \theta} d\theta < 0 \quad \text{(elastic state or unloading)},$$

$$d\bar{\varepsilon}^p = 0 \text{ for } f = 0, \quad \frac{\partial f}{\partial \bar{\sigma}} \cdot d\bar{\sigma} + \frac{\partial f}{\partial \theta} d\theta = 0 \quad \text{(neutral loading)}, \qquad (66)$$

$$d\bar{\varepsilon}^p \ne 0 \text{ for } f = 0, \quad \frac{\partial f}{\partial \bar{\sigma}} \cdot d\bar{\sigma} + \frac{\partial f}{\partial \theta} d\theta > 0 \quad \text{(plastic loading)}.$$

The direction of the plastic strain vector $d\bar{\varepsilon}^p$ can be established by following a load cycle that starts at some stress $\bar{\sigma}^*$ within the current loading surface, continues to the stress state $\bar{\sigma}$ on the yield surface where the excursion into the plastic range takes place, and then returns to $\bar{\sigma}^*$. The elastic work is recovered in the cycle, whereas the plastic work performed during the load cycle is

$$W_p = \int_{t_0}^{t} (\bar{\sigma} - \bar{\sigma}^*) \cdot d\bar{\varepsilon}^p \, dt \ge 0. \qquad (67)$$

This result suggests that

$$(\bar{\sigma} - \bar{\sigma}^*) \cdot d\bar{\varepsilon}^p \ge 0, \qquad (68)$$

where the equality is valid only for neutral loading. The relations indicate two important conclusions [42]. First, in materials that satisfy (68), the yield and loading surfaces are always convex. Second, at regular points of the loading surface, the plastic strain rate vector is always normal to the loading surface.

Thus one can write

$$d\bar{\varepsilon}^p = d\lambda \frac{\partial f}{\partial \bar{\sigma}}, \qquad (69)$$

where the magnitude $d\lambda$ needs to be determined.

During plastic straining, the yield surface must follow the motion of the loading point to satisfy (65). This can be assured by a proper evolution equation for the vector $\bar{\alpha}$ in (60). A suitable general form of a kinematic hardening law is

$$d\bar{\alpha} = d\bar{\sigma} + d\bar{\gamma}, \qquad (70)$$

where the magnitude of the vector $d\bar{\gamma}$ is specified by various rules. Well-known prescriptions were suggested by Prager and modified by Ziegler [57]. However, experimental results [45–48] indicate that in aluminum alloys

$d\bar{\varepsilon}^p$ is best approximated by selecting $d\bar{\gamma} = 0$ in (60). This hardening law was first suggested by Phillips [45, 46] and Phillips and Lee [47].

An illustration of the strain evaluation procedure for a uniformly stressed metal sample is provided by the following example [58]. Let the loading surface be represented by the Mises form (62), but with a temperature dependent yield stress $Y = Y(\theta)$, and the hardening by the Phillips law. If the latter is written in the form

$$d\bar{\alpha} = \mu \, d\bar{\sigma}, \tag{71}$$

then the consistency equation suggests that

$$
\mu = \left[\left(\frac{\partial f}{\partial \bar{\sigma}} \right)^T d\bar{\sigma} \right]^{-1} \left[\left(\frac{\partial f}{\partial \bar{\sigma}} \right)^T d\bar{\sigma} + \frac{\partial f}{\partial \theta} d\theta \right]
$$

$$
= [(\bar{\sigma} - \bar{\alpha})^T C \, d\bar{\sigma}]^{-1} [(\bar{\sigma} - \bar{\alpha})^T C \, d\bar{\sigma} - Y(\theta) Y'(\theta) \, d\theta], \tag{72}
$$

where the second expression was derived from the Mises form (62).

The plastic strain magnitude is evaluated from Ziegler's equality

$$
c d\bar{\varepsilon}_{ij}^p \frac{\partial f}{\partial \bar{\sigma}_{ij}} = d\bar{\alpha}_{ij} \frac{\partial f}{\partial \bar{\sigma}_{ij}}, \tag{73}
$$

which gives

$$
d\lambda = [c \mathbf{q}^T \mathbf{q}^*]^{-1} \left[\mathbf{q}^T d\bar{\sigma} + \frac{\partial f}{\partial \theta} d\theta \right] = \frac{1}{cQ} \left[\boldsymbol{\eta}^T d\bar{\sigma} - \frac{2Y(\theta) Y'(\theta)}{Q} d\theta \right], \tag{74}
$$

where

$$
\boldsymbol{\eta} = \frac{1}{Q} \mathbf{q}, \qquad Q = (\mathbf{q}^T \mathbf{q}^*)^{1/2},
$$

$$
q_1 = q_1^* = 2(\bar{\sigma}_{11} - \bar{\alpha}_{11}) - (\bar{\sigma}_{22} - \bar{\alpha}_{22}) - (\bar{\sigma}_{33} - \bar{\alpha}_{33}),
$$

$$
q_2 = q_2^* = -(\bar{\sigma}_{11} - \bar{\alpha}_{11}) + 2(\bar{\sigma}_{22} - \bar{\alpha}_{22}) - (\bar{\sigma}_{33} - \bar{\alpha}_{33}),
$$

$$
q_3 = q_3^* = -(\bar{\sigma}_{11} - \bar{\alpha}_{11}) - (\bar{\sigma}_{22} - \bar{\alpha}_{22}) + 2(\bar{\sigma}_{33} - \bar{\alpha}_{33}),
$$

$$
q_4 = 2q_4^* = 6(\bar{\sigma}_{21} - \bar{\alpha}_{21}), \qquad q_5 = 2q_5^* = 6(\bar{\sigma}_{31} - \bar{\alpha}_{31}),
$$

$$
q_6 = 2q_6^* = 6(\bar{\sigma}_{32} - \bar{\alpha}_{32}).
$$

The magnitudes of c and H are sometimes defined from a simple tension test; this gives

$$
c = 2H/3, \qquad H = (d\bar{\sigma} - dY)/d\bar{\varepsilon}^p,
$$

but a much better agreement with experiments follows when H is found from one of the contemporary theories of plasticity discussed, for example, in [48, 49, 59].

B.   Heterogeneous Materials

1.   INITIAL YIELDING AND HARDENING

In comparison with the analysis of elastic systems, micromechanics of elastic–plastic composites is much more difficult. Some of the earlier results apply, however, the instantaneous mechanical properties of the phases, such as stiffness and compliance are no longer known constants, instead they depend on the current stress and on past history of plastic deformation. Therefore, a plastically deforming phase that has experienced some non-uniform plastic straining is no longer homogeneous, its instantaneous properties change from point to point.

The plastic deformation process can be better understood if the medium is viewed as an aggregate of small volume elements $k = 1, 2, \ldots, N$, which subdivide both the matrix and the reinforcement phase, and in which the local strains are regarded as piecewise uniform. The refinement of such subdivision can be varied from two elements, one for each phase, to as many as desired. Since each subelement is replicated in numerous locations in the actual composite, it is appropriate to assume that local properties are similar to those that would be found in a neat polycrystalline matrix or fiber material that was exposed to the actual in situ processing sequence. Of course, in many instances such properties can be found only indirectly, but it is apparent that those that apply to the inelastic phase must fit into the same general framework as those of the phase material, described in Section III.A.

The inelastic response of such a subdivided composite to external loading will be studied in more detail with the micromechanical models in Section III.B.4. However, an example of their behavior will be presented now, as an introduction to the subject. The example is taken from the work of Lin and Dvorak [60], who examined an idealized periodic model geometry of a discontinuously reinforced fiber composite shown in Fig. 1. The fibers are cylinders of hexagonal cross section, arranged in a periodic hexagonal array in the transverse $x_2 x_3$-plane. In the longitudinal x_1-direction, the fibers are distributed also in a periodic manner. Figure 2 shows a representative volume of the composite and its subdivision into finite elements. When appropriate periodic boundary conditions are prescribed, the deformation of a large volume of the composite under macroscopically uniform overall stress can be studied with one such representative volume element. Finite element analysis of the representative volume is usually used to obtain specific results.

Suppose that the fiber is always elastic but that the matrix is an elastic–plastic material of the type described in Section III.A. The composite is initially in a stress-free state, but loading excursions of sufficient magnitude may cause plastic yielding in parts of, or in the entire matrix volume. At

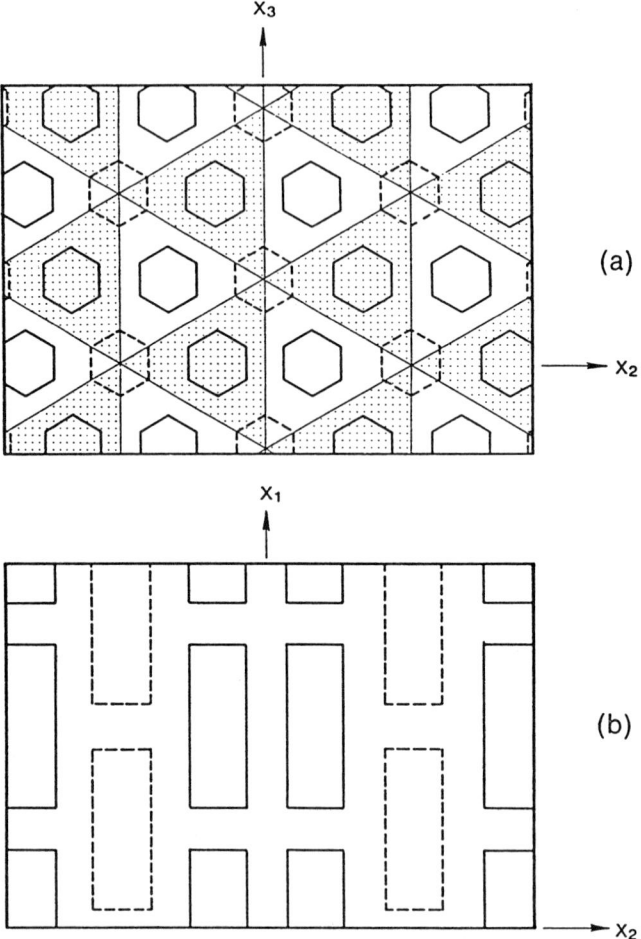

FIG. 1.   Idealized microgeometry of a whisker-reinforced composite medium, (a) transverse plane, (b) longitudinal plane [60].

current overall stress, the boundary of the elastic region can be established by loading in many directions from this stress state. Onset of overall plastic deformation can be detected as a specific deviation from linearity in the stress–strain diagram. In Fig. 3, this has been done by many radial loading excursions from the origin. Then, the overall stress was changed along the path indicated from A to B. Many excursions were again made from B to find the next loading surface. This was repeated at C, and the path CDE was completed within the last surface.

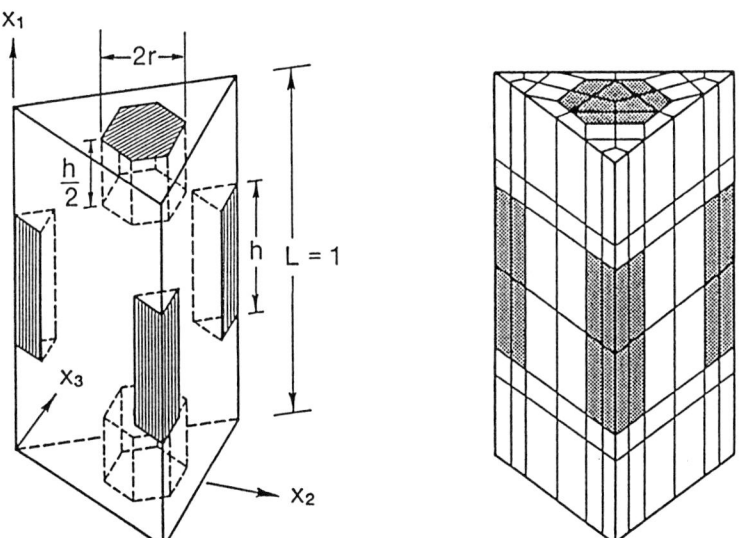

FIG. 2.   Geometry and finite element subdivision of the representative volume element [*60*].

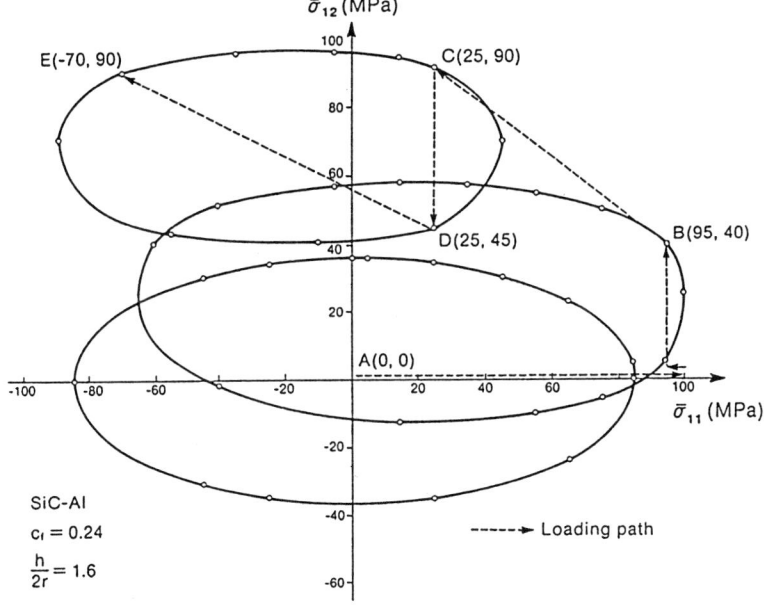

FIG. 3.   Overall initial and subsequent yield surfaces of a whisker-reinforced composite in the axial tension ($\bar{\sigma}_{11}$), longitudinal shear ($\bar{\sigma}_{12}$) plane. The subsequent yield surfaces are given at loading points B and C [*60*].

These overall surfaces reveal some aspects of plastic yielding, such as the essentially kinematic motion of the overall surface in the direction of the stress increment; this will be related to the Phillips-type kinematic hardening of the matrix. Also, there is some distortion, mostly expansion or contraction of the overall surface during plastic loading. In contrast, the yield surfaces of the kinematically hardening matrix do not change size.

A better insight into local and overall yielding can be derived from an examination of the local yield surfaces at integration points in the finite elements. To construct such local surfaces in the overall space, it is necessary to find the elastic stress concentration factors $\mathbf{B}_k$ of all elements. Each column of $\mathbf{B}_k$ is generated by a single component of the unit overall stress. Then, if the local surface of point k is given by $f_k(\boldsymbol{\sigma}_k, H_k) = 0$, in the local stresses, its image in the overall stress space is $g_k(\bar{\boldsymbol{\sigma}}, H_k) = f_k(\mathbf{B}_k\bar{\boldsymbol{\sigma}}, H_k) = 0$. With reference to (60), this can be written as

$$g_k(\bar{\boldsymbol{\sigma}} - \bar{\boldsymbol{\alpha}}_k) = f_k(\mathbf{B}_k(\bar{\boldsymbol{\sigma}} - \bar{\boldsymbol{\alpha}}_k)), \tag{75}$$

or, using (62),

$$g_k(\bar{\boldsymbol{\sigma}} - \bar{\boldsymbol{\alpha}}) = (\bar{\boldsymbol{\sigma}} - \bar{\boldsymbol{\alpha}}_k)^T \mathbf{B}_k^T \mathbf{C} \mathbf{B}_k(\bar{\boldsymbol{\sigma}} - \bar{\boldsymbol{\alpha}}_k) - Y^2 = 0. \tag{76}$$

Figure 4 shows a plane section of the cluster of local surfaces for the

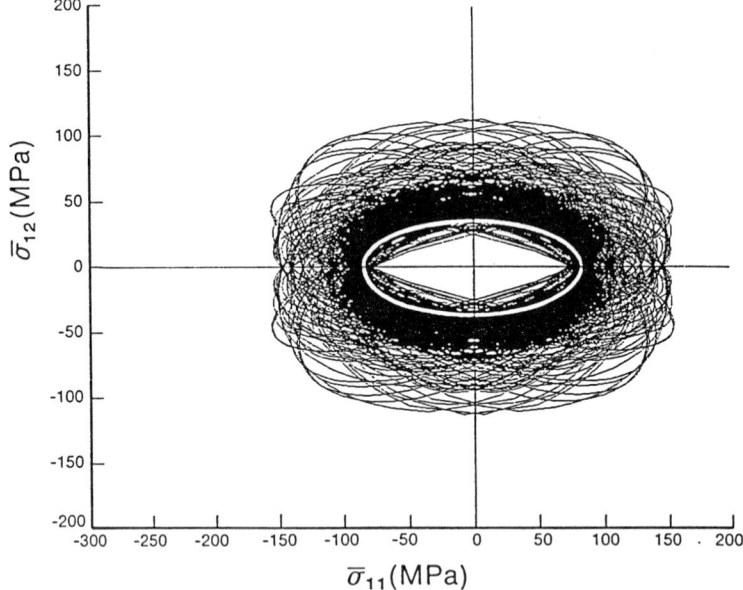

FIG. 4. Cluster of initial yield surfaces found in the overall $\bar{\sigma}_{11}\bar{\sigma}_{12}$-plane from local initial yield surfaces of the matrix phase. Overall yield surface found from deviation of the overall stress–strain response from linearity is superimposed [60].

integration points in Fig. 2, in the overall stress space for the undeformed state. The surface found from macroscopic deviations from linearity in Fig. 3 is superimposed. Figures 5 and 6 show similar clusters at points B and C, together with the corresponding surfaces from Fig. 3. Of course, the loading excursions designed to establish the points on the overall surface cause local plastic yielding and rearrangement of the clusters. In fact, even the seemingly elastic path CDE causes such rearrangement and plastic yielding. Figure 7 shows a detail of the cluster at point C. This is a section of a yield cone in the six-dimensional overall space. In the plane shown, it represents a corner; note the hinge-like arrangement which is typical after extended excursions. At the current point C, one can draw normals to the internal envelope and also evaluate the plastic strain increment. Hill [61] shows that the direction of the plastic strain must fall inside a cone of normals, this is illustrated by the plane section in Fig. 7.

The example indicates that it is not generally possible to derive the actual overall yield surface from a local yield condition. Instead, the overall surface that corresponds to measurable deviations from linearity is a locus of vertices

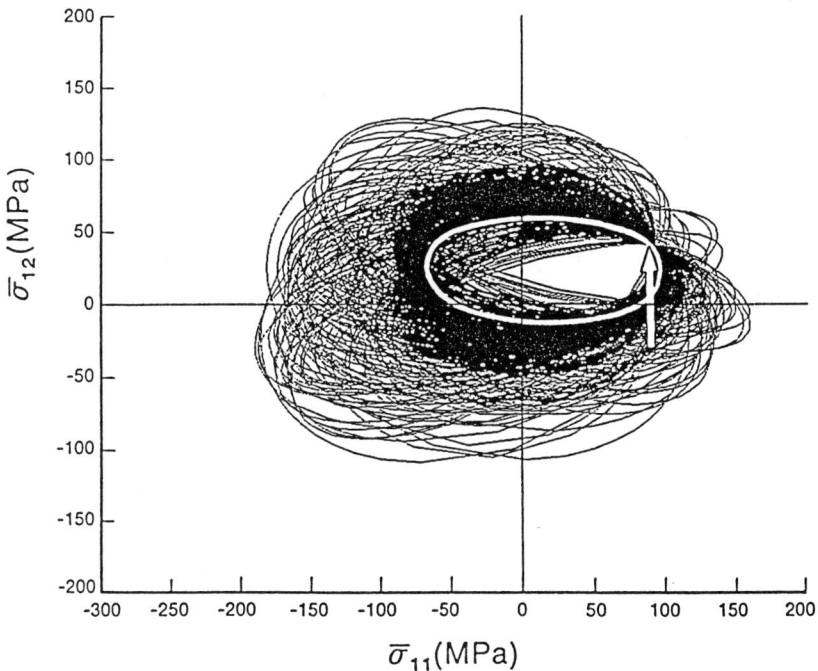

FIG. 5. Cluster of yield surfaces found at B in the overall $\bar{\sigma}_{11}\bar{\sigma}_{12}$-plane from local yield surfaces of the matrix phase. The loading path leading to B is indicated [60].

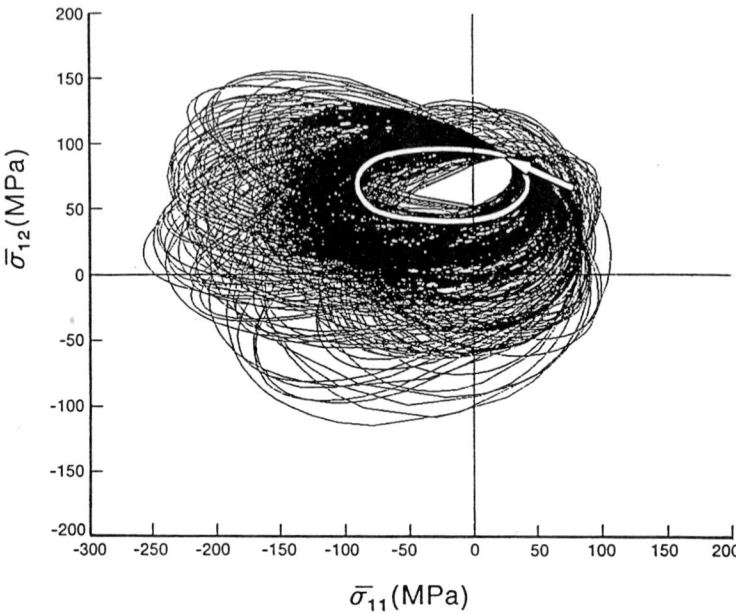

FIG. 6. Cluster of yield surfaces found at C in the overall $\bar{\sigma}_{11}\bar{\sigma}_{12}$-plane from local yield surfaces of the matrix phase [60].

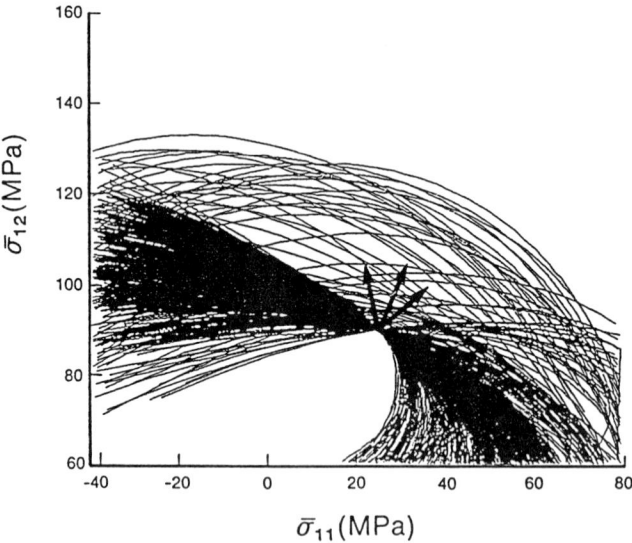

FIG. 7. Overall incremental plastic strain vector and cone of normals found at loading point C [60].

of local yield cones, with associated cones of normals that contain the plastic strain vectors.

From (75), one can read the following relation between the "radii" of local surfaces in the overall and element space,

$$(d\boldsymbol{\sigma}_k - d\boldsymbol{\alpha}_k) = \mathbf{B}_k(d\bar{\boldsymbol{\sigma}} - d\bar{\boldsymbol{\alpha}}_k). \tag{77}$$

In a kinematically hardening matrix, (70) applies to the stresses and translations in each element k. Together with (77), this specifies the direction of motion of the yield surface of each plastically deforming element k in the overall space as

$$d\bar{\boldsymbol{\alpha}}_k = d\bar{\boldsymbol{\sigma}} - \mathbf{B}_k^{-1}\, d\bar{\boldsymbol{\gamma}}_k, \tag{78}$$

or

$$d\bar{\boldsymbol{\alpha}}_k = d\bar{\boldsymbol{\sigma}}$$

for the Phillips law. This last form is particularly simple in that it does not require any information about the local stresses or translations. Of course, it still applies only to surfaces of the plastic elements; hence it does not suggest an identical translation of all surfaces in the cluster.

Under certain limited circumstances that will be described in the sequel, the subdivision of the microstructure may be restricted to the two phases and the overall surface found from (77) and (78), with k replaced by the matrix subscript β, and $\mathbf{B}_\beta$ taken from the elastic estimates in Section II.B. However, an indiscriminate application of such a coarse subdivision of the representative volume may produce entirely misleading estimates of overall yield surfaces and hardening.

2. PLASTIC CONSTITUTIVE RELATIONS

An examination of the constitutive equations for the inelastic phase, Section III.A.2, suggests that they can be reduced to the form

$$d\boldsymbol{\sigma}(\mathbf{x}) = \mathscr{L}_r[\boldsymbol{\varepsilon}(\mathbf{x}) - \boldsymbol{\beta}(\mathbf{x}), H(\mathbf{x})]\, d\boldsymbol{\varepsilon}(\mathbf{x}) + \boldsymbol{\ell}_r[H(\mathbf{x})]\, d\theta,$$

$$d\boldsymbol{\varepsilon}(\mathbf{x}) = \mathscr{M}_r[\boldsymbol{\sigma}(\mathbf{x}) - \boldsymbol{\alpha}(\mathbf{x}), H(\mathbf{x})]\, d\boldsymbol{\sigma}(\mathbf{x}) + \boldsymbol{m}_r[H(\mathbf{x})]\, d\theta, \tag{79}$$

where $\mathscr{M}_r[\boldsymbol{\sigma}(\mathbf{x}) - \boldsymbol{\alpha}(\mathbf{x}), H(\mathbf{x})]$ represents the instantaneous compliance of the inelastic phase, which can be constructed from (64), (69), and (74). The $\mathscr{L}_r[\boldsymbol{\varepsilon}(\mathbf{x}) - \boldsymbol{\beta}(\mathbf{x}), H(\mathbf{x})]$ is the instantaneous stiffness of the phase that follows from an analogous strain-space formulation. The vectors $\boldsymbol{\ell}_r$, $\boldsymbol{m}_r$ are composed from the temperature-dependent yield stress terms in (74), and from the thermal vectors $\mathbf{l}_r$, $\mathbf{m}_r$ taken from (1). These forms must replace (1) in the inelastic phase.

The phase volume averages (2) now are·

$$d\sigma_r = \frac{1}{V_r} \int_{V_r} \{\mathscr{L}_r[\varepsilon(\mathbf{x}) - \boldsymbol{\beta}(\mathbf{x}), H(\mathbf{x})] \, d\varepsilon(\mathbf{x}) + \boldsymbol{\ell}_r[H(\mathbf{x})] \, d\theta\} \, dV_r,$$

$$\tag{80}$$

$$d\varepsilon_r = \frac{1}{V_r} \int_{V_r} \{\mathscr{M}_r[\sigma(\mathbf{x}) - \boldsymbol{\alpha}(\mathbf{x}), H(\mathbf{x})] \, d\sigma(\mathbf{x}) + \boldsymbol{m}_r[H(\mathbf{x})] \, d\theta\} \, dV_r.$$

These equations replace (3), but they do not reduce to that simple form. Unless the local deformation is uniform, one cannot write the phase constitutive relations in terms of phase averages of stress and strain, because these are not connected by an instantaneous phase stiffness or compliance. Indeed, no such phase properties exist in the average sense. The phase constitutive relations must be satisfied pointwise, and this can be achieved only through integration of (79) in the phase volume along the prescribed overall stress or strain path. The fields cannot be found exactly for an actual or model geometry. An approximate evaluation is most conveniently done with the finite-element method, as indicated in the previous example.

Of course, there is a strong temptation to circumvent the difficulty associated with evaluation of the local fields. Typically, an attempt is made to reduce the actual phase constitutive relations (79) to the form

$$d\sigma_r \doteq \mathscr{L}_r(\varepsilon_r, H_r) \, d\varepsilon_r + \boldsymbol{\ell}_r(H_r) \, d\theta, \tag{81}$$

$$d\varepsilon_r \doteq \mathscr{M}_r(\sigma_r, H_r) \, d\sigma_r + \boldsymbol{m}_r(H_r) \, d\theta, \tag{81'}$$

where the instantaneous $\mathscr{L}_r$, $\boldsymbol{\ell}_r$, and $\mathscr{M}_r$, $\boldsymbol{m}_r$ relate the averages of phase stress and strain fields, as if these fields were always uniform. Such assumptions seem to be justified in modeling of the instantaneous properties of individual grains in elastic–plastic polycrystals [63]. Models of this kind regard the polycrystal as a multiphase medium, consisting of many differently oriented but otherwise identical grains, where it seems appropriate to view the deformation field as piecewise uniform. However, this view becomes untenable in two-phase systems where both the geometry and the elastic and elastic–plastic properties of the phases are entirely different. Usually, only the matrix phase deforms plastically, hence the phase properties grow even further apart in the plastic range.

A typical application of (81) is in variants of the self-consistent or Mori–Tanaka methods in plasticity. The schemes that have been proposed often employ composite cylinder or sphere inclusions in an effective medium, or other adjustments of geometry designed to improve the estimate of the local field in the matrix. We recall that in elastic analysis, these methods derive their legitimacy from proofs that show that their predictions of overall properties are bracketed by rigorous bounds. Such proofs are not available in plasticity. Therefore, there is no assurance that the estimates are reliable. Errors introduced in this manner are easily compounded by integration of the incremental forms along a prescribed overall loading path. Additional problems may be introduced by a loss of consistency and/or symmetry of the estimates. Therefore, the approximation (81) should be either avoided altogether or used only in situations where it may be indicated by comparisons with experiments and more accurate micromechanical models. In two-phase fiber composites, such situations arise in the fiber-dominated deformation of the bimodal plasticity theory discussed in the following.

3. EXACT RESULTS

In most situations, the overall response of an inelastic composite is related to the nonuniform local fields, and those must be evaluated from an analysis of a specific micromechanical model. Such models are often based on simplified assumptions that may or may not be admissible in plasticity of heterogeneous media. To assist in model formulation, we now present certain exact results that are independent of the choice of model, and therefore, must be satisfied by all admissible models. They refer to overall and local plastic strains, to thermal hardening caused by a uniform change in temperature, and to normality and convexity of the overall yield surfaces. Except as noted, the results apply not only to two-phase composites, but to representative volumes of all inelastic heterogeneous media under homogeneous boundary conditions.

a. Local and Overall Plastic Strains

Suppose that an initially stress-free aggregate is subjected to a load cycle that starts and terminates at $\bar{\sigma} = 0$, but reaches some intermediate level $\bar{\sigma} = \bar{\sigma}^0$ such that one or more phases undergoes plastic deformation. At each overall stress $\bar{\sigma}^0$, the overall plastic strain is defined as the difference between the total overall strain and the strain caused by purely elastic unloading by $-\bar{\sigma}^0$. As long as the local plastic strains are regarded in the usual continuum sense envisioned, for example by (79), a similar definition applies to each homogeneously stressed and strained material point of the inelastic phases. The local strain field in the inelastic phases consists of an elastic and plastic part, and the latter may be written as a sum of the phase volume average $\bar{\varepsilon}_r^p$ and a variable plastic field $\tilde{\varepsilon}_r^p(\mathbf{x})$. If $\sigma(\mathbf{x})$ is the local stress under current uniform overall stress, the phase strain field is

$$\varepsilon_r(\mathbf{x}) = \mathbf{M}_r \sigma_r(\mathbf{x}) + \bar{\varepsilon}_r^p + \tilde{\varepsilon}_r^p(\mathbf{x}). \tag{82}$$

After complete elastic unloading to zero overall stress, these strains become

$$\varepsilon_r(\mathbf{x}) = \mathbf{M}_r[\sigma_r(\mathbf{x}) - \mathbf{B}_r(\mathbf{x})\bar{\sigma}^0] + \bar{\varepsilon}_r^p + \tilde{\varepsilon}_r^p(\mathbf{x}), \tag{83}$$

where $\mathbf{B}_r(\mathbf{x})$ is related to the overall stress by the elastic influence function in (7), and $\mathbf{M}_r$ is the elastic phase compliance. The stress

$$\sigma_r^\rho(\mathbf{x}) = \sigma_r(\mathbf{x}) - \mathbf{B}_r(\mathbf{x})\bar{\sigma}^0 \tag{84}$$

is the elastic stress field associated with the residual strain, which together with the plastic strain $\varepsilon^P(\mathbf{x})$ creates a compatible field in the unloaded composite. The variable part of the plastic strain field has the property

$$\frac{1}{V_r} \int_{V_r} \tilde{\varepsilon}_r^p(\mathbf{x})\, dV = \mathbf{0} \Rightarrow \frac{1}{V} \int_V \tilde{\varepsilon}_r^p(\mathbf{x})\, dV = \mathbf{0}.$$

The phase volume average of the field (82) then is

$$\varepsilon_r = \mathbf{M}_r\sigma_r + \bar{\varepsilon}_r^p \quad \text{where} \quad \bar{\varepsilon}_r^p = \frac{1}{V_r} \int_{V_r} [\varepsilon_r(\mathbf{x}) - \mathbf{M}_r\sigma_r(\mathbf{x})]\, dV. \tag{85}$$

The overall plastic strain $\bar{\varepsilon}^P$ is defined as the volume average of the local strains (83) after complete elastic unloading from $\bar{\sigma}^0$:

$$\bar{\varepsilon}^P = \frac{1}{V} \int_V [\mathbf{M}_r\sigma_r^\rho(\mathbf{x}) + \bar{\varepsilon}_r^p]\, dV. \tag{86}$$

In the alternative strain space formulation (64'), the overall strain $\bar{\varepsilon}$ is regarded as the independent variable. At current $\bar{\varepsilon}$, the local stress fields are

$$\sigma_r(\mathbf{x}) = \mathbf{L}_r\varepsilon_r(\mathbf{x}) - \bar{\sigma}_r^R - \tilde{\sigma}_r^R(\mathbf{x}) \tag{87}$$

and after elastic unloading leading to $\bar{\varepsilon} = \mathbf{0}$:

$$\sigma_r(\mathbf{x}) = \mathbf{L}_r[\varepsilon_r(\mathbf{x}) - \mathbf{A}_r(\mathbf{x})\bar{\varepsilon}] - \bar{\sigma}_r^R - \tilde{\sigma}_r^R(\mathbf{x}) = \mathbf{L}_r\varepsilon_r^\rho(\mathbf{x}) - \bar{\sigma}_r^R - \tilde{\sigma}_r^R(\mathbf{x}). \tag{88}$$

The variable part of the phase relaxation stress $\tilde{\sigma}_r^R(\mathbf{x})$ has the property

$$\frac{1}{V_r} \int_{V_r} \tilde{\sigma}_r^R(\mathbf{x})\, dV = \mathbf{0} \Rightarrow \frac{1}{V} \int_V \tilde{\sigma}_r^R(\mathbf{x})\, dV = \mathbf{0}, \tag{89}$$

and the average of the field (88) thus becomes

$$\sigma_r = \mathbf{L}_r\varepsilon_r - \bar{\sigma}_r^R = \mathbf{L}_r(\varepsilon_r - \bar{\varepsilon}_r^p). \tag{90}$$

The overall relaxation stress follows as

$$\bar{\sigma}^R = \frac{1}{V} \int_V [\mathbf{L}_r\varepsilon_r^\rho(\mathbf{x}) + \bar{\sigma}_r^R]\, dV \tag{91}$$

where

$$\bar{\sigma}_r^R = \frac{1}{V_r} \int_{V_r} [\mathbf{L}_r\varepsilon_r(\mathbf{x}) - \sigma_r(\mathbf{x})]\, dV.$$

The variable parts of the local plastic strain and relaxation stress fields do not contribute to the overall strain or stress. Only the averages of the fields and the residual elastic fields make such contributions. However, the variable parts, together with the applied elastic fields, may have a very significant influence on the magnitude of plastic deformation which actually takes place in the phases, and thus on the magnitude of the averages. For example, Dvorak *et al.* [50, 62] show that extensive matrix-dominated plastic yielding may occur in an actual system under overall stress which would not satisfy a macroscopic yield condition based on average phase stresses.

We now proceed to derive a quantitative relationship between local and overall plastic strains that does not involve the residual field. A representative volume of a multiphase composite material under uniform overall stress or strain is again considered. The virtual work theorem is used; it states that the integral over a representative volume

$$\frac{1}{V} \int_V \boldsymbol{\sigma}'(\mathbf{x}) \cdot \boldsymbol{\varepsilon}''(\mathbf{x}) \, dV = \bar{\boldsymbol{\sigma}}' \cdot \bar{\boldsymbol{\varepsilon}}'' \tag{92}$$

where $\boldsymbol{\sigma}'(\mathbf{x})$ is any stress field that satisfies equations of equilibrium, with volume average (6) equal to $\bar{\boldsymbol{\sigma}}'$. Similarly, $\boldsymbol{\varepsilon}''(\mathbf{x})$ is any strain field derivable from continuous displacements, with volume averaging (6') equal to $\bar{\boldsymbol{\varepsilon}}''$ [1].

The theorem is first applied to the elastic residual stress field in (84) and to the purely elastic strain field in a composite without plastic strains. The result is

$$\frac{1}{V} \int_V \boldsymbol{\sigma}^\rho(\mathbf{x}) \cdot \mathbf{M}_r \mathbf{B}_r(\mathbf{x}) \bar{\boldsymbol{\sigma}}'' \, dV = \bar{\boldsymbol{\sigma}}^\rho \cdot \mathbf{M} \bar{\boldsymbol{\sigma}}'' = 0 \tag{93}$$

because $\bar{\boldsymbol{\sigma}}^\rho = \mathbf{0}$.

Next, the theorem is applied to the elastic stress field in a composite without plastic strains and to the total strain field in the unloaded composite (83) and (86):

$$\frac{1}{V} \int_V \mathbf{B}_r(\mathbf{x}) \bar{\boldsymbol{\sigma}}'' \cdot [\mathbf{M}_r \boldsymbol{\sigma}^\rho(\mathbf{x}) + \boldsymbol{\varepsilon}_r^p(\mathbf{x})] \, dV = \bar{\boldsymbol{\sigma}}'' \cdot \bar{\boldsymbol{\varepsilon}}^p. \tag{94}$$

With reference to (93), and to the symmetry $\mathbf{M} = \mathbf{M}^T$, one then finds the relation between local and overall plastic strains as

$$\bar{\boldsymbol{\varepsilon}}^p = \frac{1}{V} \int_{V_p} \mathbf{B}_r^T(\mathbf{x}) \boldsymbol{\varepsilon}_r^p(\mathbf{x}) \, dV \tag{95}$$

where V_p is the part of the total volume undergoing plastic deformation, where $\boldsymbol{\varepsilon}_r^p \neq \mathbf{0}$.

This is a general result valid for a heterogeneous medium with any number of inelastic phases. Related but not identical relations have been derived by Hill [61] and Rice [67].

A dual relationship between the local and overall relaxation stresses in a composite subjected to a prescribed uniform deformation path is

$$\bar{\sigma}^R = \frac{1}{V} \int_{V_p} \mathbf{A}_r^T(\mathbf{x}) \sigma_r^R(\mathbf{x}) \, dV. \tag{96}$$

In many micromechanical models, the actual local fields are replaced by their piecewise uniform approximations. For example, the self-consistent model assumes that the local fields are uniform at each phase. In the Mori–Tanaka model and in the dilute approximation, uniform fields result in the inclusions, but not in the matrix. The unit cell models discussed in Section III.B.1 and in Section III.B.4.b, typically subdivide each phase into many finite elements with uniform or piecewise uniform strain and stress fields that are related by the elastic or inelastic phase constitutive relations. During plastic loading, each such uniformly strained subdomain will have different instantaneous properties and can be regarded as a separate inelastic phase. In practice, the fields evaluated by a finite element procedure represent the best available approximations of the actual fields.

Under such circumstances, both terms in the above integrals are constant within a certain subdomain k, and the integrals may be replaced by the sums

$$\bar{\varepsilon}^p = \sum_{k_p} c_k \mathbf{B}_k^T \bar{\varepsilon}_k^p, \tag{97}$$

$$\bar{\sigma}^R = \sum_{k_p} c_k \mathbf{A}_k^T \bar{\sigma}_k^R, \tag{97'}$$

taken over all k_p in which the local plastic fields exist. Note that only in elastically homogeneous media ($\mathbf{B}_r = \mathbf{I}$) is the overall plastic strain equal to the volume average of local plastic strains.

The above results can be expanded to include the effect of a uniform change in temperature and summarized as follows.

Under a given overall stress $\bar{\sigma}$ and temperature change θ, the total overall strain is the sum of the elastic strain, the thermal strain, and the plastic strain:

$$\bar{\varepsilon} = \mathbf{M}\bar{\sigma} + \mathbf{m}\theta + \bar{\varepsilon}^p. \tag{98}$$

Likewise, under fixed overall strain and temperature change θ, the overall stress is the sum of the following terms:

$$\bar{\sigma} = \mathbf{L}\bar{\varepsilon} + \mathbf{l}\theta - \bar{\sigma}^R \tag{99}$$

with $\bar{\varepsilon}^p$ and $\bar{\sigma}^R$ related to local plastic strain by (95) to (97).

When written as relations between increments, the above relations are analogous to (64).

b. *Constitutive Relations for Two-Phase Fibrous Composites*

For each fiber composite system, there are certain loading conditions that promote the fiber-dominated deformation mode defined in Section III.B.4.c below. In this mode, the magnitude of the variable parts of the inelastic fields in (82) and (87) seems to be negligible. If this is the case, then one may simplify the analysis of such fibrous composite systems by assuming that the phase constitutive relations are satisfied by phase volume averages of the stress and strain fields, as indicated by (81). This suggests that the actual forms (82) and (87) of the local fields are now replaced by the much simpler forms (85) and (90), respectively. In other words, both the phase plastic strains and relaxation stresses are assumed to remain uniform during fiber-dominated deformation.

Now compare (85) with the phase average of (34), and (90) with the phase average of (34'). The implication is that the $\bar{\varepsilon}_r^p$ can be identified with a uniform phase transformation strain $\mathbf{\mu}_r$, and $-\bar{\sigma}_r^R$ with $\mathbf{\lambda}_r$. Similar connections exist between the overall transformation strain $\mathbf{\mu}$ in (35) and (52) and the overall plastic strain $\bar{\varepsilon}^p$ derived from (97) and also between the overall transformation stress $\mathbf{\lambda}$ in (35') and (52') and the overall relaxation stress $-\bar{\sigma}^R$ found from (97'). Therefore, (52) can be utilized to write the following result for a two-phase fiber system:

$$\bar{\varepsilon}^p = (\mathbf{K}_\alpha - \mathbf{MN}_\alpha)\bar{\varepsilon}_\alpha^p + (\mathbf{K}_\beta - \mathbf{MN}_\beta)\bar{\varepsilon}_\beta^p,$$
$$\bar{\sigma}^R = (\mathbf{LK}_\alpha - \mathbf{N}_\alpha)\bar{\varepsilon}_\alpha^p + (\mathbf{LK}_\beta - \mathbf{N}_\beta)\bar{\varepsilon}_\beta^p,$$
(100)

which is analogous to (97), but employs the overall elastic moduli instead of the phase concentration factors. Another form of (97) follows from (53):

$$\bar{\varepsilon}^p = c_\alpha \mathbf{B}_\alpha^T \bar{\varepsilon}_\alpha^p + c_\beta \mathbf{B}_\beta^T \bar{\varepsilon}_\beta^p,$$
(101)

$$\bar{\sigma}^R = c_\alpha \mathbf{A}_\alpha^T \bar{\sigma}_\alpha^R + c_\beta \mathbf{A}_\beta^T \bar{\sigma}_\beta^R,$$
(101')

where the relation employs again the elastic mechanical concentration factors. A comparison with the Levin [3] and Rosen–Hashin [4] formulae in (11) and (11[IV]) serves to establish an analogy between the uniform thermal and plastic strains.

If only one of the phases does yield, then the relation can be inverted and the phase average of the plastic strain field can be evaluated in terms of the overall plastic strain. Thus, if $r = \beta$ is the inelastic phase,

$$\bar{\varepsilon}_\beta^p = (\mathbf{K}_\beta - \mathbf{MN}_\beta)^{-1}\bar{\varepsilon}^p = \frac{1}{c_\beta}(\mathbf{B}_\beta^T)^{-1}\bar{\varepsilon}^p,$$
$$\bar{\sigma}_\beta^R = \mathbf{L}_\beta(\mathbf{LK}_\beta - \mathbf{N}_\beta)^{-1}\bar{\sigma}^R = \frac{1}{c_\beta}(\mathbf{A}_\beta^T)^{-1}\bar{\sigma}^R,$$
(102)

providing that the inverses exist.

When the local strains are assumed to be uniform in each of the phases, equations (100) and (101) relate the local and overall plastic strains in a two-phase fibrous composite.

We now derive *relations between local and overall total strains*. For the elastic composite, such relations are given by (8). Here we include the effect of a uniform temperature change θ in a similar way, but consider the plastic strains as separate eigenstrains, which by assumption are now uniform in the two phases. Recall that the effect of phase eigenstrains on the total phase strain or stress was evaluated in (58) or (59), respectively. With regard to the above connections between the uniform phase eigenstrains and plastic strains, or phase eigenstresses and relaxation stresses, and with the thermal strains taken from (8), the local and overall quantities are related as follows:

Under applied overall strain and temperature change:

$$\boldsymbol{\varepsilon}_r = \mathbf{A}_r \bar{\boldsymbol{\varepsilon}} + \mathbf{a}_r \theta + \mathbf{D}_{r\alpha} \bar{\boldsymbol{\varepsilon}}_\alpha^p + \mathbf{D}_{r\beta} \bar{\boldsymbol{\varepsilon}}_\beta^p. \tag{103}$$

Under a uniform overall stress and temperature change:

$$\boldsymbol{\sigma}_r = \mathbf{B}_r \bar{\boldsymbol{\sigma}} + \mathbf{b}_r \theta - \mathbf{F}_{r\alpha} \bar{\boldsymbol{\sigma}}_\alpha^R - \mathbf{F}_{r\beta} \bar{\boldsymbol{\sigma}}_\beta^R, \tag{104}$$

where the transformation concentration factors are given by (54) and (56). While both the overall $\bar{\boldsymbol{\varepsilon}}$ or $\bar{\boldsymbol{\sigma}}$, and the temperature change θ may have contributed to the local plastic strains, the two leading terms in the equations represent the elastic contribution to local averages; they may or may not be equal to the overall thermomechanical loads that produced the plastic strains. The relations are valid after a partial or complete unloading and in any circumstances where $\bar{\boldsymbol{\varepsilon}}$, $\bar{\boldsymbol{\sigma}}$, and θ cause no further plastic flow.

The average phase stress, which corresponds to the strain (103), follows from (34). When all contributions are included, one finds the local stresses in a constrained composite as:

$$\boldsymbol{\sigma}_r = \mathbf{L}_r \boldsymbol{\varepsilon}_r + \mathbf{l}_r \theta - \bar{\boldsymbol{\sigma}}_r^R = \mathbf{L}_r (\boldsymbol{\varepsilon}_r - \bar{\boldsymbol{\varepsilon}}_r^p) + \mathbf{l}_r \theta. \tag{105}$$

Alternatively, the constrained composite may be viewed as medium loaded by the overall stress $\bar{\boldsymbol{\sigma}}$, given by (99). Then, (104) and (101′) may be utilized to find another form of (105):

$$\boldsymbol{\sigma}_r = \mathbf{B}_r \mathbf{L} \bar{\boldsymbol{\varepsilon}} + (\mathbf{b}_r + \mathbf{B}_r \mathbf{l}) \theta + (c_\alpha \mathbf{B}_r \mathbf{A}_\alpha^T - \mathbf{F}_{r\alpha}) \bar{\boldsymbol{\sigma}}_\alpha^R + (c_\beta \mathbf{B}_r \mathbf{A}_\beta^T - \mathbf{F}_{r\beta}) \bar{\boldsymbol{\sigma}}_\beta^R. \tag{106}$$

The identity $\bar{\boldsymbol{\sigma}}_r^R = \mathbf{L}_r \bar{\boldsymbol{\varepsilon}}_r^p$ may be introduced if desired.

Next, we find the local strains in a composite under uniform overall stress, which coexist with the local stresses (104). According to (34), the result is:

$$\boldsymbol{\varepsilon}_r = \mathbf{M}_r \boldsymbol{\sigma}_r + \mathbf{m}_r \theta + \bar{\boldsymbol{\varepsilon}}_r^p. \tag{107}$$

Again, the overall uniform stress may be regarded as a consequence of the applied overall strain (98) that has the component (101′) derived from the local plastic strains. When this is utilized in (103), the local strain (107) assumes the form:

$$\varepsilon_r = \mathbf{A}_r \mathbf{M} \bar{\sigma} + (\mathbf{a}_r + \mathbf{A}_r \mathbf{m})\theta + (c_\alpha \mathbf{A}_r \mathbf{B}_\alpha^T + \mathbf{D}_{r\alpha})\bar{\varepsilon}_\alpha^p + (c_\beta \mathbf{A}_r \mathbf{B}_\beta^T + \mathbf{D}_{r\beta})\bar{\varepsilon}_\beta^p. \quad (108)$$

The above pairs of expressions for local stresses and strains, together with (9), (10), (85), and (90), provide the following general connections between the eigenstress and eigenstrain concentration factors:

$$\mathbf{L}_r \mathbf{A}_r = \mathbf{B}_r \mathbf{L}$$

$$\mathbf{F}_{rr} = \mathbf{L}_r[\mathbf{I} - c_r \mathbf{A}_r \mathbf{B}_r^T - \mathbf{D}_{rr}]\mathbf{M}_r \qquad (r = \alpha, \beta) \qquad (109)$$

$$\mathbf{F}_{r\rho} = -\mathbf{L}_r[c_\rho \mathbf{A}_r \mathbf{B}_\rho^T - \mathbf{D}_{r\rho}]\mathbf{M}_\rho \qquad (r\rho = \alpha\beta \text{ or } \beta\alpha).$$

We remark that only (100) and (102) refer specifically to fiber systems, while all the other results from (101) to (109) apply to any two-phase composite for which the various concentration factors can be found. Exact connections between the mechanical and transformation concentration factors have been derived in [40], hence it is only necessary to find the elastic $\mathbf{A}_r$ and $\mathbf{B}_r$; this should be possible for most geometries of practical interest. For completeness, we reproduce the expressions that apply to any two-phase composite:

$$\mathbf{D}_{r\alpha} = (\mathbf{I} - \mathbf{A}_r)(\mathbf{L}_\alpha - \mathbf{L}_\beta)^{-1}\mathbf{L}_\alpha, \qquad \mathbf{D}_{r\beta} = -(\mathbf{I} - \mathbf{A}_r)(\mathbf{L}_\alpha - \mathbf{L}_\beta)^{-1}\mathbf{L}_\beta,$$

$$\mathbf{F}_{r\alpha} = (\mathbf{I} - \mathbf{B}_r)(\mathbf{M}_\alpha - \mathbf{M}_\beta)^{-1}\mathbf{M}_\alpha, \qquad \mathbf{F}_{r\beta} = -(\mathbf{I} - \mathbf{B}_r)(\mathbf{M}_\alpha - \mathbf{M}_\beta)^{-1}\mathbf{M}_\beta. \quad (110)$$

For a fiber composite, these can be shown to represent another form of (54).

We now proceed to derive *macroscopic constitutive relations* for those two-phase composites that admit the approximate relations (81) for the phases. To this end, we utilize (81′):

$$d\varepsilon_r = \mathcal{M}_r \, d\sigma_r + \mathbf{m}_r \, d\theta \qquad (81')$$

and the incremental forms of (104) and (107):

$$d\sigma_r = \mathbf{B}_r \, d\bar{\sigma} + \mathbf{b}_r \, d\theta - \mathbf{F}_{r\alpha}\mathbf{L}_\alpha \, d\bar{\varepsilon}_\alpha^p - \mathbf{F}_{r\beta}\mathbf{L}_\beta \, d\bar{\varepsilon}_\beta^p \qquad (111)$$

$$d\varepsilon_r = \mathbf{M}_r \, d\sigma_r + \mathbf{m}_r \, d\theta + d\varepsilon_r^p. \qquad (112)$$

From the first and third equation, one finds the average phase plastic strain:

$$d\bar{\varepsilon}_r^p = (\mathcal{M}_r - \mathbf{M}_r) \, d\sigma_r + (\mathbf{m}_r - \mathbf{m}_r) \, d\theta, \qquad (113)$$

which we redefine to read as

$$d\bar{\varepsilon}_r^p = \mathcal{G}_r \, d\sigma_r + \mathbf{g}_r \, d\theta.$$

This is used in (112) to eliminate the local plastic strains and to write the following two equations for the unknown local stresses:

$$d\boldsymbol{\sigma}_r = \mathbf{B}_r\, d\bar{\boldsymbol{\sigma}} + \mathbf{b}_r\, d\theta$$
$$- \mathbf{F}_{r\alpha}\mathbf{L}_\alpha(\mathscr{G}_\alpha\, d\boldsymbol{\sigma}_\alpha + \boldsymbol{g}_\alpha\, d\theta) - \mathbf{F}_{r\beta}\mathbf{L}_\beta(\mathscr{G}_\beta\, d\boldsymbol{\sigma}_\beta + \boldsymbol{g}_\beta\, d\theta) \tag{114}$$

for $r = \alpha, \beta$.

The solution is:

$$d\boldsymbol{\sigma}_r = \mathscr{B}_r\, d\bar{\boldsymbol{\sigma}} + \boldsymbol{\ell}_r\, d\theta, \tag{115}$$

where $\mathscr{B}_r$ and $\boldsymbol{\ell}_r$ are the instantaneous stress concentration factors given by

$$\mathscr{B}_\alpha = \left[\mathbf{I} - \mathbf{F}_{\alpha\alpha}\mathbf{L}_\alpha\mathscr{G}_\alpha - \frac{c_\alpha}{c_\beta}\mathbf{F}_{\alpha\beta}\mathbf{L}_\beta\mathscr{G}_\beta \right]^{-1}\left[\mathbf{B}_\alpha - \frac{1}{c_\beta}\mathbf{F}_{\alpha\beta}\mathbf{L}_\beta\mathscr{G}_\beta \right],$$

$$\boldsymbol{\ell}_\alpha = \left[\mathbf{I} - \mathbf{F}_{\alpha\alpha}\mathbf{L}_\alpha\mathscr{G}_\alpha - \frac{c_\alpha}{c_\beta}\mathbf{F}_{\alpha\beta}\mathbf{L}_\beta\mathscr{G}_\beta \right]^{-1}[\mathbf{b}_\alpha - \mathbf{F}_{\alpha\alpha}\mathbf{L}_\alpha\boldsymbol{g}_\alpha - \mathbf{F}_{\alpha\beta}\mathbf{L}_\beta\boldsymbol{g}_\beta], \tag{116}$$

$$\mathscr{B}_\beta = \left[\mathbf{I} + \mathbf{F}_{\beta\beta}\mathbf{L}_\beta\mathscr{G}_\beta - \frac{c_\beta}{c_\alpha}\mathbf{F}_{\beta\alpha}\mathbf{L}_\alpha\mathscr{G}_\alpha \right]^{-1}\left[\mathbf{B}_\beta - \frac{1}{c_\alpha}\mathbf{F}_{\beta\alpha}\mathbf{L}_\alpha\mathscr{G}_\alpha \right],$$

$$\boldsymbol{\ell}_\beta = \left[\mathbf{I} + \mathbf{F}_{\beta\beta}\mathbf{L}_\beta\mathscr{G}_\beta - \frac{c_\beta}{c_\alpha}\mathbf{F}_{\beta\alpha}\mathbf{L}_\alpha\mathscr{G}_\alpha \right]^{-1}[\mathbf{b}_\beta - \mathbf{F}_{\beta\alpha}\mathbf{L}_\alpha\boldsymbol{g}_\alpha - \mathbf{F}_{\beta\beta}\mathbf{L}_\beta\boldsymbol{g}_\beta]. \tag{117}$$

This applies when both phases experience plastic straining. If one phase remains elastic, say the α phase, then $\mathscr{G}_\alpha = \mathbf{0}$ and $\boldsymbol{g}_\alpha = \mathbf{0}$.

The strain increments in the phases now follow by substitution of the above stress increments into (81'):

$$d\boldsymbol{\varepsilon}_r = \mathscr{M}_r\,\mathscr{B}_r\, d\bar{\boldsymbol{\sigma}} + (\mathscr{M}_r\boldsymbol{\ell}_r + \boldsymbol{m}_r)\, d\theta. \tag{118}$$

Finally, the overall constitutive relation for the total strains is obtained from (9') as

$$d\bar{\boldsymbol{\varepsilon}} = \mathscr{M}d\bar{\boldsymbol{\sigma}} + \boldsymbol{m}\, d\theta \tag{119}$$

where

$$\mathscr{M} = \sum_{r=\alpha,\beta} c_r\mathscr{M}_r\mathscr{B}_r \qquad \boldsymbol{m} = \sum_{r=\alpha,\beta} c_r(\mathscr{M}_r\boldsymbol{\ell}_r + \boldsymbol{m}_r). \tag{120}$$

are analogous to the elastic forms in (10) and (11).

Another possible derivation of these results may utilize the general formula (101). Substitute (113) into (101) and write the overall instantaneous compliances as:

$$\mathscr{M} = \mathbf{M} + c_\alpha\mathbf{B}_\alpha^T\mathscr{G}_\alpha\mathscr{B}_\alpha + c_\beta\mathbf{B}_\beta^T\mathscr{G}_\beta\mathscr{B}_\beta$$
$$\boldsymbol{m} = \mathbf{m} + c_\alpha\mathbf{B}_\alpha^T(\mathscr{G}_\alpha\boldsymbol{\ell}_\alpha + \boldsymbol{g}_\alpha) + c_\beta\mathbf{B}_\beta^T(\mathscr{G}_\beta\boldsymbol{\ell}_\beta + \boldsymbol{g}_\beta), \tag{121}$$

where $\mathbf{M}$ and $\mathbf{m}$ are elastic, and the remaining terms are inelastic contributions that vanish for each elastic phase, when the particular $\mathscr{G}_r = \mathbf{0}$.

In principle, the above results may be extended to a multiphase composite medium. The only requirement is to evaluate the additional transformation concentration factor terms, up to $-\mathbf{F}_{rn}\mathbf{L}_n\varepsilon_n^p$ in (110). Once those are found, then the solution of n equations (114), if it exists, provides the instantaneous stress concentration factors for all n phases. Both (119) and (120) apply to any number of phases if additional phase terms are attached.

For a two-phase medium, the above results may be simplified. Recall that the phase transformation concentration factors are related to the mechanical concentration factors by (110). Therefore, in a two-phase medium (116) and (117) may be written without reference to the transformation concentration factors. The result may be derived in several ways. Here we recall the identities following (10) and restate them for the instantaneous concentration factors:

$$c_\alpha \mathscr{B}_\alpha + c_\beta \mathscr{B}_\beta = \mathbf{I} \qquad c_\alpha b_\alpha + c_\beta b_\beta = \mathbf{0}. \tag{122}$$

This is verified by the derivation leading to (116) and (117).

Next, compare the forms (120) and (121), and use (122) to eliminate one pair of the instantaneous concentration factors. Then, in turn, solve for the remaining instantaneous quantities to obtain

$$\mathscr{B}_\alpha = \frac{1}{c_\alpha}[\mathbf{M}_\alpha - \mathbf{M}_\beta - (\mathbf{B}_\alpha^T - \mathbf{I})\mathscr{G}_\alpha + (\mathbf{B}_\beta^T - \mathbf{I})\mathscr{G}_\beta]^{-1}[\mathbf{M} - \mathbf{M}_\beta + (\mathbf{B}_\beta^T - \mathbf{I})\mathscr{G}_\beta]$$

$$b_\alpha = \frac{1}{c_\alpha}[\mathbf{M}_\alpha - \mathbf{M}_\beta - (\mathbf{B}_\alpha^T - \mathbf{I})\mathscr{G}_\alpha + (\mathbf{B}_\beta^T - \mathbf{I})\mathscr{G}_\beta]^{-1} \tag{123}$$

$$\times [-c_\alpha(\mathbf{B}_\alpha^T - \mathbf{I})g_\alpha - c_\beta(\mathbf{B}_\beta^T - \mathbf{I})g_\beta + c_\alpha\mathbf{m}_\alpha + c_\beta\mathbf{m}_\beta - \mathbf{m}]$$

for any two-phase system. Both are symmetric with respect to the exchange of α and β subscripts, and thus (122) is satisfied. It can be verified that this agrees with (116) and (117) if (110) is taken into account.

These results represent an explicit macroscopic constitutive relation for any two-phase composite in which the response of the phases is approximated by (81). Note that no material model has been used in the derivation. The only information about the microstructure is reflected in the elastic stress and strain concentration factors $\mathbf{A}_r$ and $\mathbf{B}_r$.

In all two-phase systems, the transformation concentration factors can be expressed in terms of $\mathbf{A}_r$ and $\mathbf{B}_r$; this is indicated by (54) and (110). If this route is chosen, then the self-consistent or Mori–Tanaka models may be adopted for convenient evaluation of $\mathbf{A}_r$ and $\mathbf{B}_r$. However, if the analysis proceeds from an independent evaluation of the overall elastic properties $\mathbf{L}$

and $\mathbf{M}$, e.g., via the Hashin–Shtrikman bounds (33), the concentration factors are derived from (10). For a fiber system, the factors also follow from (53) and (56). Then, the above constitutive relation is independent of the details of microstructural geometry and, therefore, is exact for any two-phase system that admits (81).

In summary, the above procedure gives closed-form expressions for the instantaneous overall properties in terms of instantaneous phase properties and overall elastic stiffness or compliance. In contrast, other available approaches usually evaluate the instantaneous concentration factors from approximate solutions of inelastic inclusion problems that often require extensive numerical computations. The estimates of $\mathcal{M}$ are usually implicit, and the plastic strains derived from these estimates may violate (95) to (97).

c. Thermal Hardening

In an elastic composite, a uniform change in temperature will influence the local fields; this can be evaluated from (7) and (12). Even in the absence of plastic loading, such changes in local stresses and strains will affect the yield and relaxation surfaces in the overall stress or strain space. Under such circumstances, it is convenient to retain the representation of the yield surfaces suggested by (75) and (76) and to include the effect of temperature through additional parameters. For example, with reference to the illustration of local yielding in Section III.B.1, and Figs. 1 to 7, consider the effect of a small change $\Delta\theta$ on the overall yield surfaces of individual subelements k in the elastic–plastic matrix. Suppose that the overall stress, if any, is adjusted such that no plastic yielding is caused by the $\Delta\theta$. The average stress change in a subelement k is equal to $\Delta\boldsymbol{\sigma}_k = \mathbf{b}_k \Delta\theta$. There is no plastic loading, hence $d\boldsymbol{\alpha}_k = \mathbf{0}$, and $\Delta\boldsymbol{\sigma}_k$ is the only change in the local stress. Then, it follows from (77) that the translation of the overall yield surface of subelement k is equal to

$$\Delta\bar{\boldsymbol{\alpha}}_k = -\mathbf{B}_k^{-1}\Delta\boldsymbol{\sigma}_k. \tag{124}$$

The implication is that the change $\Delta\theta$ will cause a rigid body translation of all subelement yield surfaces in the overall space. This effect may be referred to as thermal hardening. Of course, it is present both in the elastic composite and during plastic deformation; in the latter case the translation in (101) must be added to the $\bar{\boldsymbol{\alpha}}_k$ during each loading step. Depending on the respective loading directions, the thermal change may accelerate or retard plastic deformation.

The result given is quite transparent, but not particularly convenient, as it involves the stress concentration factors. A more useful description follows from the decomposition procedure of Section II.C.1. Recall that this procedure admits eigenstrains in the phases and prescribes auxiliary overall stresses that make the strain field uniform in the entire aggregate, while the stresses become piecewise uniform. Moreover, if both the homogeneous and particular solutions (40) are combined, one can create a one-parameter family of such uniform fields. When each of the phases is isotropic in the transverse plane, i.e., isotropic or transversely isotropic, it is possible to adjust the parameter in such a way that the uniform auxiliary strain field is isotropic in the aggregate.

These considerations suggest the following strategy for evaluation of the effects of uniform changes in temperature on plastic deformation in composites with elastic–brittle fibers. In contrast to the example given, consider now a general plastic loading step from some current reference state in which a temperature change $d\theta$ is applied simultaneously with an overall stress $d\bar{\sigma}$. First, apply the temperature change and the auxiliary surface tractions that create the above isotropic uniform strain field. In an isotropic metal matrix that is plastically incompressible, such combined loading will cause only an isotropic stress increment but no plastic yielding; the local stress increment vector is parallel to the axis of the cylindrical local yield surface (75) or (76). Next, remove the auxiliary tractions and add the overall stress increment, if any, that has been prescribed together with the change $d\theta$. This modifies the current stress increment and the entire overall mechanical loading path. The effect of temperature is represented by a uniform and isotropic strain field together with a certain mechanical loading, which in superposition with the applied mechanical loads may cause plastic deformation of the aggregate.

This approach to thermal hardening was first outlined by Dvorak [7], without the benefit of the results in Section II.C.1. Here we utilize (40) and seek the magnitude of dS_T that guarantees an isotropic stress $d\hat{\sigma}_\beta$ in the metal matrix $(r = \beta)$, which is assumed to be elastically isotropic and plastically incompressible. The thermal strains in the phases are uniform phase eigenstrains given by

$$d\mathbf{\mu}_r = [\alpha_r, \beta_r, \beta_r, 0, 0, 0]^T \, d\theta, \tag{125}$$

where α_r and β_r denote the linear longitudinal and transverse coefficients of thermal expansion of the phases.

Substitute (125) into (40$_2$), and with the help of (42) to (46), evaluate the nonvanishing components of $d\hat{\sigma}_r$. The result is:

$$d\hat{\sigma}_1^\alpha = q(l_\alpha \Delta l - n_\alpha \Delta k) \, dS_T + qk_\alpha E_\alpha (l_\beta \Delta \alpha + 2k_\beta \Delta \beta) \, d\theta, \tag{126}$$

$$d\hat{\sigma}_1^\beta = q(l_\beta \Delta l - n_\beta \Delta k) \, dS_T + qk_\beta E_\beta (l_\alpha \Delta \alpha + 2k_\alpha \Delta \beta) \, d\theta, \tag{127}$$

$$d\hat{\sigma}_2^\alpha = d\hat{\sigma}_2^\beta = d\hat{\sigma}_3^\alpha = d\hat{\sigma}_3^\beta = dS_T. \tag{128}$$

Require that $d\hat{\sigma}_1^\beta = d\hat{\sigma}_2^\beta = d\hat{\sigma}_3^\beta$ and find

$$dS_T = \{qk_\beta E_\beta(l_\alpha\Delta\alpha + 2k_\alpha\Delta\beta)[1 - q(l_\beta\Delta l - n_\beta\Delta k)]^{-1}\}\, d\theta. \qquad (129)$$

The overall auxiliary stress $d\hat{\sigma}$ has the nonvanishing components

$$d\hat{\sigma}_1 = c_\alpha\, d\hat{\sigma}_1^\alpha + c_\beta\, d\hat{\sigma}_1^\beta = dS_A, \qquad d\hat{\sigma}_2 = d\hat{\sigma}_3 = dS_T. \qquad (130)$$

Together with $d\theta$, this overall stress creates a uniform strain field in the aggregate. It also guarantees that the stress increment in the matrix is isotropic and thus causes no yielding.

Finally, the auxiliary stress must be removed. This is a mechanical loading step that also should include the actual stress $d\bar{\sigma}$ that was prescribed together with the $d\theta$. After this step, the stress fields in the phases are

$$\begin{aligned} d\boldsymbol{\sigma}_\alpha(\mathbf{x}) &= d\hat{\boldsymbol{\sigma}}_\alpha + \mathscr{B}_\alpha(\mathbf{x})(d\bar{\boldsymbol{\sigma}} - d\hat{\boldsymbol{\sigma}}), \\ d\boldsymbol{\sigma}_\beta(\mathbf{x}) &= d\hat{\boldsymbol{\sigma}}_\beta + \mathscr{B}_\beta(\mathbf{x})(d\bar{\boldsymbol{\sigma}} - d\hat{\boldsymbol{\sigma}}). \end{aligned} \qquad (131)$$

Inasmuch as plastic straining may be caused in the matrix during this step, the $\mathscr{B}_r(\mathbf{x})$ denote the instantaneous stress influence functions (116), (117) derived for a particular model geometry from an appropriate integration of (79) along the modified loading path. The implication is that the effect of temperature on local fields can be represented by a modification of the mechanical loading path from $d\bar{\boldsymbol{\sigma}}$ to $(d\bar{\boldsymbol{\sigma}} - d\hat{\boldsymbol{\sigma}})$, where $d\hat{\boldsymbol{\sigma}}$ depends on $d\theta$ as indicated by (126) to (130).

Similar conclusions apply to evaluation of the overall strain. In particular, the dS_T in (129) is substituted into (42$_3$) and the $d\hat{\boldsymbol{\varepsilon}}_0$ into (40$_3$); this gives the overall strain $d\hat{\boldsymbol{\varepsilon}}$ caused by $d\hat{\boldsymbol{\sigma}}$ and $d\theta$. Then, the strain caused by the mechanical loading $(d\bar{\boldsymbol{\sigma}} - d\hat{\boldsymbol{\sigma}})$ is added. The result is

$$d\bar{\boldsymbol{\varepsilon}} = d\hat{\boldsymbol{\varepsilon}} + \mathscr{M}(d\bar{\boldsymbol{\sigma}} - d\hat{\boldsymbol{\sigma}}), \qquad (132)$$

where $\mathscr{M}$ is the instantaneous compliance of the aggregate (122).

During an elastic unloading step, the results (131) and (132) convert to their elastic counterparts, i.e., the $\mathscr{B}_r(\mathbf{x})$ are replaced by $\mathbf{B}_r(\mathbf{x})$, and $\mathscr{M}$ by $\mathbf{M}$.

Recall that such elastic loading by $\Delta\theta$ alone ($\Delta\bar{\boldsymbol{\sigma}} = \mathbf{0}$) was specified in the example leading to (124). We now reconstruct (124) in the following way. The change $\Delta\theta$ is applied together with $\Delta\hat{\boldsymbol{\sigma}}$, selected such that $\Delta\hat{\boldsymbol{\sigma}}_\beta$ is isotropic; i.e., (129) is used to evaluate ΔS_T. By definition, the Mises yield condition (62) or (76) does not depend on the hydrostatic stress component; hence $\Delta\hat{\boldsymbol{\sigma}}_\beta$ renders $\Delta\boldsymbol{\sigma}_k = \mathbf{0}$, and the $\Delta\theta$ is accounted for by application of the overall stress $-\Delta\hat{\boldsymbol{\sigma}}$, which does not vanish. As before, $\Delta\boldsymbol{\alpha}_k = \mathbf{0}$ because there is no plastic loading. Accordingly, (77) now becomes

$$\mathbf{0} = \mathbf{B}_k(\Delta\hat{\boldsymbol{\sigma}} - \Delta\bar{\boldsymbol{\alpha}}_k). \qquad (133)$$

Therefore, instead of (124), we now obtain

$$\Delta \bar{\pmb{\alpha}}_k = \Delta \hat{\pmb{\sigma}}. \tag{134}$$

This is equivalent to (124), but it is now apparent that during elastic deformation, a temperature change $\Delta\theta$ causes a rigid-body translation of *all* subelement yield surfaces in the overall stress space by the same amount $\Delta\hat{\pmb{\sigma}}$.

These results lead to several conclusions. First, unlike most polycrystals, composite materials may experience macroscopic plastic straining due to a sufficiently large change in temperature. The temperature change θ_Y, which will cause the onset of local yielding in an initially stress-free composite, can be evaluated from (134). The stress $\Delta\hat{\pmb{\sigma}}$ is a function of temperature; (130) suggests that in the system considered, it is an axisymmetric overall stress consisting of an axial normal stress dS_A and transverse hydrostatic stress dS_T. The temperature needed to generate a certain dS_T follows from (129), and the axial components from (126), (127). Then, if the subelement yield surface is known in the overall stress space, or if the overall yield surface is derived from physical or numerically simulated experiments, the $d\hat{\pmb{\sigma}}$ is integrated along the thermal loading path until $\theta = \theta_Y$, where $\hat{\pmb{\sigma}}$ reaches the respective yield surface. Of course, in reality the yield surface of a stress-free composite undergoes a rigid-body translation equal to its diameter in the $\hat{\pmb{\sigma}}$-direction in overall stress space. In the absence of an overall mechanical stress, the origin is the loading point, and plastic yielding starts when the overall surface comes into contact with this point. Similar conclusions apply to prestrained systems where the center of the yield surface is at some initial position distinct from the origin. A somewhat different treatment of such effects and specific examples can be found in [64–66].

In applications, it is useful to know that the effect of combined thermal and mechanical loads on plastic deformation of a fiber composite can be evaluated by prescribing only mechanical loading, but along a modified path. The uniform auxiliary fields must be added to the results. If the matrix yield stress depends on temperature, then it must be changed in each mechanical loading step that corresponds to a specific temperature change. With such adjustments, an existing model that was designed to predict overall response of a fiber composite under mechanical loading can be easily modified to accommodate both mechanical and thermal loads [7].

d. Normality and Convexity

Additional connections between local and overall behavior of composite materials can be established for certain work relations and for the direction of the plastic strain vector. The results summarized here were obtained mostly by Hill [61].

Consider the direction of the overall plastic strain vector. Suppose that the composite is subdivided into many small subelements, as in the example in Section III.B.1 above. Assume that each inelastic phase and therefore each subelement conforms with the Drucker postulate (68). Locally, the plastic strain rate vector coincides with the outside normal to the yield surface at the current local stress σ_k. Any elastic stress increment $d\sigma_k^*$ must be directed into the local surface, for example, if one takes $d\sigma^* \approx (\sigma^* - \sigma)$, then (68) indicates that

$$d\sigma_k^*(d\varepsilon_k - \mathbf{M}_r \, d\sigma_k) \leq 0. \tag{135}$$

After some algebra, this can be recast into the form:

$$d\sigma_k^* \, d\varepsilon_k - d\sigma_k \, d\varepsilon_k^* \leq 0, \tag{136}$$

where $d\sigma_k^*$ and $d\varepsilon_k^*$ are elastic, while $d\sigma_k$ and $d\varepsilon_k$ are arbitrary.

Note that each of these terms is a product of an equilibrium stress field with a strain field derivable from continuous displacements. This opens the way to an application of (92) and to the result

$$d\bar{\sigma}^* \, d\bar{\varepsilon} - d\bar{\sigma} \, d\bar{\varepsilon}^* \leq 0. \tag{137}$$

The implication is that the plastic part of the overall strain rate lies within the cone of normals associated with the yield cone at the current vertex. Figure 7 shows an example of such a configuration of local yield surfaces, where the yield cone could be inscribed as an internal envelope of the local surfaces at the current loading point.

This property does not necessarily guarantee normality to a yield surface evaluated from numerical or physical experiments, such as those shown in Figs. 3 to 6, or in Fig. 25. Indeed, such surfaces are merely loci of vertices of adjacent yield cones, and at each loading point, normality limits the plastic rate vector only to the cone of normals.

Finally, consider the relation between the products of stress and plastic strain increments. Appeal again to the stability postulate (68) in the form

$$d\sigma_k(d\varepsilon_k - \mathbf{M}_r \, d\sigma_k) \geq 0, \tag{138}$$

where the equality holds in the elastic subelements. This can be integrated over V to yield

$$d\bar{\sigma} \, d\bar{\varepsilon} \geq \frac{1}{V} \int_V d\sigma_k(\mathbf{M}_r \, d\sigma_k) \, dV. \tag{139}$$

If the actual stress field in the inelastic aggregate is regarded as an admissible field in equilibrium with $d\bar{\sigma}$, and the strain field $\mathbf{M}_r \, d\bar{\sigma}_k$ compatible with the strain $\mathbf{M} \, d\bar{\sigma}$, then the principle of minimum complementary energy in elasticity suggests that when $\mathbf{M}_r$ is positive-definite

$$d\bar{\sigma}(\mathbf{M}\,d\bar{\sigma}) < \frac{1}{V}\int_V d\sigma_k(\mathbf{M}_r\,d\sigma_k)\,dV, \tag{140}$$

where the left-hand side represents the energy of the actual field.

From the last two equations and from (92), one obtains the inequalities

$$d\bar{\sigma}(d\bar{\varepsilon} - \mathbf{M}\,d\bar{\sigma}) > \frac{1}{V}\int_V d\sigma_k(d\varepsilon_k - \mathbf{M}_r\,d\sigma_k)\,dV \geq 0, \tag{141}$$

under changing overall load $d\bar{\sigma} \neq \mathbf{0}$. The sharp inequality guarantees that the overall plastic strain rate may vanish only in the absence of all local rates. It also suggests that in the absence of local strain hardening, when $d\sigma_r = \mathbf{0}$ everywhere, the overall load must still increase. This is sometimes referred to as constraint hardening, since it arises from mechanical interactions between the phases.

Many additional aspects of the general structure of plasticity of heterogeneous media were discussed by Hill and Rice [*69, 70*] and Rice [*67, 68*].

4. MICROMECHANICAL MODELS

The discussion of elasticity and plasticity of heterogeneous media suggests that overall elastic properties can be estimated or bounded with relative ease and that the simplicity of elastic analysis is derived from the homogeneity of the phases during deformation. The elastic properties are known constants, hence volume averaging (2) of local fields (7) gives the relations (8), which provide the desired result (10). The estimates of $\mathbf{A}_r$ or $\mathbf{B}_r$ then follow from well-known approximate solutions of elastic inclusion problems, such as the self-consistent or Mori–Tanaka methods. Direct evaluation of bounds on elastic moduli can be made according to Section II.B.4.

When at least one phase deforms plastically, its homogeneity is lost. Local instantaneous stiffnesses and compliances are no longer known constants. Instead, they depend on current stress and on past history of plastic deformation, and as such they are functions of local coordinates. Although Sections III.B.2 and III.B.3 present many exact results, they do not address the influence of specific microstructural geometry on overall behavior. This can be accomplished only through evaluation of local fields along an incremental loading path. When the approximation (81) is no longer acceptable, evaluation of local fields becomes a necessary part of inelastic modeling of composite materials.

Before the advent of micromechanics, problems of this kind were sometimes approached by introduction of certain assumptions about the overall behavior of fibrous composites that were motivated by micromechanical considerations. For example, in the model by Mulhern *et al.* [*51*] and Spencer [*71*], the fibers were taken to be inextensible and the composite plastically incompressible; and with these restrictions, the aggregate was regarded as a homogeneous medium. The early micromechanical models often employed simplified phase geometries in order to introduce actual mechanical properties of the phases. A natural goal was to adjust the geometry of the microstructure in such a way that the local fields became piecewise uniform. This was accomplished, for example, by the self-consistent method [*72*], or by assuming specific simple geometries of the matrix and fiber [*73–76*].

Each of these approximate models had certain advantages as well as drawbacks, but at least one of the early models, the vanishing-fiber-diameter model of Dvorak and Bahei-El-Din [*74*] permitted a simplified analysis of metal matrix composite structures. This model reduces the effect of the fiber to a unidirectional elastic constraint on the elastic–plastic matrix that is otherwise free to deform uniformly under uniform overall stress or strain. The model will not be reviewed here, but it is useful to recall its extension to plasticity of laminated plates [*75*], its implementation in a general purpose finite element program for structural analysis [*77*], and applications in the evaluation of stress fields at holes and notches in laminated composite plates [*78, 79*]. More recent extensions to thermoplasticity were discussed by Bahei-El-Din [*58*] and structural analysis applications in [*80*]. A strain–space form of the model [*56*] has been recently implemented into the ABAQUS program.

As an example of more recent work, we now describe two different approaches to plasticity analysis of fibrous composite media; one that provides bounds on certain instantaneous overall properties together with estimates of local fields, and one that takes advantage of certain newly recognized deformation mechanisms of fibrous composites.

a.   Bounds on Overall Instantaneous Properties

Models of this kind typically utilize a particular geometry of the fibrous medium and derive bounds on instantaneous overall properties from estimates of local fields and minimum principles of plasticity [*44, 81–83*]. Here we briefly summarize the derivation introduced by Dvorak and Teply [*84, 85*]; an analogous model was later described by Accorsi and Nemat-Nasser [*86*].

In the minimum principle of plasticity for strain rates, one considers a representative volume V of a composite material that is subjected to overall uniform strain, applied along an incremental path that leads to the current strain point $\bar{\varepsilon}$. It is assumed that the current state is represented by known actual local stress and strain fields in the phases and that these fields satisfy the local constitutive equations, which are also assumed as known in the entire representative volume V at the current point $\bar{\varepsilon}$ of the overall deformation path. At this current reference state, a uniform strain increment $d\bar{\varepsilon}$ is applied through certain displacements $d\bar{u}$, prescribed on the entire surface S of V. This creates local as well as overall stress changes that need to be determined. An exact solution is often beyond reach, but an approximate solution can be found with the help of suitably chosen trial functions. A kinematically admissible field $d\varepsilon^*$ derived from a continuous velocity field is selected in V such that its volume average $\{d\varepsilon^*\}_V = d\bar{\varepsilon}^* = d\bar{\varepsilon}$, is compatible with the surface displacements $d\bar{u}^* = d\bar{u}$ on $S = S_u$. This field is also supposed to satisfy the actual local constitutive equations in the current reference state, so that a certain stress field $d\sigma^*$ can be found from the $d\varepsilon^*$. Under such circumstances, the fields $d\sigma^*$, $d\varepsilon^*$, and du^* represent a kinematically admissible set. The energy changes that would occur in the actual state (no asterisks), and those that take place in the admissible state are related through the minumum principle for strain rates

$$\int_V \tfrac{1}{2} d\varepsilon \, d\sigma \, dV - \int_{S_p} d\bar{p} \, d\bar{u} \, dS \le \int_V \tfrac{1}{2} d\varepsilon^* \, d\sigma^* \, dV - \int_{S_p} d\bar{p} \, d\bar{u}^* \, dS, \quad (142)$$

where $S_p = S - S_u$. Body forces are neglected. Recall that displacements are prescribed on the entire surface $S = S_u$, hence the surface integrals vanish. The volume integrals can be written in terms of volume averages and work averages $\bar{\sigma}\bar{\varepsilon} = \{\sigma\varepsilon\}_V$. The result is

$$\tfrac{1}{2} d\bar{\varepsilon}\mathbf{L} \, d\bar{\varepsilon} \le \tfrac{1}{2} d\bar{\varepsilon}\mathbf{L}_\varepsilon \, d\bar{\varepsilon}. \quad (143)$$

If one prescribes the boundary conditions in terms of surface tractions rather than displacements, such that $d\bar{\sigma}^* = d\bar{\sigma}$ and $d\bar{p}^* = d\bar{p}$ on $S = S_p$, then one can obtain the inequality

$$-\tfrac{1}{2} d\bar{\sigma}\mathbf{M} \, d\bar{\sigma} \le -\tfrac{1}{2} d\bar{\sigma}\mathbf{M}_\varepsilon \, d\bar{\sigma}. \quad (144)$$

The $\mathbf{L}$ and $\mathbf{M}$ represent the actual instantaneous overall stiffness and compliance. The $\mathbf{L}_\varepsilon$ and $\mathbf{M}_\varepsilon$ are their approximations computed from the admissible field $d\bar{\varepsilon}^*$.

In a similar way, one can specify a statically admissible stress field in the domain and use the minimum principle for stress rates to obtain lower bounds on energy rates. The final result can be summarized as

$$d\bar{\varepsilon}\mathbf{L}_\varepsilon \, d\bar{\varepsilon} \ge d\bar{\varepsilon}\mathbf{L} \, d\bar{\varepsilon} \ge d\bar{\varepsilon}\mathbf{L}_\sigma \, d\bar{\varepsilon} \ge 0 \quad (145)$$

and

$$d\bar{\sigma}\mathbf{M}_\sigma \, d\bar{\sigma} \ge d\bar{\sigma}\mathbf{M} \, d\bar{\sigma} \ge d\bar{\sigma}\mathbf{M}_\varepsilon \, d\bar{\sigma} \ge 0, \quad (146)$$

where $\mathbf{L}_\sigma$ and $\mathbf{M}_\sigma$ are the approximate values of instantaneous overall properties computed from the admissible stress field.

Since each of these terms is a positive-definite quadratic form, all the $\mathbf{L}$ and $\mathbf{M}$ matrices are also positive-definite. This property leads to the inequalities

$$d\bar{\varepsilon}(\mathbf{L}_\varepsilon - \mathbf{L}) \, d\bar{\varepsilon} \geq 0, \qquad d\bar{\varepsilon}(\mathbf{L} - \mathbf{L}_\sigma) \, d\bar{\varepsilon} \geq 0, \tag{147}$$

where the equality signs apply only to the exact solution. Let

$$\mathbf{L}_\varepsilon - \mathbf{L} = \mathbf{H}, \qquad \mathbf{L} - \mathbf{L}_\sigma = \mathbf{D}, \tag{148}$$

and observe that all these matrices are positive-definite. In typical applications, these are (6×6) matrices with six eigenvalues, diagonal terms, and leading principal minors that are all positive. If the eigenvalues are arranged in nonincreasing order, then according to the monotonicity theorem for eigenvalues of symmetric matrices, the ordered eigenvalues of the above matrices, denoted here by lower-case kernel letters, are related by

$$l_i^\varepsilon \geq l_i + h, \qquad l_i \geq l_i^\sigma + d, \tag{149}$$

where

$$h_1 \geq h \geq h_n, \qquad d_1 \geq d \geq d_n,$$

and the ordered eigenvalues satisfy the inequalities

$$l_1^\varepsilon \geq l_2^\varepsilon \geq \cdots \geq l_6^\varepsilon > 0, \qquad h_1 \geq h_2 \geq \cdots \geq h_6 > 0, \text{ etc.}$$

This indicates that $h_n > 0$, $d_n > 0$, hence it follows that

$$l_i^\varepsilon \geq l_i \geq l_i^\sigma. \tag{150}$$

If the ordered eigenvalues are arranged in a nonincreasing order as diagonal terms, then one can write the bounds as

$$\text{diag } l_i^\varepsilon \geq \text{diag } l_i \geq \text{diag } l_i^\sigma, \quad i = 1, 2, \dots 6. \tag{151}$$

A similar procedure can be applied to (146) to find analogous bounds on the ordered eigenvalues of $\mathbf{M}$,

$$\text{diag } m_i^\sigma \geq \text{diag } m_i \geq \text{diag } m_i^\varepsilon. \tag{152}$$

One also can find bounds on the diagonal terms of $\mathbf{L}$ and $\mathbf{M}$. Recall that these terms are also positive, and it follows from (147) that

$$L_{kk}^\varepsilon \geq L_{kk} \geq L_{kk}^\sigma, \qquad M_{kk}^\sigma \geq M_{kk} \geq M_{kk}^\varepsilon, \quad \text{(no sum on } k\text{).} \tag{153}$$

However, no close bounds can be found for the off-diagonal terms of $\mathbf{L}$ and $\mathbf{M}$. That can be done only on the basis of known connections between the terms, such as those that exist, for example, in transversely isotropic and other elastic media.

If the bounds are used in an incremental evaluation of the overall properties of an elastic–plastic composite, e.g., by a finite-element analysis of the representative volume, then they must be qualified in the following way. We recall that the minimum principles compare the energy changes of the actual and admissible states from a current reference state that corresponds to the actual solution of the problem. This is assumed to be known, together with the actual local properties. However, in an incremental numerical solution, this condition is not met. Except in the elastic state, the local properties in each current state are not known exactly, they are known only in terms of the finite element approximations. The current state does not represent an admissible set, and it is not possible to find admissible local fields for the next plastic step. Therefore, the bounds do not apply to the actual composite system. Instead, they apply to a system in which the local properties have been replaced by those computed, say, in the finite element solution.

In a typical implementation, approximate upper bounds are computed from the displacement formulation of the finite element method, and approximate lower bounds are obtained from the hybrid formulation. Thus one follows a sequence of upper-bound solutions that also approximate the local properties, or a sequence of lower-bound solutions that give different approximations of the local properties. It remains to be established if either procedure is convergent.

b. Periodic Array Models

Evaluation of bounds on instantaneous overall properties of composite aggregates is best performed for a specific model material that approximates the microstructural geometry of the actual system. Although most microstructures are random, periodic distributions of the reinforcement are often used in the development of model materials. The advantage of this approach is that it may divide the composite into identical unit cells of smallest possible size, which are repeated throughout the representative volume. The cells are chosen so that one can prescribe for each cell certain periodic boundary conditions that correspond to uniform overall strain or stress states. The periodic representation appears to be justified in composites reinforced by continuous large-diameter ($\sim 150 \ \mu$m) fibers, such as boron or silicon carbide, in which a nearly periodic distribution of the fibers is assured in manufacture. Models of this type are also useful in other fibrous systems, and in particulate composites, particularly at higher volume concentrations when interaction between phases becomes significant.

The model described here is the periodic hexagonal array (PHA) model [*84, 85*]. Figure 8 shows the typical cross section of the model material in

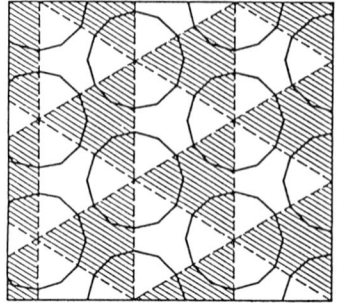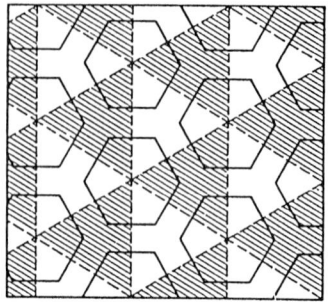

FIG. 8.   Transverse cross-sections of periodic hexagonal array models of fibrous composites with hexagonal and dodecagonal cylindrical fibers. Reprinted with permission from *J. Mech. Phys. Solids* **36**, 29, J. L. Teply and G. J. Dvorak, © 1988, Pergamon Press plc.

the transverse plane. The medium consists of a matrix reinforced by aligned fibers of identical cross section, which is approximated by a regular $n \times 6$-sided polygon. As shown in the figure, the microstructure is divided into two sets of identical triangular prisms, with vertices in the fiber axes; these are selected as the unit cells.

Under uniform overall stress or strain, the deformation of the composite can be compared to that of a homogeneous effective medium that has the same overall properties. In particular, it is possible to identify a set of *contact points* in both the periodic and effective media, such that these points undergo exactly the same displacements when either medium is subjected to a given uniform overall deformation. For example, the previous choice of unit cells suggests that to each point x_0 in a given cell, there correspond points x in all other cells of the same (shaded or unshaded) type, such that the local stresses and strains are equal at all such points. In the x_i system of Fig. 9, the coordinates of x are

$$x = x_0 + c, \tag{154}$$

where $c = c/2(i\sqrt{3},\ j,\ 2s/c)$ and the values of i and j must be selected by an appropriate combination of the following:

$i = 0, j = \pm 2n,$ for translation parallel to x_2

$i = j = \pm n,$ for translation parallel to $V_1 V_3$ direction

$i = -j = \pm n,$ for translation parallel to $V_1 V_2$ direction,

where n is an integer. Clearly, the relative displacements of all points (154) must be identical to those of the effective homogeneous medium, such points represent one set of possible contact points. Note that the fiber centers belong to this set.

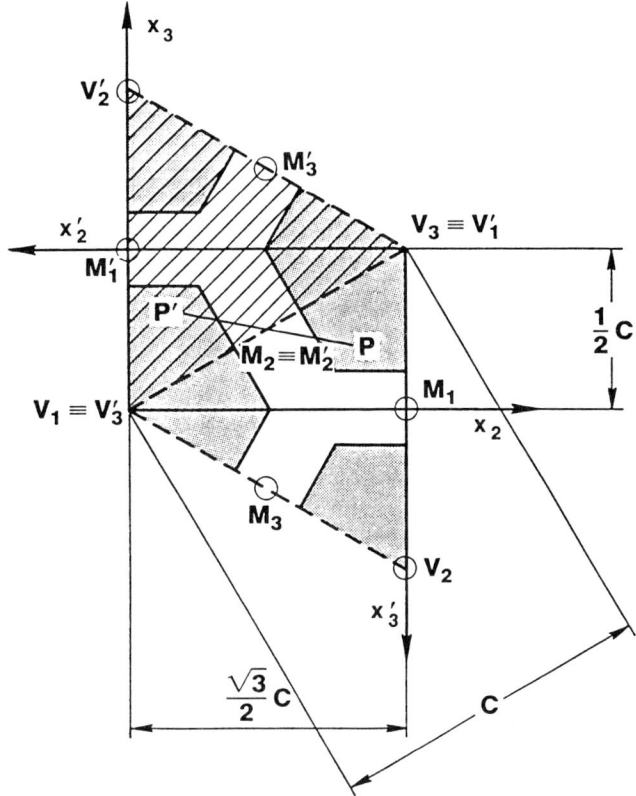

FIG. 9. Two adjacent unit cells and their local coordinate systems. Reprinted with permission from *J. Mech. Phys. Solids* **36**, 29, J. L. Teply and G. J. Dvorak, © 1988, Pergamon Press plc.

Next, one can show that the shaded and unshaded prisms are equivalent. In particular, the transformation

$$\mathbf{x}' = -\delta\mathbf{x} + \mathbf{c}_0 \qquad (155)$$

(where δ is Kronecker's symbol and $\mathbf{c}_0$ is a particular value of $\mathbf{c}$) converts the shaded prisms into the unshaded ones and vice versa. Both the overall and local stresses and strains remain invariant under this transformation, and the surface tractions and displacements are identical in the $\mathbf{x}$ and $\mathbf{x}'$ coordinate systems.

Finally, the periodic displacement boundary conditions must be specified for the unit cells. For a uniform overall strain increment $\Delta\bar{\varepsilon}$, Teply and Dvorak [*85*] show that, with reference to Fig. 9,

$$\Delta\mathbf{u}_{V'} - \Delta\mathbf{u}_V = \Delta\mathbf{u}'_V - \Delta\mathbf{u}'_{V'} = \Delta\bar{\varepsilon}\mathbf{c}_0, \qquad (156)$$

where V and V' indicate pairs $V_1 V'_1$, $V_2 V'_2$, and so forth. Also, they find that

$$\Delta \mathbf{u}_M = \tfrac{1}{2}(\Delta \mathbf{u}_P + \Delta \mathbf{u}_{P'}) \tag{157}$$

for any pair of points P and P' selected on the boundary between two adjacent cells (Fig. 9). When $P \equiv V_3$ and $P' \equiv V_1$,

$$\Delta \mathbf{u}_M = \tfrac{1}{2}(\Delta \mathbf{u}_V + \Delta \mathbf{u}_{V'}) = \tfrac{1}{2}\Delta \bar{\varepsilon} \mathbf{c}_0. \tag{158}$$

Hence,

$$\Delta \mathbf{u}_M - \Delta \mathbf{u}_V = \tfrac{1}{2}\Delta \bar{\varepsilon} \mathbf{c}_0, \qquad \Delta \mathbf{u}'_M - \Delta \mathbf{u}'_V = \tfrac{1}{2}\Delta \bar{\varepsilon} \mathbf{c}'_0. \tag{159}$$

Therefore, the displacements of V_i and M_i, $(i = 1, 2, 3)$, of the unit cell are related to $\Delta \bar{\varepsilon}$ in the same way as the displacements of the same points

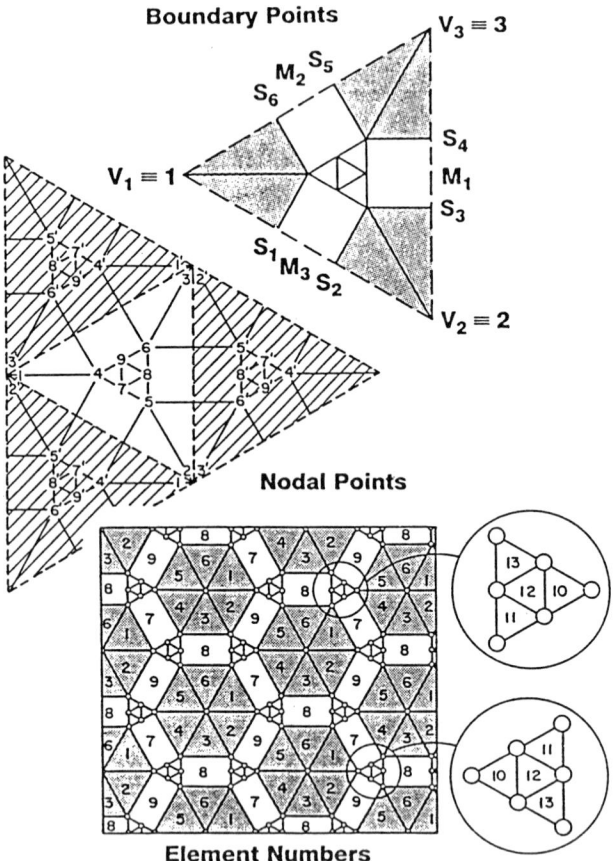

FIG. 10.   Unit cell and the finite element mesh used in evaluation of upper bounds. Reprinted with permission from *J. Mech. Phys. Solids* **36**, 29, J. L. Teply and G. J. Dvorak, © 1988, Pergamon Press plc.

located in the homogeneous effective medium. Accordingly, both sets represent contact points between the two media under uniform overall strain.

The subdivision of the unit cells into subelements is usually motivated by the particular purpose of the calculation. If only distant bounds on overall properties are needed, then one may prefer to choose the coarsest possible mesh that, however, still provides the number of degrees of freedom needed for plastic deformation. On the other hand, if one also wants to obtain some insight into the local fields, then a much more refined mesh is required.

Figure 10 shows the mesh used in evaluation of the upper-bound solutions for the PHA model. The number of subelements was determined with regard to the considerations in the previous section, and is suitable only for bounding of overall properties. Note that elements 7, 8, and 9 are shared by two unit cells. The actual upper-bound solution employs the displacement formulation of the finite element method, where the admissible strain-rate field is derived from a continuous, piecewise linear displacement field, prescribed in *V.* Figure 11 shows the actual solution domain and support conditions of the unit cell.

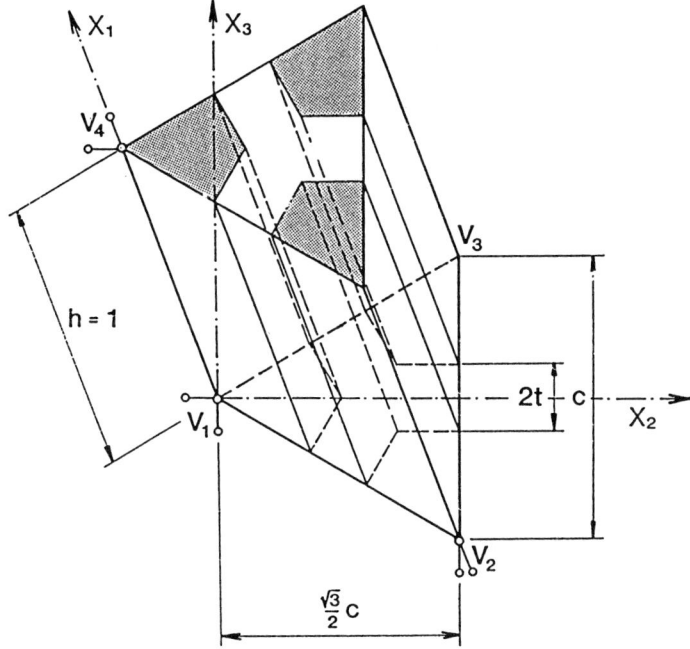

FIG. 11. Dimensions and support conditions of the unit cell in upper-bound evaluation. Reprinted with permission from *J. Mech. Phys. Solids* **36**, 29, J. L. Teply and G. J. Dvorak, © 1988, Pergamon Press plc.

A complementary lower-bound solution was also obtained. A somewhat coarser mesh was used in this case, with elements 10–13 in Fig. 10 replaced by a single element. The equilibrium or hybrid formulation was used; the admissible stress field was specified as uniform in each element. Continuity of the field was satisfied by boundary tractions applied at nodal points selected at midside points of element boundaries. These tractions were balanced with Lagrange multipliers, identified with nodal displacements [87, 88, 89].

Figure 12 shows an example of the results found with the bounding approach implementation in the PHA model. A collection of stress–strain

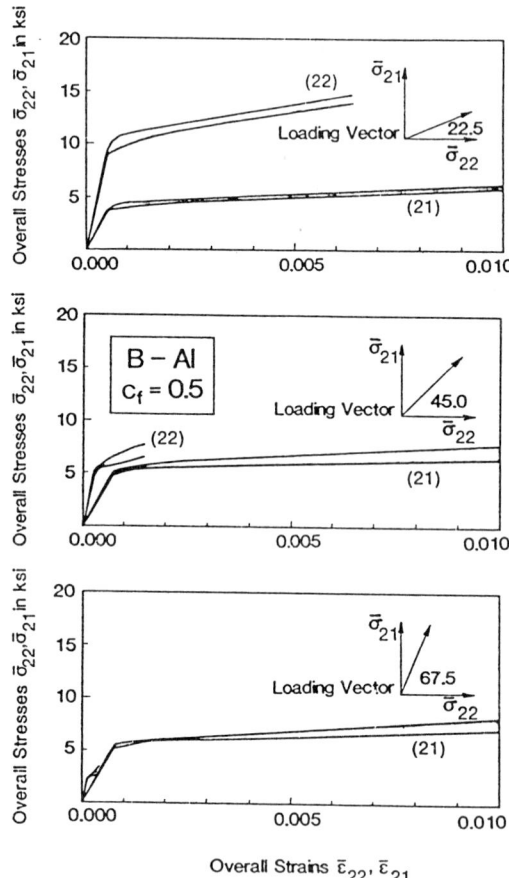

FIG. 12. Approximate bounds on overall response under proportional loading. Reprinted with permission from J. Mech. Phys. Solids **36**, 29, J. L. Teply and G. J. Dvorak, © 1988, Pergamon Press plc.

curves was computed for a unidirectional fibrous material subjected to combined transverse tension and longitudinal shear stresses $\bar{\sigma}_{22}$ and $\bar{\sigma}_{21}$. A linearly hardening aluminum matrix and an elastic boron fiber were used in the unit cell. For the three proportional loading paths used in the evaluation, two stress–strain curves in close proximity were computed, one from the upper bound and one from the lower bound procedure. Note that both transverse normal and longitudinal shear strains appear in the first loading case. However, the second and third paths apparently promote preferential straining in shear. This turns out to be a demonstration of a so-called matrix deformation mode of fibrous composites, which is examined as a part of the bimodal plasticity theory in the following.

Local stress fields can be evaluated with more refined meshes in the PHA model unit cell. In the examples that follow, the cell was subdivided into 87 elements; the local stress fields were found and stress contours were plotted in the domain [90]. Figure 13 shows such fields for a boron–aluminum composite under transverse tension and longitudinal shear, at two overall stress levels. Figure 14 shows contours of local normal stress caused in the composite by uniform thermal change. At the lower levels of overall stress or temperature change, the matrix is strained only elastically, but it becomes fully plastic at the higher levels. Large gradients are present in each case.

The local fields can be utilized in a comparison of predictions of overall instantaneous properties by different micromechanical procedures. Of particular interest is the question of accuracy of those approaches that rely on averages of local fields, such as the self-consistent and the Mori–Tanaka methods. No attempt was actually made to reproduce either of the two techniques. Instead, the overall properties were found from averages of the fields found by the refined finite element analysis of the PHA unit cell. At many steps of each loading path, the local stresses in each element were integrated over the volume of each phase, as in (4). Then the specific forms of the constitutive equation (3) and (64) to (74), originally prescribed for the elastic and plastic strain increments in the matrix subelements, were applied to find the average total strain that would have been caused in the matrix by the average of the matrix stress field. A similar strain average was also evaluated in the elastic fiber. Finally, these phase strain averages were added as in (9′) to arrive at the total overall strain. Figure 15 compares the overall response of the PHA model with that evaluated from the above averages of local fields. The agreement is quite poor in the B/Al system, but better in the T50–Gr/Al system. Similar comparisons were made for longitudinal shear loading of the two systems; the two methods gave reasonably close predictions, but the response computed from the averages was somewhat stiffer. Figure 16 shows the response under uniform thermal change. Again, the agreement of the two methods is not very satisfactory in the B/Al

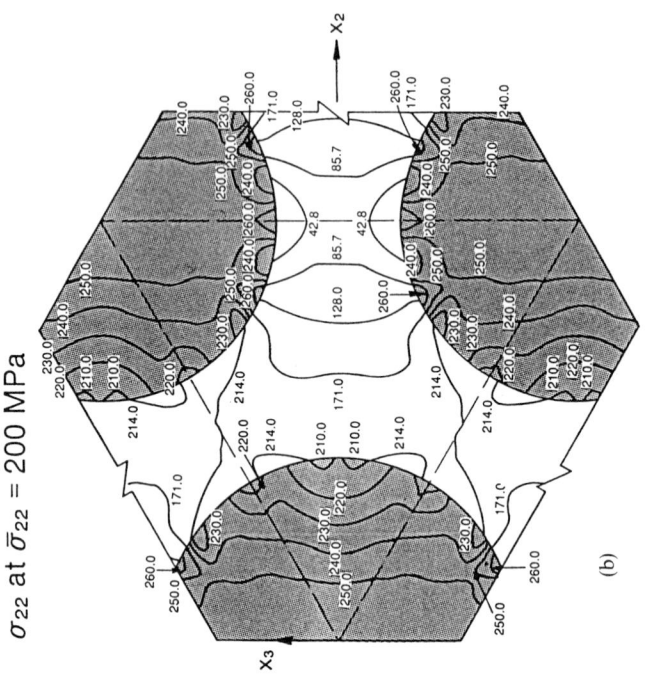

σ_{22} at $\bar{\sigma}_{22}$ = 200 MPa

(b)

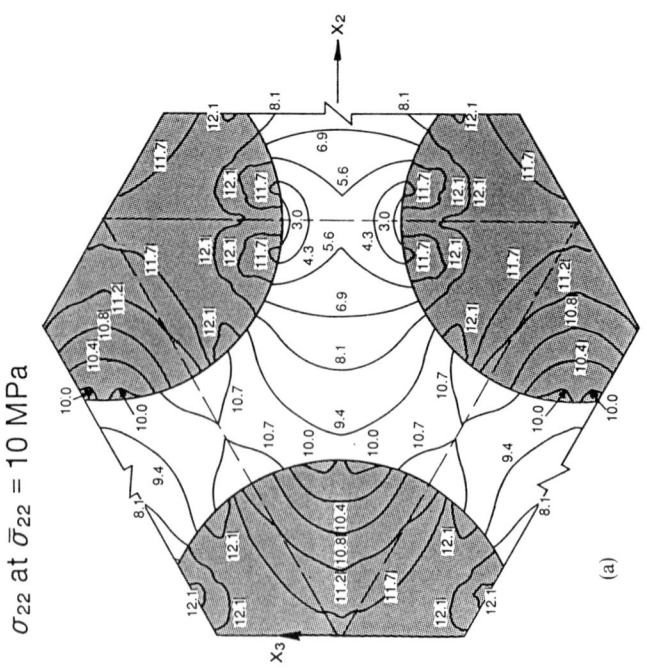

σ_{22} at $\bar{\sigma}_{22}$ = 10 MPa

(a)

56

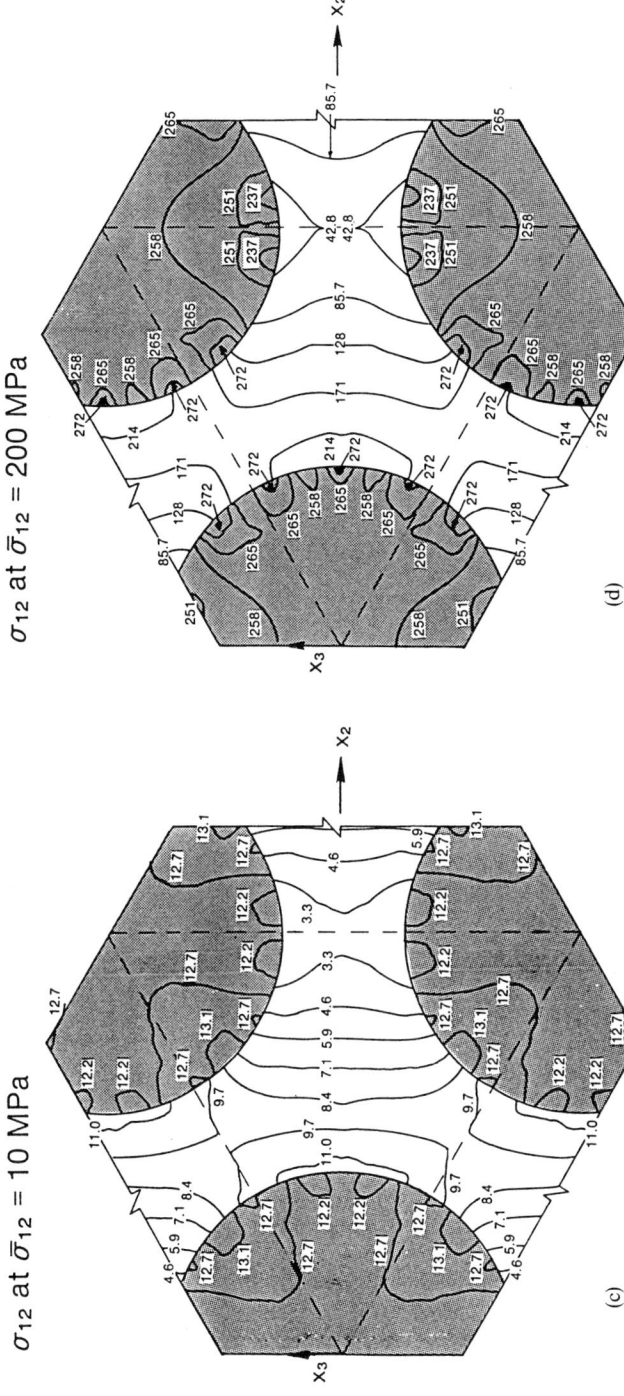

σ_{12} at $\bar{\sigma}_{12}$ = 200 MPa

σ_{12} at $\bar{\sigma}_{12}$ = 10 MPa

(c)

(d)

Fig. 13. Local stress contours in a B/Al composite subjected to mechanical load [90]. (a) Local transverse normal stress contours (σ_{22}) at $\bar{\sigma}_{22}$ = 10 MPa. Elastic state. (b) Same contours at $\bar{\sigma}_{22}$ = 200 MPa. Fully plastic state. (c) Local longitudinal shear stress contours (σ_{12}) at $\bar{\sigma}_{12}$ = 10 MPa. Elastic state. (d) Same contours at $\bar{\sigma}_{12}$ = 200 MPa. Fully plastic state.

57

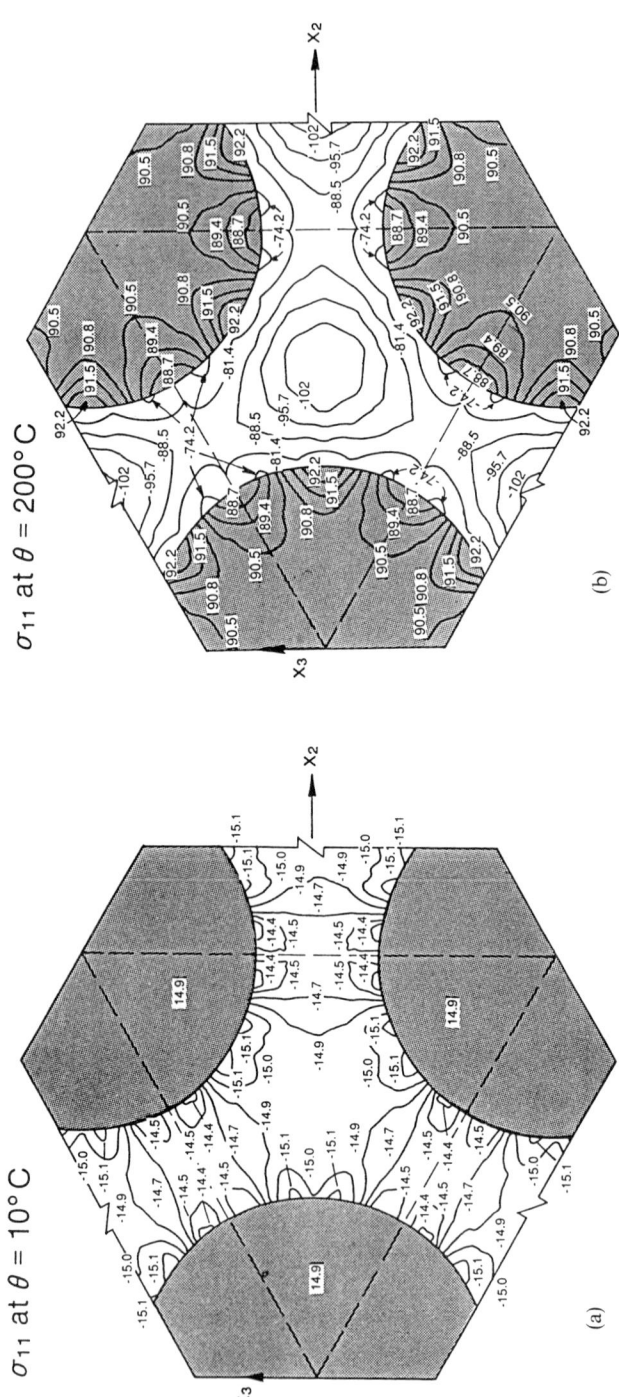

σ_{11} at $\theta = 200°\text{C}$

(b)

σ_{11} at $\theta = 10°\text{C}$

(a)

58

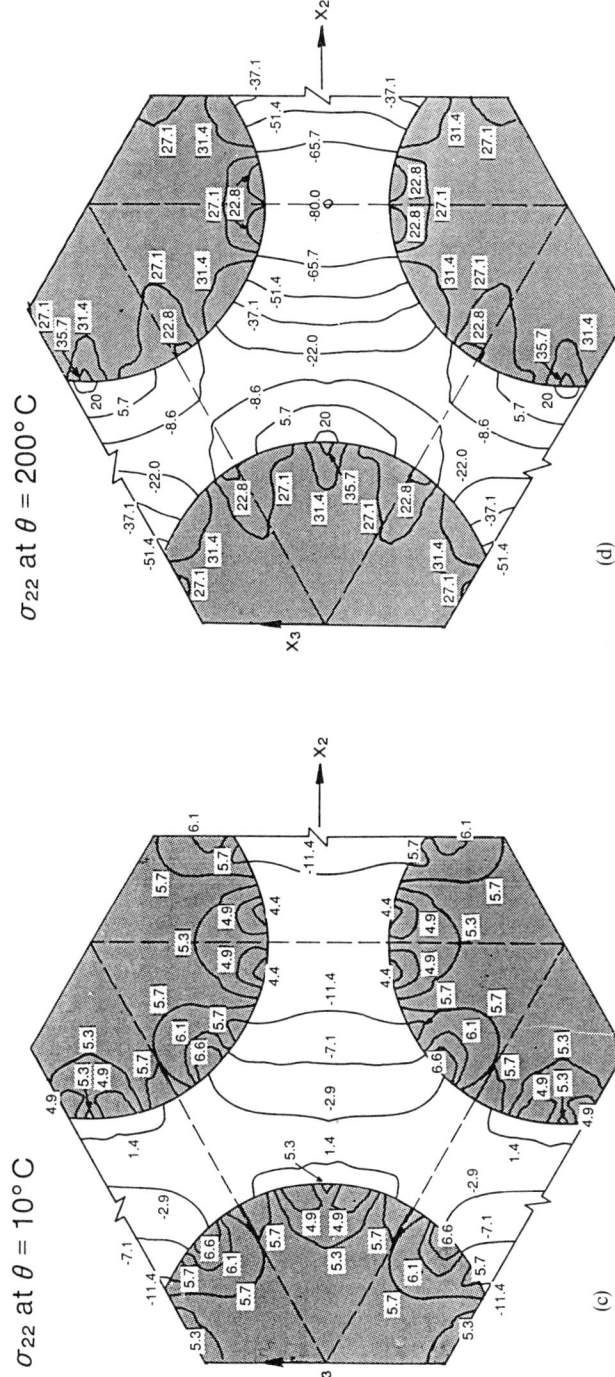

FIG. 14. Local stress contours in a B/Al composite subjected to a uniform thermal change [90]. (a) Axial stress (σ_{11}) at $\Delta\theta = 10°$C. Elastic state. (b) Axial stress (σ_{11}) and $\Delta\theta = 200°$C. Fully plastic state. (c) Transverse normal stress (σ_{22}) at $\Delta\theta = 10°$C. Elastic state. (d) Transverse normal stress (σ_{22}) at $\Delta\theta = 200°$C. Fully plastic state.

59

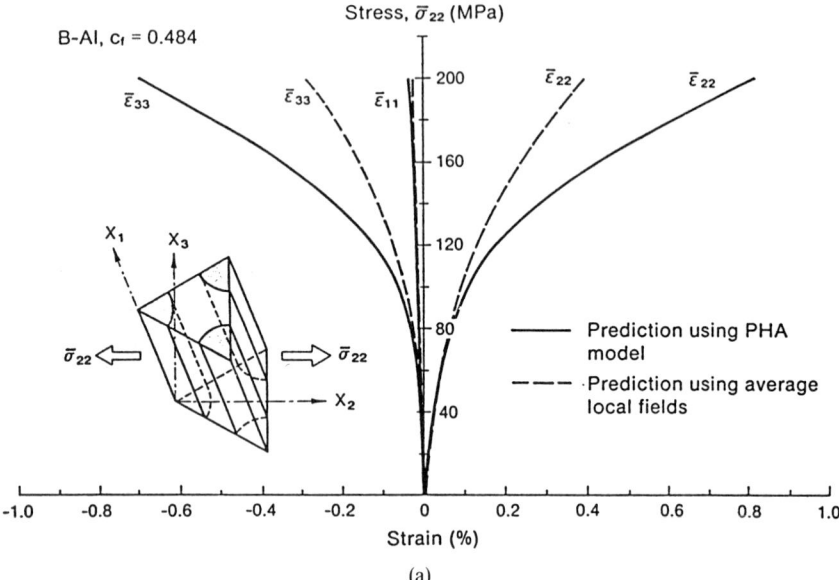

(a)

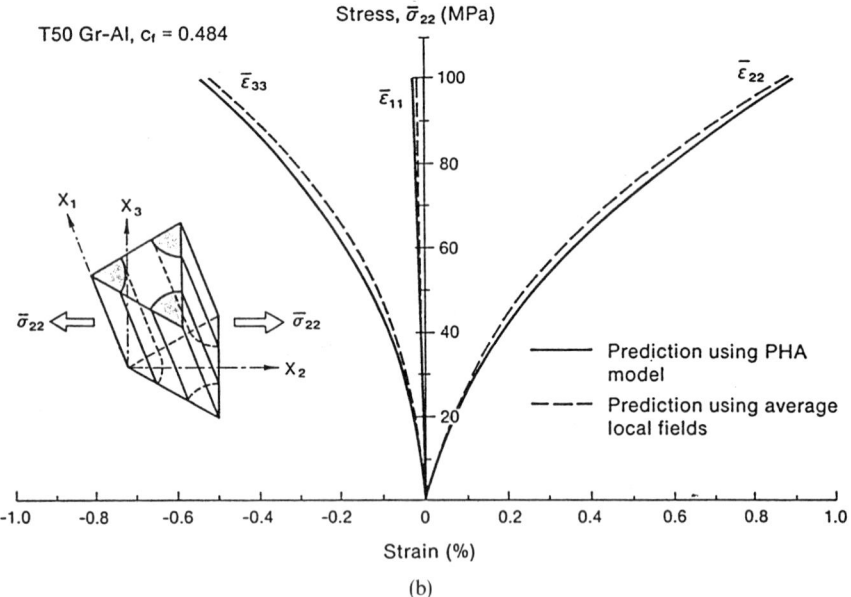

(b)

FIG. 15. Overall response under transverse tension $\bar{\sigma}_{22}$ [90]. (a) Of a B/Al system. (b) Of a Gr/Al system.

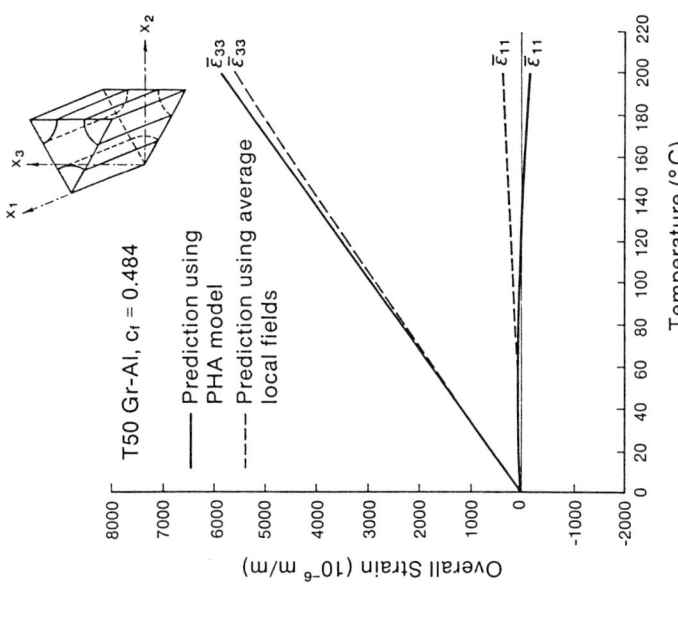

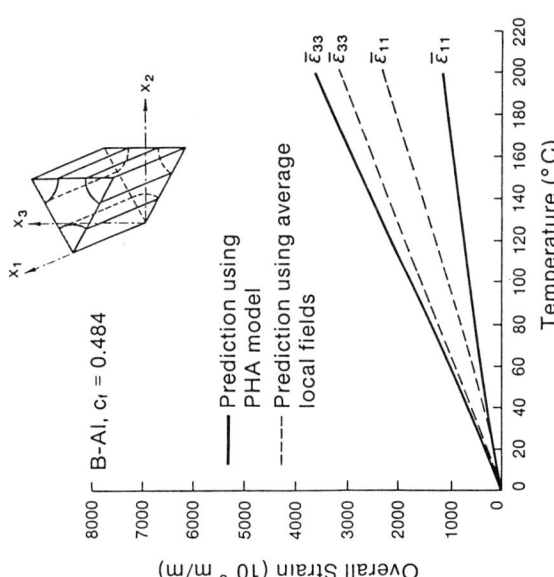

FIG. 16.   Predictions of overall strain–temperature response. Comparison of PHA model results with those derived from average local fields [90].

system. There appears to be a better agreement in the T50–Gr/Al case, but in fact, the prediction of the axial strain by the averaging approach is entirely misleading, both in magnitude and in sign.

This example can serve as a salutory reminder of the possibly large errors that can be introduced by averaging; similar discrepancies would be found in analogous comparisons with the self-consistent or Mori–Tanaka schemes. Several other examples of such errors appear in the following.

We note that the PHA model has been implemented as a UMAT routine in the ABAQUS program [91] and can thus be used in structural applications. Also, the model opens an avenue to numerical experimentation, which is useful in the development of simpler constitutive theories and in interpretation of physical experiments. It is shown in the following that the PHA model agrees well with experimental results, provided that an accurate description of actual phase properties is specified. Finally, the PHA model serves as a check of accuracy of other approaches, such as the averaging scheme discussed earlier.

c. Bimodal Plasticity Theory

Another approach to modeling of overall behavior of fibrous composites was motivated, in part, by the experimental results described in Section III.B.5, [50, 62]. The experiments suggest that a unidirectional fibrous layer under plane stress may exhibit two distinct deformation modes that affect only the inelastic response. If the overall stress increments have a large normal component in the fiber direction, then the material tends to deform in the fiber-dominated mode (FDM). In this mode, the composite appears to deform as described in Section III.B.3.b. In any event, the segments of the overall yield surface that correspond to this mode are well approximated if the matrix stress concentration factors derived from self-consistent estimates of local stresses. To this end, one finds the estimate of the elastic stress concentration factors in the matrix and relates the local averages to the overall stress. At yield, the overall stresses assume the magnitudes required to satisfy the Mises yield condition (62) by the stress averages in the matrix. This procedure leads to a single surface, given by (75) with $k = r = \beta$, for the matrix phase. The elastic concentration factor $\mathbf{B}_\beta$ in (75) is given by (21$_2$) for the self-consistent method and by the inverse of the coefficient matrix in (27$_1$) for the Mori–Tanaka method. Overall hardening follows from (77), $d\bar{\boldsymbol{\alpha}} = d\bar{\boldsymbol{\sigma}}$.

In the matrix-dominated mode (MDM), the theory postulates that the composite can be regarded as an elastic–plastic continuum in which plastic straining occurs in the form of smooth shearing deformations in the matrix, on certain hypothetical slip planes that are parallel to the fiber axis, and in certain preferred slip directions on these planes. This mode ignores the actual microstructural geometry. Instead, the matrix is regarded as a homogeneous medium, and the fibers that occupy a finite volume fraction are assumed to constrain the matrix deformation to the slip planes. There is a similarity with continuum slip models of single-crystal plasticity, which ignore the discrete dislocation substructure of the crystal and postulate smooth shearing to take place on certain slip systems [92].

The MDM mode is illustrated by Fig. 17, which shows the admissible slip planes. The actual slip directions are determined by the requirement that the resolved shear stress reaches a maximum on the active slip systems. It is possible to show that there are always two conjugate systems that satisfy this requirement on planes parallel to the fiber axis, which is parallel to x_1. The corresponding MDM yield surface for an overall state of plane stress appears in Fig. 18. It is an infinite cylinder of the oval cross section shown, with generators parallel to the $\bar{\sigma}_{11}$ axis; hence the $\bar{\sigma}_{11}$ stress does not influence the onset of MDM yielding.

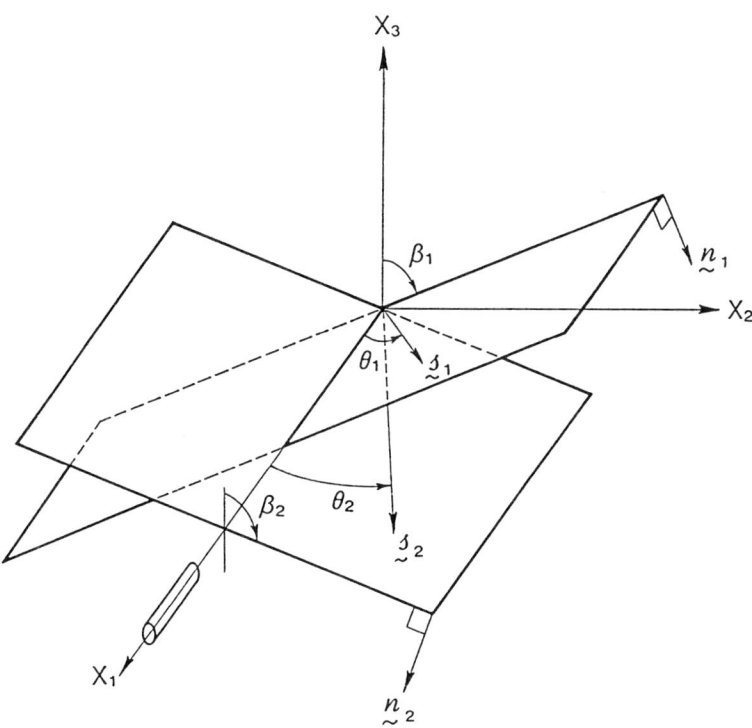

FIG. 17. Geometry of the two conjugate slip systems of the matrix-dominated mode (MDM). Reprinted with permission from Springer-Verlag, G. J. Dvorak and Y. A. Bahei-El-Din, *Acta Mechanica* **69**, 219 (1987).

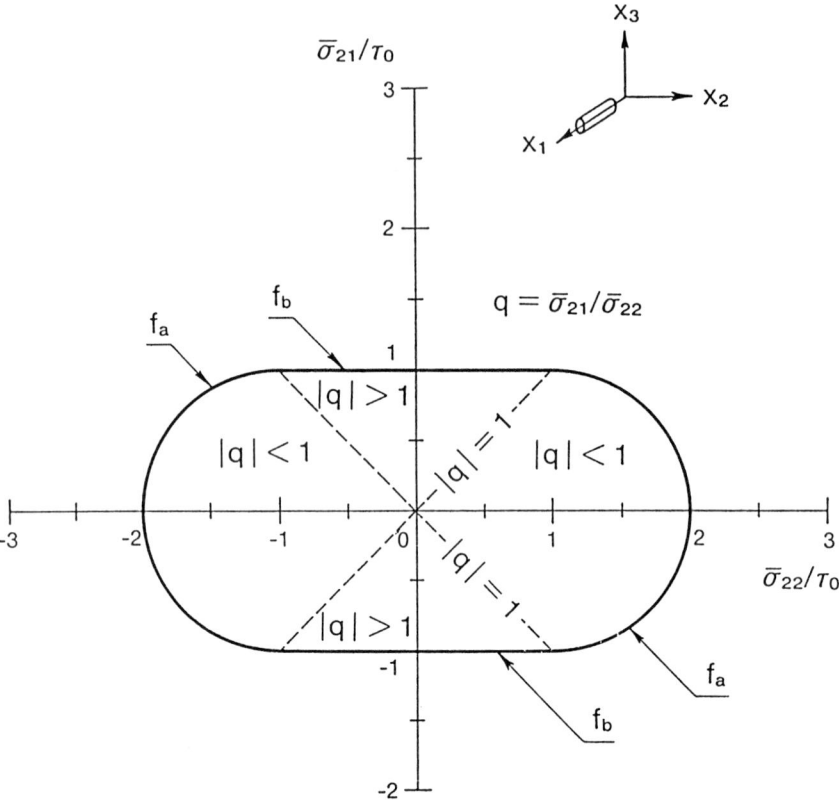

FIG. 18. Transverse cross section of the cylindrical overall yield surface of the matrix-dominated mode. Reprinted with permission from Springer-Verlag, G. J. Dvorak and Y. A. Bahei-El-Din, *Acta Mechanica* **69**, 219 (1987).

The surface consists of two segments defined by the equations

$$f_a \equiv (\bar{\sigma}_{21} - \bar{\alpha}_{21})^2/\tau_0 + [(\bar{\sigma}_{22} - \bar{\alpha}_{22})/\tau_0 \mp 1]^2 - 1 = 0 \qquad \text{for} \quad |q| \le 1,$$

$$f_b \equiv (\bar{\sigma}_{21} - \bar{\alpha}_{21})^2 - \tau_0 = 0 \qquad \text{for} \quad |q| \ge 1,$$

$$(160)$$

where $q = (\bar{\sigma}_{21} - \bar{\alpha}_{21})/(\bar{\sigma}_{22} - \bar{\alpha}_{22})$, and $\bar{\alpha}_{ij} = 0$ for an initial yield surface.

Figures 19 and 20 show superimposed yield surfaces for the two modes and for two material systems. Figure 19 represents the section by the transverse normal stress $\bar{\sigma}_{22}$ and longitudinal shear $\bar{\sigma}_{21}$ plane, Fig. 20 by the axial normal stress $\bar{\sigma}_{11}$ and longitudinal shear plane. In both figures, the solid line indicates the MDM yield surface, which is not affected by phase elastic properties and is therefore unique for all systems. The various

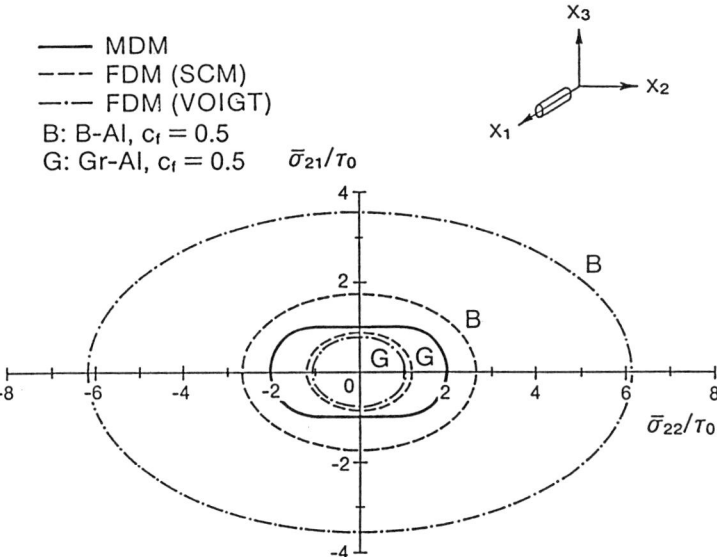

FIG. 19. Initial yield surfaces in the transverse tension ($\bar\sigma_{22}$) and longitudinal shear ($\bar\sigma_{21}$) plane. Comparison of the fiber-dominated (FDM) and matrix-dominated (MDM) yield modes in B/Al and Gr/Al composites. Reprinted with permission from Springer-Verlag, G. J. Dvorak and Y. A. Bahei-El-Din, *Acta Mechanica* **69**, 219 (1987).

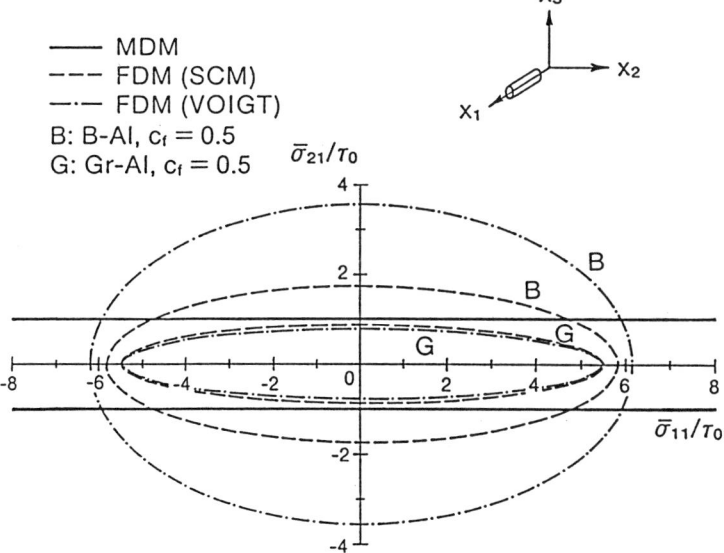

FIG. 20. Initial yield surfaces in the axial tension ($\bar\sigma_{11}$) and longitudinal shear ($\bar\sigma_{21}$) plane. Comparison of fiber-dominated (FDM) and matrix-dominated (MDM) yield modes in B/Al and Gr/Al composite systems. Reprinted with permission from Springer-Verlag, G. J. Dvorak and Y. A. Bahei-El-Din, *Acta Mechanica* **69**, 219 (1987).

ellipses represent the FDM surfaces. Those were found for the B/Al and the T-50 Gr/Al systems and are designated by letters B and G, respectively. The phase volume fractions were taken as equal to 0.5 in both systems. One set of the FDM surfaces was found using the self-consistent estimate of local stresses (SCM), and another set with the Voigt approximation, cf. (30).

Note that the B/Al system yield surfaces have both MDM and FDM segments, the actual yield surface is the internal envelope of the segments. Within this envelope, the FDM segments can be regarded as end caps on the MDM oval cylinder. In the Gr/Al system, the FDM surfaces are always within the MDM surface, hence the former is the active mode. Since only phase elastic properties are involved, one may ask which property makes a particular mode more or less prominent. In the example shown, the matrix is isotropic and its properties are identical in both systems. The B fiber is also isotropic, but the Gr fiber is transversely isotropic. Elastic constants used in the evaluation appear in Table I.

The axial fiber moduli are very similar, but there is a large difference in the longitudinal shear modulus G_A, which is smaller in the graphite fiber than in the matrix and an order of magnitude smaller than the corresponding fiber modulus. The implication is that the MDM deformation is preferred in systems that have fibers of large shear stiffness, and the FDM modes are preferred in fibers that are compliant in shear. One may speculate that the stiff fiber tends to prevent plastic shearing on planes that intersect the fiber axis, and therefore encourages the slip pattern of Fig. 17. The more compliant fiber cannot do that, and thus the FDM mode may prevail.

Viewed from a different perspective, the results in Figs. 19 to 22 can be taken as a serious warning against indiscriminate use of averaging methods in plasticity of fibrous composites. Indeed, only the yield surface of the fiber mode may possibly be approximated by the self-consistent estimate of the local stresses, and then only under plane stress. Figure 14 tends to confirm this; note the agreement between the PHA and the averaging predictions in the Gr/Al, which appear in Figs. 15 and 16. However, even in the seemingly simple case of uniform thermal change, averaging methods typically predict

TABLE I
ELASTIC PROPERTIES OF SELECTED MATRIX AND FIBER MATERIALS

	E_A [GPa]	G_A [GPa]	ν_A	E_T [GPa]	G_T [GPa]
6061 Aluminum	72.5	27.2	0.33	72.5	27.2
Boron	400.0	166.8	0.20	400.0	166.8
T-50 Graphite	386.4	15.2	0.41	7.6	2.6

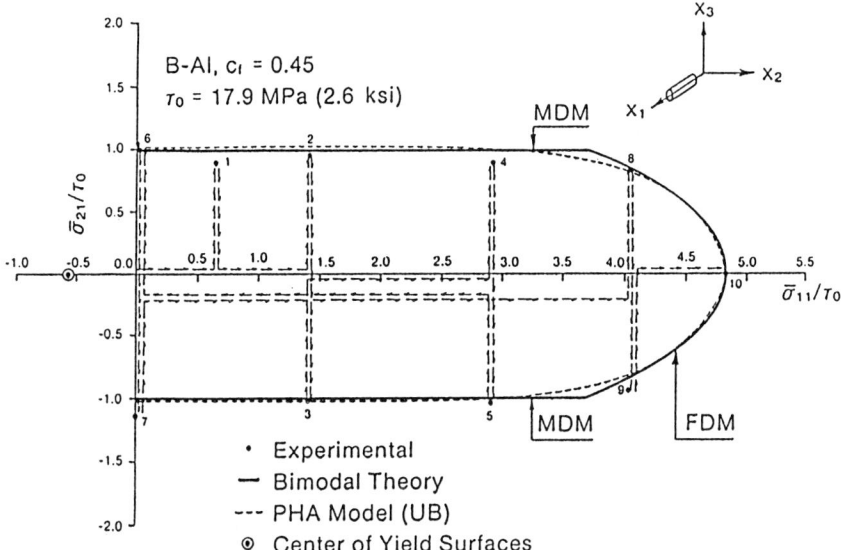

FIG. 21.   Initial yield surface of a B/Al composite in the longitudinal plane. Comparison of experimental results with yield surfaces derived from the bimodal plasticity theory and the PHA model. Reprinted with permission from Springer-Verlag, G. J. Dvorak and Y. A. Bahei-El-Din, *Acta Mechanica* **69**, 219 (1987).

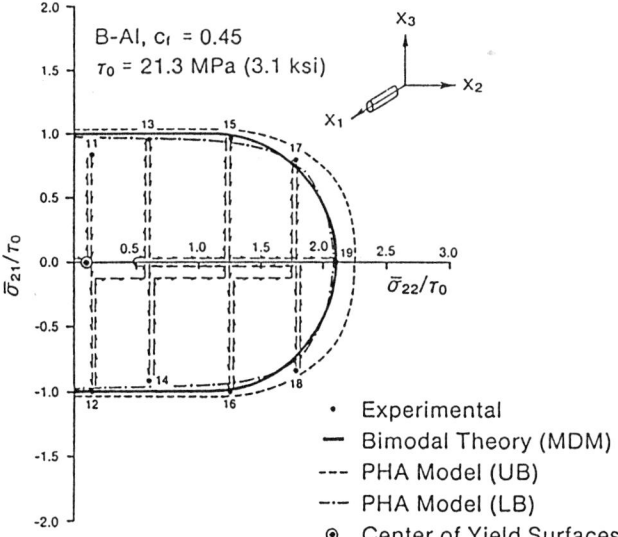

FIG. 22.   Initial yield surface of the B/Al system in the transverse stress ($\bar{\sigma}_{22}$) and longitudinal shear ($\bar{\sigma}_{21}$) plane. Comparison of experimental results with surfaces derived from the PHA model, and from the matrix-dominated deformation mode (MDM). Reprinted with permission from Springer-Verlag, G. J. Dvorak and Y. A. Bahei-El-Din, *Acta Mechanica* **69**, 219 (1987).

an open cylindrical yield surface in the principal overall stress space [72, 75], whereas the actual overall yield surface is closed [64, 66]. Such disagreements in yield surface predictions then imply further problems in the evaluation of subsequent overall surfaces and plastic strains; this is illustrated in Fig. 16. The conclusion that emerges from these and other examples is that one may use the self-consistent or the Mori–Tanaka method with some confidence only in exceptional circumstances, which are currently limited to the plane stress FDM deformation.

We now turn our attention to the evaluation of plastic strains. Note first that the yield condition (160) suggests that the yield stress is independent of the slip system involved, which is confirmed by the experimental results that follow. Of course, that would be true in any case on the flat branches that engage only a single system. However, comparisons of the MDM results with the PHA model, and experimental measurements of flow stress, indicate that the slip direction, and perhaps even the direction of the stress increment, both influence the instantaneous overall properties in the plastic range if loading involves any of the semicircular segments of the MDM yield surface. The relevant connections have not yet been fully established, but work currently in progress indicates that good comparisons of predicted plastic strains with those measured in carefully conducted experiments are possible with the PHA model.

This is actually a part of a broader problem that remains unresolved in plasticity of both homogeneous and heterogeneous media, namely the formulation of reliable predictions of plastic strains during loading along an arbitrary path. A promising approach is offered by the contemporary multiple surface theories that are discussed, for example, in [48, 49, 59], but a definite treatment of this subject must await further research.

5. COMPARISON WITH EXPERIMENTAL RESULTS

Whereas the theoretical aspects of elastic–plastic behavior of fibrous composites have been investigated rather extensively, only limited attention has been given to experimental investigations of the actual behavior of composite systems. To be useful in verification of a theory, the experiments must reflect overall response under multiaxial incremental loading. So far, only the recent study by Dvorak et al. [50] has been designed with this purpose in mind.

The work was conducted on thin-walled B/Al tubes, which were reinforced by continuous fibers aligned in the axial direction. Similar matrix tubes were tested as well. The tubes were loaded incrementally by combinations of an axial force, internal pressure, and torque; the in-plane strains were measured and recorded. The purpose of the investigation was to establish initial and subsequent yield surfaces, and plastic strain magnitudes and directions, for several different loading programs.

Figures 21 and 22 show the initial yield surfaces in two stress planes. The coordinates have been normalized with respect to the composite yield stress τ_0 in longitudinal shear. The experimental points are connected by the arrows, which retrace the actual loading sequence. The solid line shows the MDM and FDM yield surface segments determined for the current τ_0. The dashed line is the prediction of the PHA model, with the same τ_0 and yield stress definition used in the experiments; the definition, we recall, relies on a back-extrapolation from few initial plastic steps. Both upper (UB) and lower (LB) bounds were computed in Fig. 22, but only the upper bound in Fig. 21. The agreement of the two predictions with experiments is satisfactory, and the existence of the flat segments of the surface is clearly confirmed. The positions of the centers of the surfaces do not coincide with the origin, which is believed to be caused by residual thermal stresses left after cooling from the annealing temperature. Note also that the values of τ_0 are different in the two figures. Such variations have been also observed in subsequent surfaces, both in the composite and in similar matrix specimens.

Many subsequent yield surfaces were evaluated. They were found to be of similar shape, but the size, i.e., the current τ_0, changed somewhat along the path. Similar changes were observed in the matrix surfaces. Figure 23 illustrates some of the results. The loading path connects points 87 and 88, then continues from 99 to 100, then from 113 to 114, from 136 to 127, and finally from 132 to 133. The intervening points define the loading surfaces established at the breaks in continuous loading. Note the similarity in shape, and also the adherence to the Phillips hardening rule (77, 78). An exception to the latter was found, however, for loading at the flat segments (Fig. 24). The normal stress component, which is parallel to the flat segment, exerts no influence on the translation, but reasserts itself if the loading vector is on the semicircular branch, *cf.* path 132–133 in Fig. 23b. A possible explanation of this behavior appears in [50]. The translation of the MDM surface is given, with reference to (160),

$$d\bar{\alpha}_{21} = d\bar{\sigma}_{21}, \qquad d\bar{\alpha}_{22} = d\bar{\sigma}_{22} \qquad \text{for} \quad f_a \equiv 0, |q| \leq 1$$
$$d\bar{\alpha}_{21} = d\bar{\sigma}_{21} \qquad\qquad\qquad \text{for} \quad f_b \equiv 0, |q| \geq 1.$$

$$(161)$$

The strain increment vectors were found to follow the normality requirement along some but not all loading directions. An interpretation of the results is still in progress.

Figure 25 is another illustration of the agreement between the two theories

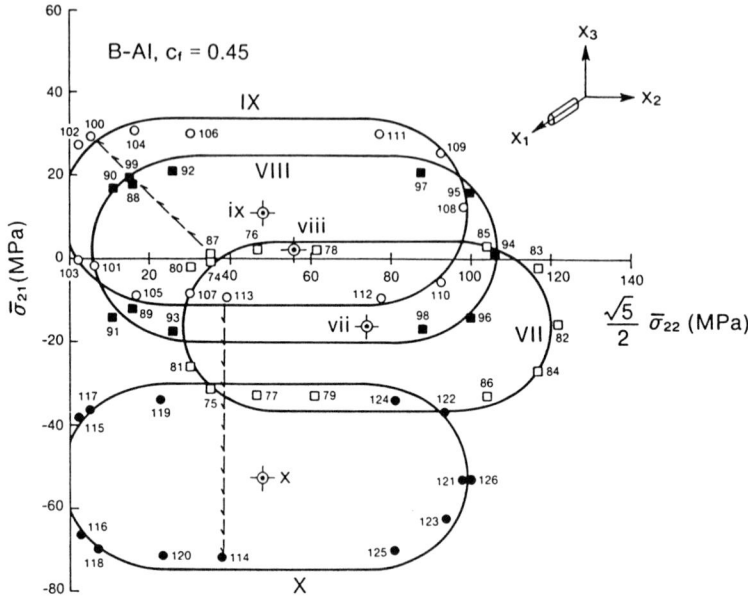

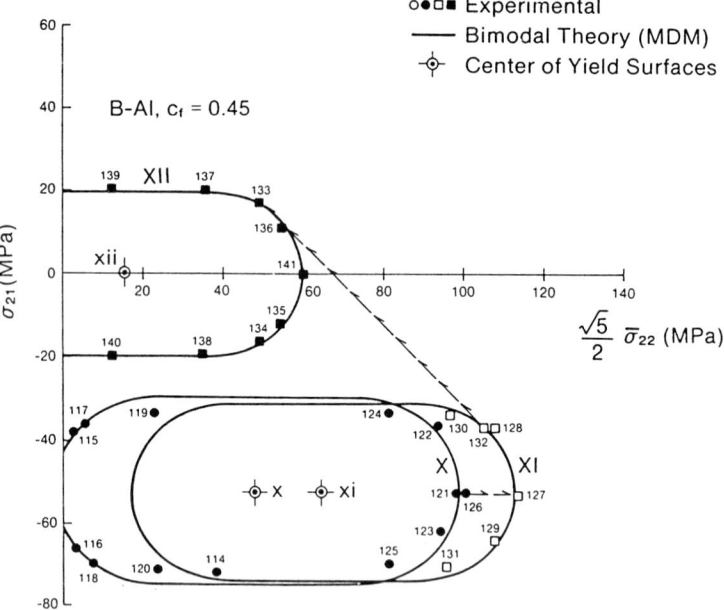

FIG. 23. Subsequent yield surfaces of the B/Al composite. Reprinted with permission from *J. Mech. Phys Solids* **36**, 655, G. J. Dvorak, Y. A. Bahei-El-Din, Y. Macheret, and C. H. Liu, © 1988, Pergamon Press plc.

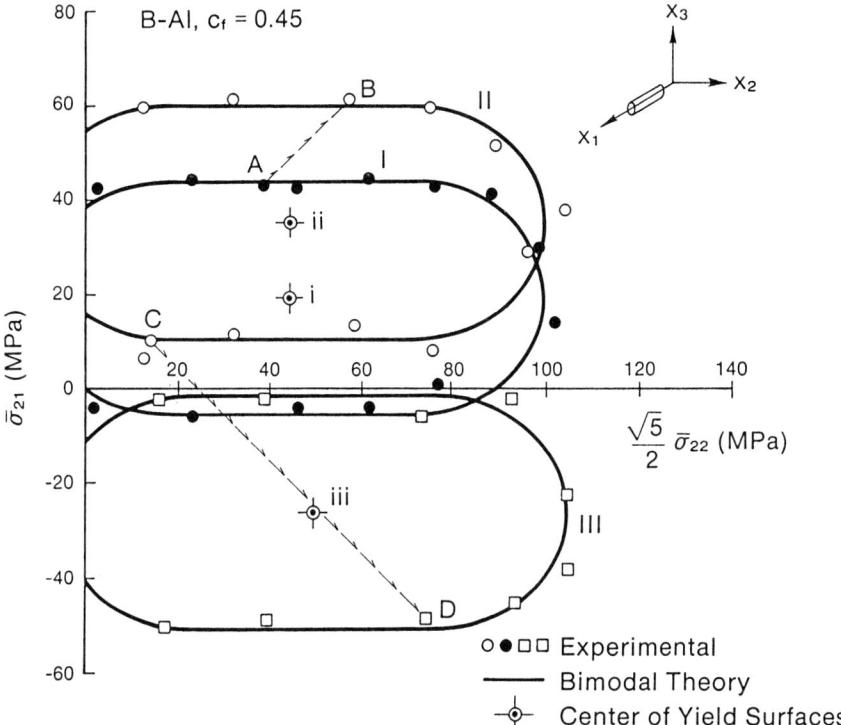

FIG. 24.   Subsequent yield surfaces of the B/Al composite. Load is applied at the flat segments of the surface. Reprinted with permission from *J. Mech. Phys Solids* **36**, 655, G. J. Dvorak, Y. A. Bahei-El-Din, Y. Macheret, and C. H. Liu, © 1988, Pergamon Press plc.

and experiments. For a cyclic loading path *ABCDEA*, which was retraced several times, we show the subsequent yield surfaces found by the different approaches. PHA-computed directions of the plastic strain vectors are shown as well. These are mostly normal to the current surface, whereas the experimentally measured plastic strains often have an additional longitudinal shear component.

IV.   Conclusion

Space limitations prevent discussion of other aspects of plasticity of composites, and also of the applications of the above theories to practical problems. We note here related studies of fatigue behavior of B/Al laminates [*93, 94*] that indicate that the onset and evolution of damage is closely related

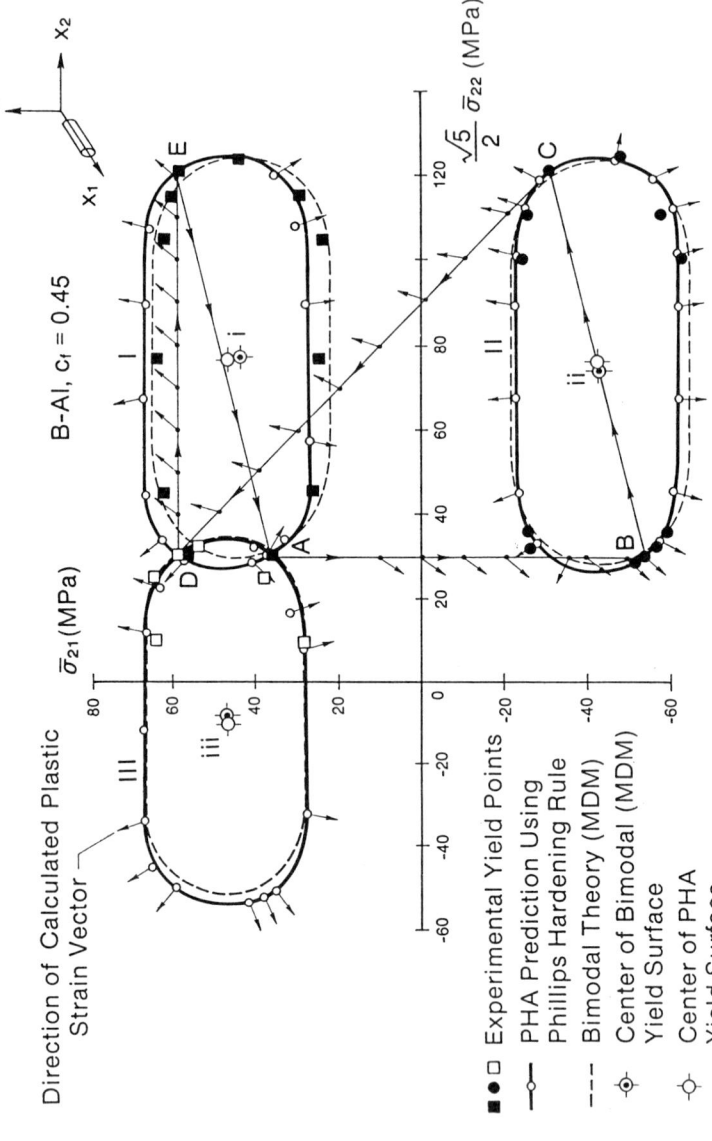

FIG. 25. Yield surfaces for a cyclic loading program *ABCDEA*. Comparison of experimental results with PHA and bimodal theory (MDM) predictions.

to shakedown of the laminate. The damage process is dominated by growth of low-cycle fatigue cracks in the matrix, on planes that are parallel to the fiber direction in each off-axis ply. Cracks perpendicular to the loading direction grow in the matrix of the axial $0°$ plies. Cyclic plastic straining of the matrix drives the cracks, and when it terminates by shakedown, the damage process reaches a saturation state. Laminate stiffness loss caused by damage saturation at a constant load amplitude has been predicted on these grounds and found to be in agreement with experimental measurements.

Another recent application of the theory has been in predictions of fracture strength of notched unidirectional B/Al plates [95, 96]. The matrix-dominated mode of deformation was identified there with discrete plastic zones that were observed to grow from notch tips, in the fiber direction, under increasing load. When plasticity was assumed to be limited to the matrix mode, the definition of zone geometry led to reliable predictions of local stresses ahead of the notch and to fracture strength estimates for the plates.

Current work that utilizes the PHA model includes a study in dimensional stability of Gr/Al $\pm \theta$ laminates [91]. There we found that plastic deformation may improve dimensional stability in comparison with a purely elastic response in certain layups. The interaction between mechanical and thermal loads was also examined; particular combinations may enhance or impair dimensional stability.

Plasticity of particle- or whisker-reinforced composite materials is closely related to the present subject. In comparison to fibrous composites, the microstructural geometry of such systems is much more complex, and that makes modeling more difficult. The averaging techniques discussed earlier, such as the self-consistent and the Mori–Tanaka methods, are often likely to be employed in model development. Examples can be found in the work of Hutchinson [97, 98], Duva [99], Weng [100], Tandon and Weng [101], and McMeeking [102]. Experimental or computational verification is apparently not available in the literature.

Acknowledgments

Financial support for this work was provided by the Office of Naval Research. Professors Yehia A. Bahei-El-Din, Mark Shephard and Jerry Lin, Dr. Jan L. Teply, and graduate students, R. Shah and J. F. Wu, contributed to the recent work on plasticity of composite materials described herein.

Appendix

This is a brief summary of the relevant details of the notation used in the chapter. The constitutive relations of an elastic, transversely isotropic solid, with the axis of rotational symmetry x_1, has the form

$$
\begin{Bmatrix} \varepsilon_1 \\ \varepsilon_2 \\ \varepsilon_3 \\ \varepsilon_4 \\ \varepsilon_5 \\ \varepsilon_6 \end{Bmatrix} = \begin{bmatrix} 1/E_L & -v_L/E_L & -v_L/E_L & 0 & 0 & 0 \\ & 1/E_T & -v_T/E_T & 0 & 0 & 0 \\ & & 1/E_T & 0 & 0 & 0 \\ & & & 1/G_T & 0 & 0 \\ & & & & 1/G_L & 0 \\ \text{SYM.} & & & & & 1/G_L \end{bmatrix} \begin{Bmatrix} \sigma_1 \\ \sigma_2 \\ \sigma_3 \\ \sigma_4 \\ \sigma_5 \\ \sigma_6 \end{Bmatrix}
$$

$$
\begin{Bmatrix} \sigma_1 \\ \sigma_2 \\ \sigma_3 \\ \sigma_4 \\ \sigma_5 \\ \sigma_6 \end{Bmatrix} = \begin{bmatrix} n & l & l & 0 & 0 & 0 \\ & (k+m) & (k-m) & 0 & 0 & 0 \\ & & (k+m) & 0 & 0 & 0 \\ & & & m & 0 & 0 \\ & & & & p & 0 \\ \text{SYM.} & & & & & p \end{bmatrix} \begin{Bmatrix} \varepsilon_1 \\ \varepsilon_2 \\ \varepsilon_3 \\ \varepsilon_4 \\ \varepsilon_5 \\ \varepsilon_6 \end{Bmatrix}.
$$

The moduli E_L, G_L refer to straining in the longitudinal direction, and E_T, G_T in the transverse plane. Poisson's ratios are defined as $v_L = -\varepsilon_2/\varepsilon_1$, $v_T = -\varepsilon_3/\varepsilon_2$ under uniaxial tension σ_1, σ_2, respectively.

Hill's moduli k, l, n, m, and p are related to the moduli in the compliance matrix by the relations

$$ k = [1/G_T - 4/E_T + 4v_L^2/E_L]^{-1}, $$

$$ l = 2kv_L, $$

$$ n = E_L + 4kv_L^2 = E_L + l^2/k, $$

$$ m = G_T, \qquad p = G_L. $$

Additional useful connections are

$$ E_T = 2(1 + v_T)G_T = 4km/(k + qm), $$

$$ v_T = (k - qm)/(k + qm), $$

$$ q = 1 + (4kv_L^2)/E_L. $$

If the solid is isotropic, with bulk modulus K and shear modulus G, then

$$ k = G/(1 - 2v), \qquad l = K - 2G/3, $$

$$ n = K + 4G/3, \qquad m = p = G. $$

References

1. R. Hill, *J. Mech. Phys. Solids*, **11**, 357 (1963).
2. R. Hill, *J. Mech. Phys. Solids*, **13**, 213 (1965).
3. V. M. Levin, *Mekh. Tverd. Tela.*, **2**, 88 (1967).
4. B. W. Rosen and Z. Hashin, *Int. J. Engng. Sci.*, **8**, 157 (1970).
5. N. Laws, *J. Mech. Phys. Solids*, **21**, 9 (1973).
6. Y. Takao and M. Taya, *J. Appl. Mech.*, **52**, 806 (1985).
7. G. J. Dvorak, *J. Appl. Mech.*, **53**, 737 (1986).
8. Y. Benveniste and G. J. Dvorak, in "Micromechanics and Inhomogeneity—The Toshio Mura Anniversary Volume," p. 65, Springer-Verlag, New York, 1989.
9. J. D. Eshelby, *Proc. Roy. Soc. London*, **A241**, 376 (1957).
10. J. D. Eshelby, in "Progress in Solid Mechanics" (I. N. Sneddon and R. Hill, eds.), Vol. II, p. 87, North-Holland Publ., Amsterdam, 1961.
11. M. Kinoshita and T. Mura, *Physica Status Solidi A*, **5**, 759 (1971).
12. T. Mura, "Micromechanics of Defects in Solids," Martinus Nijhoff Publ., The Hague, 1982.
13. L. J. Walpole, *Proc. Roy. Soc. London*, **A300**, 270 (1967).
14. L. J. Walpole, *J. Mech. Phys. Solids*, **17**, 235 (1969).
15. F. Ghahremani, *Mech. Res. Commun.*, **4**, 89 (1977).
16. R. Hill, *J. Mech. Phys. Solids*, **12**, 199 (1964).
17. A. V. Hershey, *J. Appl. Mech.*, **21**, 236 (1954).
18. B. Budiansky, *J. Mech. Phys. Solids*, **13**, 223 (1965).
19. R. Hill, *J. Mech. Phys. Solids*, **13**, 189 (1965).
20. T. Mori and K. Tanaka, *Acta Metall.*, **21**, 571 (1973).
21. Y. Benveniste, *Mech. Mater.*, **6**, 147 (1987).
22. A. N. Norris, *J. Appl. Mech.*, **56**, 83 (1989).
23. G. J. Weng, *Int. J. Engng. Sci.*, **22**, 8945 (1984).
24. T. Chen and G. J. Dvorak, "On the Overall Elastic Moduli of Composite Materials by the Mori–Tanaka Scheme," to be published.
25. Z. Hashin, and S. Shtrikman, *J. Mech. Phys. Solids*, **10**, 335 (1962).
26. Z. Hashin, and S. Shtrikman, *J. Mech. Phys. Solids*, **10**, 343 (1962).
27. Z. Hashin, and S. Shtrikman, *J. Mech. Phys. Solids*, **11**, 127 (1963).
28. Z. Hashin, *J. Mech. Phys. Solids*, **13**, 119 (1965).
29. Z. Hashin, and B. W. Rosen, *J. Appl. Mech.*, **31**, 223 (1964).
30. R. Hill, in "Progress in Applied Mechanics, The Prager Anniversary Volume," Vol. 99, Macmillan, New York, 1963.
31. L. J. Walpole, *J. Mech. Phys. Solids*, **14**, 151 (1966).
32. L. J. Walpole, *J. Mech. Phys. Solids*, **14**, 289 (1966).
33. J. R. Willis, *J. Mech. Phys. Solids*, **25**, 185 (1977).
34. J. R. Willis, in "Advances in Applied Mechanics," Vol. 21, p. 1. Academic Press, New York, 1981.
35. G. W. Milton, and R. V. Kohn, *J. Mech. Phys. Solids*, **36**, 597 (1988).
36. R. M. Christensen, "Mechanics of Composite Materials," Wiley, New York, 1979.
37. L. J. Walpole, in "Advances in Applied Mechanics," Vol. 21, p. 169. Academic Press, New York, 1981.
38. L. J. Walpole, in "Fundamentals of Deformation and Fracture" (B. A. Bilby, K. J. Miller, and J. R. Willis, eds.), p. 91. Cambridge University Press, Cambridge, 1985.
39. Z. Hashin, *J. Appl. Mech.*, **50**, 481 (1983).
40. G. J. Dvorak, "On Uniform Fields in Heterogeneous Media," to be published in *Proc. Roy. Soc. London*, 1990.

41. R. Hill, "The Mathematical Theory of Plasticity," Oxford University Press, Oxford, 1950.
42. D. C. Drucker, in "Proc. 1st U. S. Natl. Congress Appl. Mech.," p. 487. ASME, New York, 1952.
43. P. M. Naghdi, in "Proc. Second Symp. Naval Structural Mechanics" (E. H. Lee and P. S. Symonds, eds.), p. 121. Pergamon Press, New York, 1960.
44. J. B. Martin, "Plasticity: Fundamentals and General Results." The MIT Press, Cambridge, Mass., 1975.
45. A. Phillips, C. S. Liu, and J. W. Justusson, *Acta Mechanica*, **14**, 119 (1972).
46. A. Phillips and H. Moon, *Acta Mechanica*, **27**, 91 (1977).
47. A. Phillips and C. W. Lee, *Int. J. Solids Structures*, **15**, 715 (1979).
48. D. L. McDowell, *J. Appl. Mech.*, **54**, 323 (1987).
49. J. L. Chaboche, *Int. J. Plasticity*, **2**, 149 (1986).
50. G. J. Dvorak, Y. A. Bahei-El-Din, Y. Macheret, and C. H. Liu, *J. Mech. Phys. Solids*, **36**, 655 (1988).
51. J. F. Mulhern, T. G. Rogers, and A. J. M. Spencer, *Proc. Roy. Soc. London*, **A301**, 473 (1967).
52. A. E. Green and J. E. Atkins, "Large Elastic Deformation," Oxford University Press, Oxford, 1960.
53. A. J. M. Spencer, "Theory of Invariants" in *Continuum Physics*, Vol. I (A. C. Eringen, ed.), p. 239. Academic Press, New York, 1971.
54. J. Casey and P. M. Naghdi, *J. Appl. Mech.*, **50**, 350 (1983).
55. P. J. Yoder and W. D. Iwan, *J. Appl. Mech.*, **48**, 773 (1981).
56. C. J. Wung and G. J. Dvorak, *Int. J. Plasticity*, **1**, 125 (1985).
57. H. Ziegler, *Q. Appl. Math.*, **17**, 55 (1959).
58. Y. A. Bahei-El-Din, "Plasticity Analysis of Fibrous Composite Laminates under Thermomechanical Loads," to appear in *ASTM-STP*.
59. Y. F. Dafalias and E. P. Popov, *J. Appl. Mech.*, **43**, 645 (1976).
60. J. Lin and G. J. Dvorak, "Local and Overall Elastic–Plastic Response of Particle-Reinforced Metal Matrix Composites," to be published.
61. R. Hill, *J. Mech. Phys. Solids*, **15**, 79 (1967).
62. G. J. Dvorak and Y. A. Bahei-El-Din, *Acta Mechanica*, **69**, 219 (1987).
63. R. Hill, *J. Mech. Phys. Solids*, **13**, 89 (1965).
64. G. J. Dvorak, M. S. M. Rao, and J. Q. Tarn, *J. Comp. Mater.*, **7**, 194 (1973).
65. G. J. Dvorak and M. S. M. Rao, and J. Q. Tarn, *J. Appl. Mech.*, **41**, 249 (1974).
66. G. J. Dvorak and M. S. M. Rao, *J. Appl. Mech.*, **43**, 619 (1976).
67. J. R. Rice, *J. Appl. Mech.*, **37**, 728 (1970).
68. J. R. Rice, *J. Mech. Phys. Solids*, **19**, 433 (1971).
69. R. Hill and J. R. Rice, *J. Mech. Phys. Solids*, **20**, 401 (1972).
70. R. Hill and J. R. Rice, *SIAM J. Appl. Math.*, **25**, 448 (1973).
71. A. J. M. Spencer, "Deformation of Fiber-Reinforced Materials," Oxford University Press, Oxford, 1972.
72. G. J. Dvorak and Y. A. Bahei-El-Din, *J. Mech. Phys. Solids*, **27**, 51 (1979).
73. G. J. Dvorak and Y. A. Bahei-El-Din, in "Proceedings of ARO/NSF Research Workshop on Mechanics of Composite Materials," (G. J. Dvorak, ed.), pp. 32–54, Duke University, Durham, N.C., 1978.
74. G. J. Dvorak and Y. A. Bahei-El-Din, *J. Appl. Mech.*, **49**, 327 (1982).
75. Y. A. Bahei-El-Din and G. J. Dvorak, *J. Appl. Mech.*, **49**, 740 (1982).
76. J. Aboudi, *Int. J. Engng Sci.*, **20**, 605 (1982).
77. Y. A. Bahei-El-Din, G. J. Dvorak, and S. Utku, *Computers and Structures*, **13**, 321 (1981).
78. Y. A. Bahei-El-Din and G. J. Dvorak, *J. Appl. Mech.*, **47**, 827 (1980).
79. W. S. Johnson, C. A. Bigelow, and Y. A. Bahei-El-Din, NASA Technical Paper 2187, 1983.

80. A. J. Svobodnik, H. J. Bohm and F. G. Rammerstorfer, in "Advances in Plasticity 1989" (A. S. Khan and M. Tokuda, eds.), p. 137, Pergamon Press, New York, 1989.

81. H. J. Greenberg, *Q. Appl. Math.*, **7**, 85 (1949).

82. D. C. Drucker, in "Calculus of Variations and its Applications, Proceedings of Symposia in Applied Mathematics," Vol. VIII, p. 7, McGraw Hill, New York, 1958.

83. P. G. Hodge, Jr., in "Engineering Plasticity," (J. Heyman and F. A. Leckie, eds.), p. 237, Cambridge University Press, Cambridge, 1968.

84. G. J. Dvorak and J. L. Teply, in "Plasticity Today: Modelling, Methods and Applications, W. Olszak Memorial Volume" (A. Sawczuck and V. Bianchi, eds.), p. 623, Elsevier, London, 1985.

85. J. L. Teply and G. J. Dvorak, *J. Mech. Phys. Solids*, **36**, 29 (1988).

86. M. L. Accorsi and S. Nemat-Nasser, *Mechanics of Materials*, **5**, 209 (1986).

87. B. Fraeijs De Veubeke, "Stress Analysis" (O. C. Zienkiewicz and G. S. Hollister, eds.), Ch. 9. Wiley, New York, 1965.

88. O. C. Zienkiewicz, "The Finite Element Method," McGraw-Hill, London, 1977.

89. J. L. Teply, Ph.D. Thesis, University of Utah, Salt Lake City, 1984.

90. Y. A. Bahei-El-Din, G. J. Dvorak, J. Lin, R. S. Shah, and J. F. Wu, "Local Fields and Overall Response of Fibrous and Particulate Metal Matrix Composites," ALCOA Laboratories Report, 1987.

91. J. F. Wu, M. S. Shephard, G. J. Dvorak, and Y. A. Bahei-El-Din, *Composites Science and Technology*, **35**, 347 (1989).

92. J. F. W. Bishop and R. Hill, *Phil. Mag.*, **42**, 414 (1951).

93. G. J. Dvorak and W. S. Johnson, *Int. J. Fracture*, **16**, 585 (1980).

94. G. J. Dvorak and E. C. J. Wung, in "Strain Localization and Size Effect Due to Cracking and Damage" (J. Mazars and Z. P. Bazant, eds.), Elsevier Applied Science Publishers, London, 1989.

95. G. J. Dvorak, Y. A. Bahei-El-Din, and L. C. Bank, *Engng. Fracture Mech.*, **34**, 87 (1989).

96. Y. A. Bahei-El-Din, G. J. Dvorak, and J. F. Wu, *Engrg. Fracture Mech.*, **34**, 105 (1989).

97. J. W. Hutchinson, *Proc. Roy. Soc. London*, **A319**, 247 (1970).

98. J. W. Hutchinson, *Proc. Roy. Soc. London*, **A348**, 101 (1976).

99. J. M. Duva, *J. Engng. Materials and Technology*, **106**, 317 (1984).

100. G. Weng, *Intl. J. Plasticity*, **1**, 275 (1985).

101. G. P. Tandon and G. Weng, *J. Appl. Mech.*, **55**, 126 (1988).

102. R. M. McMeeking, in "Mechanics of Material Behavior—The Daniel C. Drucker Anniversary Volume" (G. J. Dvorak and R. T. Shield, eds.), p. 275, Elsevier Science Publishers, Amsterdam, 1984.

103. Y. Benveniste, G. J. Dvorak, and T. Chen, *Mech. of Mater.*, **305**, 8 (1989).

Strengthening of Metal Matrix Composites Due to Dislocation Generation Through CTE Mismatch

R. J. ARSENAULT

Metallurgical Materials Laboratory
University of Maryland
College Park, Maryland

I.   Introduction

Numerous strengthening mechanisms have been proposed for composites (see Chapter 1). Therefore, a legitimate question may be raised: Why consider dislocation generation as an alternate mechanism? If one is considering continuous filament metal matrix composites (MMCs), then the need for considering dislocation generation is not that important. This does not mean that dislocation generation due to differences in the coefficients of thermal expansion of the filaments and matrix (ΔCTE) does not produce dislocations, it is just that the effect of these dislocations on the strength of the composite has not been determined and is probably not large. However, if one wishes to explain the observed strengthening of discontinuous metal matrix composites (DMMCs), then dislocation generation becomes extremely important. In this chapter, the reasons dislocation generation is important will be presented along with a discussion of the mechanisms of dislocation generation. This will be followed by a discussion of strengthening mechanisms due to the presence of dislocations and small subgrain size generated by thermal stresses.

II. Reasons for and Mechanisms of Dislocation Generation

In this section, a case for the importance of dislocation generation in a discussion of strengthening mechanisms of metal matrix composite will be made. Experimental evidence that dislocations are generated will be discussed in the next section.

The need to consider dislocation generation by the ΔCTE effect became evident when attempts to explain the observed strengthening of DMMCs by rule-of-mixtures (ROM) models failed. Let us briefly list the potential strengthening mechanisms:

- classical composite strengthening (load transfer),
- dispersion strengthening,
- residual elastic stresses,
- differences in texture,
- high dislocation densities due to dislocation generation as a result of differences in coefficients of thermal expansion, and
- small subgrain size as a result of the generation of a high dislocation density.

Evidence for or against each mechanism will be presented in the following.

A. Classical Composite Strengthening

The basic elements of the classical continuum composite strengthening are discussed in detail in the chapter by Dvorak. However, another continuum treatment, based on the Eshelby model [1], was carried out by Taya and Arsenault [2]. In Fig. 1, the ratio of the yield stress of a SiC_w composite (σ_{yc}) to the yield stress of the matrix material (σ_{ym}) is plotted as a function of V_w. The solid line indicates the experimental results, and the dashed line is the result of the Taya and Arsenault analysis. In the Taya and Arsenault analysis, the residual stress effect was not considered. It is obvious from the

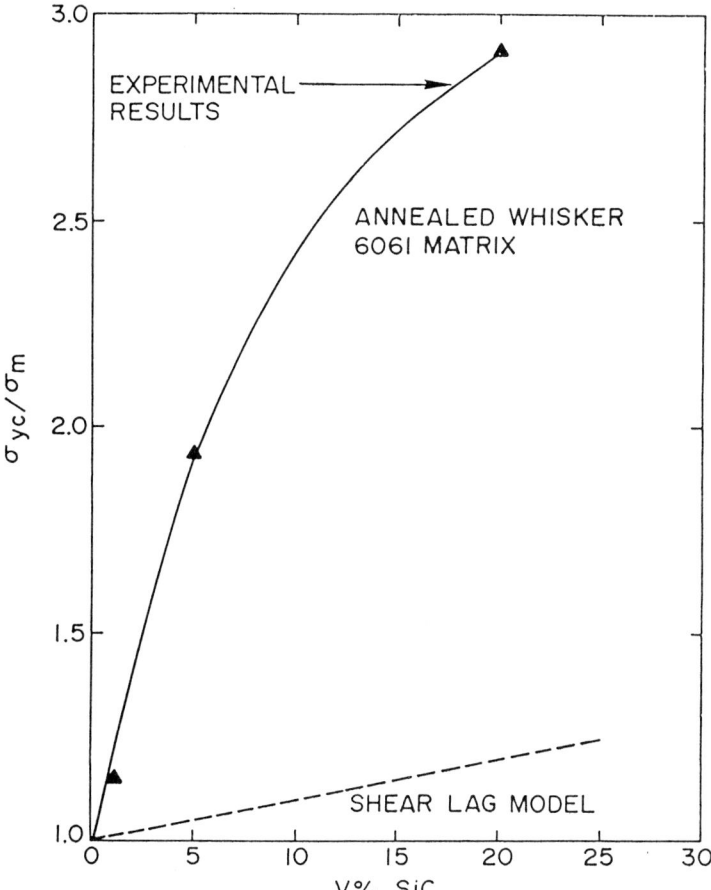

FIG. 1. The ratio of the yield stress of the composite (σ_{yc}) to that of the metal matrix (σ_{ym}) versus volume fraction of SiC whiskers.

plot that the Eshelby model is not capable of accounting for the observed strengthening. If the modified shear-lag model is considered (the *dashed* line in Fig. 1), the agreement between the predicted result and the experimental observed strengthening is even poorer.

B. Dispersion Strengthening

Another possible explanation of the observed strengthening of SiC/Al is based on the Orowan theory, which enables one to estimate the stress required for a dislocation to bypass a particle. The Orowan theory predicts

$$\sigma = \frac{2\mu\mathbf{b}}{\lambda},$$

where $\mathbf{b}$ is the Burgers vector and λ is the interparticle spacing. The particle spacing is the primary parameter controlling this stress. Taking an average particle spacing of 2 μm yields an approximate increase in composite strength of 5 MPa. This increase is too small compared with the 60 to 110-MPa-range increases observed experimentally [3].

C. Residual Elastic Stress

An analytical model by Taya and Arsenault [4] based on an ellipsoidal-shaped SiC particle in an Al matrix was developed that predicts that a tensile thermal residual stress should exist in the matrix for a whisker with a length-to-diameter ratio of 1.8. Furthermore, the longitudinal residual stress should be higher than the transverse residual stress. The experimental data obtained from a neutron diffraction analysis are shown in Table I. These neutron diffraction results do indicate that the matrix is in tension and that the longitudinal residual stress is higher than the transverse residual stress. However, the neutron data indicate a higher value of residual stress than predicted residual stress.

The model of Arsenault and Taya [4] also predicts that the yield stress in compression should equal the yield stress in tension if the SiC is in the form of spheres. However, in the case of spherical SiC composites, the experimentally measured tensile stress is higher than that of the compressive yield stress, whereas the difference predicted by the model is zero.

On the other hand, the model of Taya and Arsenault successfully predicts the differences in the tensile and compressive yield stress due to the thermal residual stresses observed in whisker composites where $\sigma_{yc}^C > \sigma_{yc}^T$. However, the model is completely incapable of predicting the absolute magnitude of the

TABLE I

TENSILE THERMAL RESIDUAL STRESS
NEUTRON DIFFRACTION MEASUREMENTS

Material	Longitudinal (MPa)	Transverse (MPa)
20 vol-% whisker SiC 1100 matrix	136–110	83–57

increase in yield stress, since it does not have the capacity to predict the matrix strengthening.

The effect of residual stress on the mechanical properties of DMMCs was further demonstrated in the work by Arsenault and Wu [5], who have shown that there is a difference in the compressive and tensile yield stress ($\sigma_{yc}^C - \sigma_{yc}^T$). The data shown in Fig. 2 are plotted as $\sigma_{yc}^C - \sigma_{yc}^T$ versus volume fraction of whisker SiC for a 6061 Al alloy matrix given a variety of heat treatments.

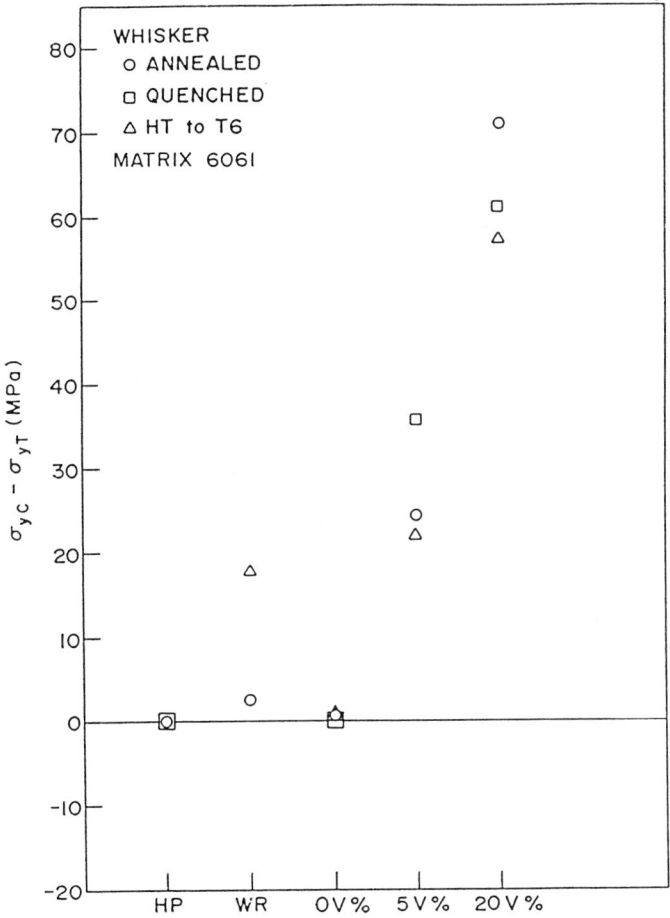

FIG. 2. A plot of the difference between the compressive yield stress and the tensile yield stress for various volume fractions of whisker composite materials (matrix, 6061 aluminum alloy); ○, annealed; □, quenched; △, heat-treated to T6 condition. Also plotted is the difference in yield stress for high-purity aluminum and wrought 6061 aluminum alloys.

In the case where matrix strengthening is not such a predominant factor, i.e., for continuous filament composites, the model is excellent in predicting the differences between tension and compression and the absolute values, as shown in Fig. 3.

Another manifestation of the residual stress is the asymmetrical behavior of the Bauschinger effect [5]. In homogeneous materials, i.e., unreinforced Al, the magnitude of the Bauschinger effect is independent of the direction of initial loading. However, in the case of SiC/Al discontinuous composites, the magnitude of the Bauschinger effect is strongly dependent upon initial loading. If the sample is initially tested in tension followed by reverse loading compression, the Bauschinger effect (Bauschinger strain) is small. If, however,

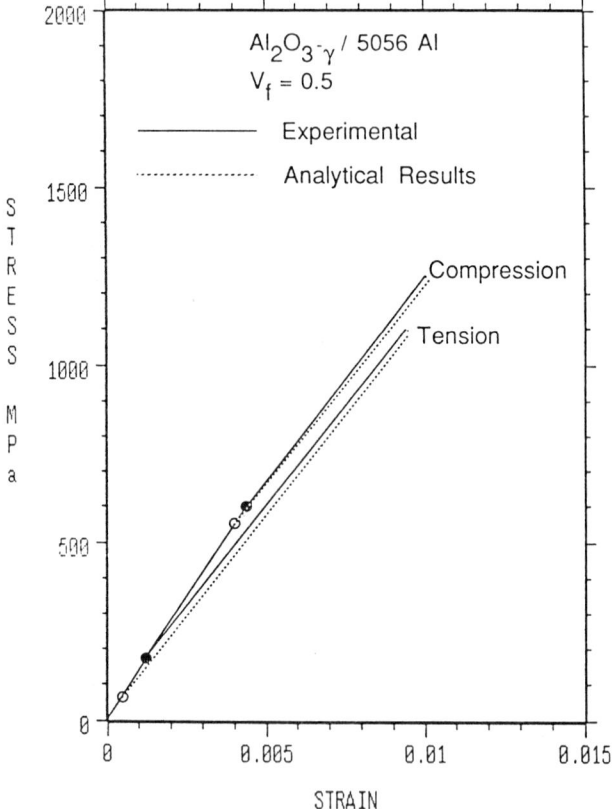

FIG. 3. Tension and compression stress–strain curves of continuous Al_2O_3 fiber/5056 Al composite with $V_f = 0.5$. The experimental and theoretical results are denoted by solid and dashed curves, respectively. The solid and open circles denote the yield stress of the experimental and theoretical results, respectively.

the test is initially begun in compression followed by tension, then there is
a large Bauschinger strain (infinity in some cases); see Fig. 4. The difference
is due to tensile thermal residual stress due to ΔCTE.

D. Texture

A texture investigation was undertaken by Arsenault [6], and from a
comparison of the data, it is apparent that there is little difference in the
texture of 99.99% Al, 0 vol-%, 6061 Al alloy and 20 vol-% 6061 Al alloy

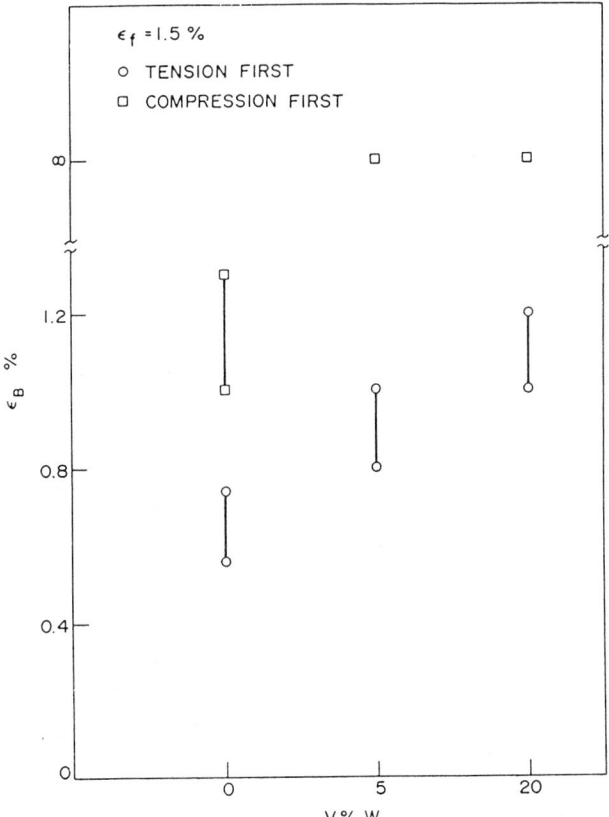

FIG. 4. A plot of the Bauschinger strain versus the volume fraction of whiskers in a
composite with a 6061 aluminum alloy matrix for both tension-first (○) and compression-first
(□) testing ($\varepsilon_F = 1.5\%$).

matrix composite, as shown in Fig. 5. Therefore, differences in texture between the matrix material and the composite cannot account for any real difference in strength between the matrix and the composite.

E. High Dislocation Density

1. INTRODUCTION

From a consideration of the possible strengthening mechanisms outlined in the previous section, it is obvious that they are not capable of explaining the observed strengthening of DMMCs. Therefore, other mechanisms were proposed [7, 8] to explain the observed strengthening. The most likely was the strengthening due to a high dislocation density within the matrix as a function of the ΔCTE between the reinforcement and the matrix. The fact that dislocations could be generated from the interface of a misfitting particle and a ductile matrix was studied by Ashby and Johnson [9] as early as the late 1960s. In the composite systems, the generation of dislocations due to misfit strain was also studied by several workers. In the investigation of the W/Cu system, Chawla and Metzger [10] observed, using etch pitting techniques, that the dislocation density is much higher at the W/Cu interface than in the bulk matrix, although the ΔCTE is only 4:1. Weatherly [11] stated that in the silica copper system, multiple dislocation tangles were actually observed around silica due to the ΔCTE. Similar results were also reported by Ashby et al. [12] and by Williams and Garmong in the Ni/W eutectic composite system [13].

Based on differences in the experimental increase observed in the strength of the composite and that predicted by classical mechanisms, Arsenault and Fisher [7] proposed that the increased strength observed in SiC/Al composites could be accounted for by postulating a high dislocation density in the aluminum matrix. An experimental simulation of dislocation generation in the SiC/Al system by analyzing the generation of slip lines around a SiC cylinder in an aluminum disk due to thermal cycling has been demonstrated by Flom and Arsenault [14]. However, high dislocation densities could be due to factors other than differences in thermal contraction. Another possibility is dislocation generation due to plastic deformation during materials processing that could be trapped during the annealing process by SiC particles. If large differences (6:1) in coefficients of thermal expansion are the primary cause for high dislocation density, dislocation generation should be seen upon cooling from annealing temperatures in an in situ high-voltage electron microscope (HVEM) experiment, as suggested by Arsenault and Fisher [7].

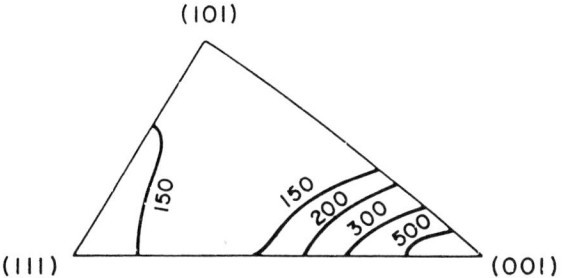

TEXTURE FOR EXTRUDED & ANNEALED
99.99 % Al

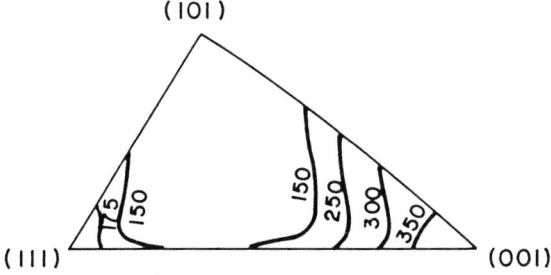

TEXTURE FOR EXTRUDED & ANNEALED
OV% SiC 1100 MATRIX

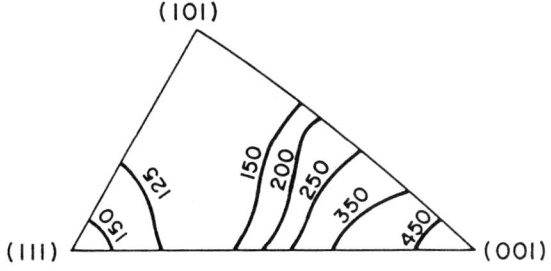

TEXTURE FOR EXTRUDED & ANNEALED
20V% SiC 1100 MATRIX

FIG. 5. The texture of two control Al samples and a 20 vol-% SiC whisker composite
in the annealed condition as determined by an x-ray technique.

An extensive investigation was undertaken by Vogelsang *et al.* [8] involving an in situ high-voltage electron transmission microscopy of the heating and cooling of SiC/Al composites. It was shown conclusively that a very high dislocation density could be produced in the Al matrix upon cooling the composite from 773 K to 300 K.

It should be pointed out that while fruitful results have been achieved from the aspect of dislocation behaviors, another dimension, which is based purely on classical mechanics, is quite active. Several workers, such as Hoffman [15], Garmong [16], Dvorak *et al.* [17], Mehan [18], and most recently, Nardone and Prewo [19], have explained the strengthening phenomenon due to the existence of reinforcement by analyzing the load transfer to the reinforcement. However, one of the shortcomings of these models is that none of them recognizes the microstructural change and its associated property change in the ductile matrix due to the presence of reinforcement as compared with matrix material in its pure form.

2. THEORETICAL MODEL FOR DMMCS

A simple model based on prismatic punching was explored [20] as a mechanism of generating dislocations due to thermal strains. In principle, this mechanism is applicable for both DMMCs and continuous metal matrix composites (CMMCs). However, in the present development, only DMMCs will be considered. Assumptions were made that both SiC and aluminum were elastically isotropic and that the SiC reinforcement was in the form of parallelepiped particles. Prismatic punching was assumed to occur equally on all faces of the particles, as shown in Fig. 6.

Consider the reinforcement particles with dimensions of height t_1, width t_2, and thickness t_3; the misfit strain due to the difference of CTE between aluminum and SiC is

$$\varepsilon = \varepsilon_i = \frac{\Delta t_i}{t_i} = \frac{\Delta CTE \times \Delta T}{2} \qquad (i = 1, 2, 3), \qquad (1)$$

where ΔT is the temperature difference and the total number of dislocation loops may be expressed as

$$N_i = t_{ii} \frac{\varepsilon}{\mathbf{b}}, \qquad (2)$$

where N_i is the number of prismatic loops punched in the ith dimension, ε is the misfit strain, $\mathbf{b}$ is the length of the Burgers vector. The factors, t_{ii}, form a specific case defined as the contribution tensor, t_{ij}, the contribution of

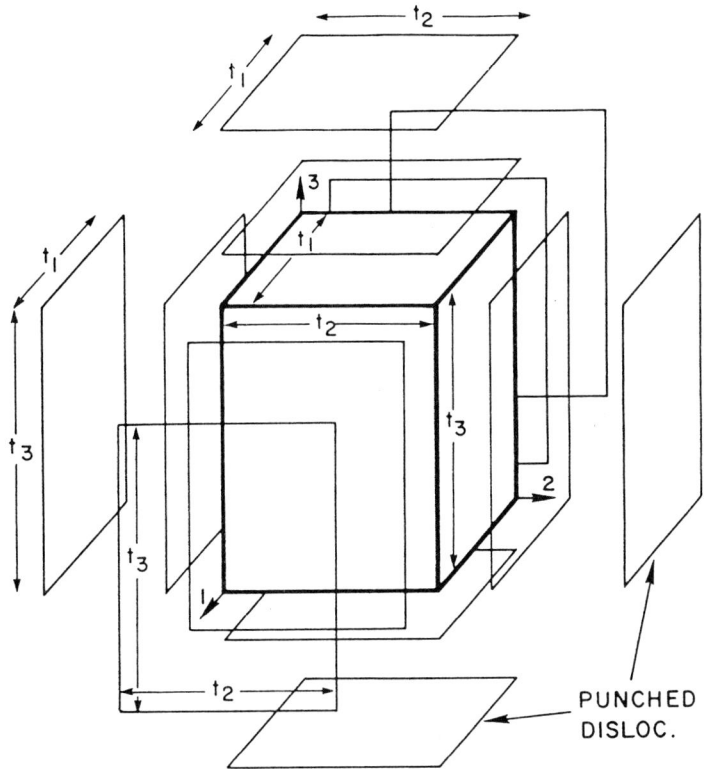

FIG. 6.   A schematic diagram of the particle and several punched dislocations.

the particle height in the ith dimension in punching a dislocation loop in the ith direction where in this case i is equal to j. It is believed that by taking account of dislocation back stresses and dislocation interaction, t_{ii} can be written in a general form,

$$t_{ii} = f(t_i, k^*, k, v, v^*), \tag{3}$$

where k^* and k are the bulk modulus of the particle and the matrix, respectively, v^* and v are Poisson's ratios for the particle and matrix, respectively, and t_i is the actual dimension in the ith dimension of the particle.

With the help of the contribution tensor, the total length of dislocation loops punched out in the ith direction (i) is given by

$$1_i = \frac{2\varepsilon}{\mathbf{b}} t_{ij}\delta_{ij}t_{ik}(1 - \delta_{ik}), \tag{4}$$

where δ_{ij} and δ_{ik} are the Kronecker deltas. The Einstein suffix notation is used here.

From a consideration of experimental data, further simplifications can be made. The speed of dislocation motion was at least as fast as the rate of dislocation generation [8]. This would result from relatively low friction stress and, therefore, result in relatively low back stress. These two factors are then neglected in the subsequent derivations. It is further assumed that the misfit strain is so relaxed that the residual elastic strain results from misfit of the magnitude, which is less than one Burgers vector; and the interaction stresses among different particles and those between punched dislocations from one particle with the dislocation punched out by another platelet are neglected.

Based on these considerations, the proceeding formulations can be further simplified. On the assumption of rigid expansion,

$$t_{ii} = t_i. \tag{5}$$

Then

$$N_i = \left[t_i \frac{\varepsilon}{\mathbf{b}} \right], \qquad (i = 1, 2, 3) \tag{6}$$

where N_i is an integer, and the square brackets represent the closest smaller integer taken from the real number enclosed. Since N_i is a large number,

$$N_i \cong t_i \frac{\varepsilon}{\mathbf{b}} = \frac{\Delta t_i}{\mathbf{b}}, \tag{6'}$$

and it is also reasonable to assume that

$$t_{ij} = t_{kj} = t_j.$$

Therefore, Eq. (4) can be expanded in the simple form

$$1_1 = \frac{2\varepsilon}{\mathbf{b}} t_1 (t_2 + t_3),$$

$$1_2 = \frac{2\varepsilon}{\mathbf{b}} t_2 (t_1 + t_3), \tag{7}$$

$$1_3 = \frac{2\varepsilon}{\mathbf{b}} t_3 (t_1 + t_2),$$

which may also be directly extracted from Fig. 6. Thus, the total length of the dislocations generated by one particle is

$$L = \sum 1_i = 1_1 + 1_2 + 1_3. \tag{8}$$

Without considering the dislocation back stresses and dislocation–particle interaction, the arrangement of the particles will not affect the dislocation density ρ due to the difference between the thermal expansion coefficients of matrix and reinforcement. The number of particles in the unit volume is

$$n = \frac{V_p}{t_1 t_2 t_3}, \tag{9}$$

where n is the number of particles in unit volume and V_p is the volume fraction of particles. Therefore, the total length of the dislocations generated by all particles will be

$$L_L = nL = (l_1 + l_2 + l_3)\frac{V_p}{t_1 t_2 t_3}, \tag{10}$$

and the dislocation density generated by the particles in the matrix is

$$\rho = \frac{L_L}{1 - V_p} = \frac{4V_p \varepsilon}{b(1 - V_p)}\left(\frac{1}{t_1} + \frac{1}{t_2} + \frac{1}{t_3}\right). \tag{11}$$

Rearranging Eq. (11), the following is obtained

$$\rho = \frac{2V_p \varepsilon}{b(1 - V_p)}\frac{S}{v'} = ks, \tag{12}$$

where k is a constant, S is the total surface area of a particle, s is the total surface area per unit volume of SiC (that is the specific particle surface area), and v' is the volume of the particle. This means that from Eqs. (11) and (12) for the same volume fraction, in general, the smaller the particle size or the larger the surface area per unit volume of reinforcement material, the *higher* a dislocation density will be produced.

Reconsidering Eq. (12), and using Eq. (9), Eq. (12) will become

$$\rho = \frac{2\varepsilon A_n}{b}S = cS, \tag{12a}$$

where A_n is the number of particles per unit matrix volume. If c is treated as a constant, and by further considering original dislocation density in the matrix (although it is small $\rho_0 \times 8.0 \times 10^{12}\,\text{m}^{-2}$ [8]),

$$\rho_{\text{total}}(S) = \rho_0 + cS. \tag{12b}$$

This has a form of linear approximation of Taylor's series of expansion of $\rho_{\text{total}}(S)$.

Let us consider the cases of composites with a whisker ratio of aspect ratio $(R = D/d)$ of 0.5, and a platelet of $R = 2$, where D is the diameter and d is the thickness (length) of the particle. The dislocation densities for both cases are

$$\rho_w = \frac{10V_p\varepsilon}{b(1 - V_p)}\frac{1}{t},\tag{13}$$

$$\rho_p = \frac{8V_p\varepsilon}{b(1 - V_p)}\frac{1}{t},\tag{14}$$

with t representing the smallest dimension, while ρ_w and ρ_p stand for the dislocation densities in the whisker and platelet composites, respectively. It can be seen that the prismatic dislocation density is higher in a whisker composite than in a platelet composite of the same volume fraction (Fig. 7). In general, the dislocation density ρ due to punching can be written as

$$\rho = \frac{BV_p\varepsilon}{b(1 - V_p)}\frac{1}{t},\tag{15}$$

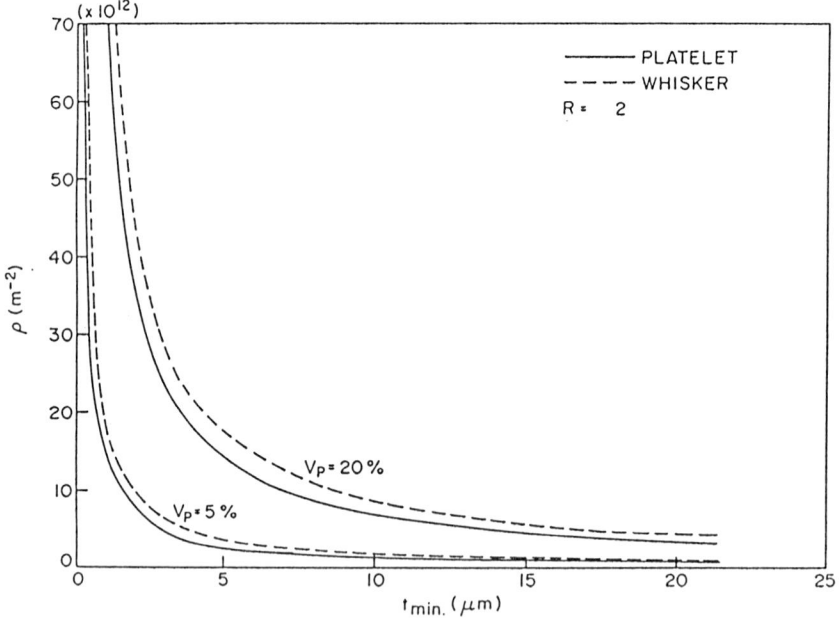

FIG. 7. Calculated dislocation density ρ due to prismatic punching as a function of minimum particle thickness, t, for both whiskers (– – –) and platelets (———) of the same volume fraction, V_p, according to Eq. (9).

where B is a geometric constant that is theoretically between 4 ($R = \infty$) and 12 ($R = 1$).

This analysis is of the lower-bound determination of the dislocation density, and in principle, the upper bound should be easily calculated under the same constraints. An attempt was made by Shi and Arsenault [21]. The stress field about the reinforcement due to the thermal mismatch was calculated by a finite-element method (FEM), and then this stress was relaxed by dislocation generation. Two additional constraints were included in the calculation. In order to prevent voids from forming upon relaxation, geometrical dislocations are necessary. Depending upon the orientations of the slip systems with respect to the reinforcement, the efficiency of the mismatch relation changes. The net result of just these two considerations resulted in extremely high dislocation densities.

The densities obtained approached infinity. These results were obviously wrong; a three-dimensional analysis is required that takes into account creep, subgrain formation, and the interaction stresses between dislocation generated by neighboring particles. Unfortunately, the computer time required to complete this is extremely long.

In summary, the observed dislocation densities in DMMCs can be accounted for. The lower-bound calculated densities are lower than the observed densities, but the attempted upper-bound calculated densities are well above the observed densities. Therefore, any concerns that the thermal strains or stresses are insufficient to produce the observed dislocation density can be completely eliminated.

3. Continuous MMCs

The importance of dislocation generation in terms of strengthening of CMMCs has been completely ignored. The probable reason is that in the use of the classical continuum models (ROM) strengthening, a change in the matrix strength due to increase in dislocation density has only a small effect on the overall strength of the composite.

However, three points should be raised. First, dislocation generation due to ΔCTE has been observed in carbon/Al composites [22]. Secondly, the prismatic punching model can be easily applied to CMMCs to predict the lower bound of dislocation density. Finally, Taya and Mori [23] have argued that if the filament exceeded a given length, dislocation generation would not occur.

It appears that dislocation generation due to ΔCTE of continuous filament would be a fruitful area of theoretical investigation, since the current theoretical prediction of Taya and Mori [23] is in conflict with the experimental results.

F. Small Subgrain Size

The observation of a much smaller subgran size in DMMC [7] has led Arsenault [6] to propose that subgrain strengthening could be a significant contributing factor to the strength of the composite. The initial observations of the small subgrain size was in DMMC in which the interparticle spacing was small, and the argument was put forward that the small subgrain size in the DMMC was due to the small interparticle spacing. However, the subgrain size about a large (250 μm) particle was also found to be small [20]. Therefore, the interparticle spacing does not control the subgrain size.

The small subgrain size is due to high density of dislocations generated by the thermal stress. The details of how the subgrains are formed are not obvious. However, the reason the subgrain boundaries form is that this is a lower-energy dislocation configuration.

II. Strengthening Mechanisms

A. Dislocation-Density Strengthening

If an increase of dislocation density is believed to contribute to strengthening, and the increased density is uniformly distributed in the matrix (which it is not; the heterogeneity of dislocation density will be discussed later), the following equation can be used to relate dislocation density to strength.

$$\Delta\sigma = \beta\mu\mathbf{b}\rho^{1/2}, \tag{16}$$

where $\Delta\sigma$ is the increase in yield strength, ρ is the increase in dislocation density over that of the matrix density, and β is a geometric constant. Hansen [24] obtained a value for β of 1.25 for aluminum, which will be used here. Substituting Eq. (15) into Eq. (16) gives

$$\Delta\sigma = \beta\mu\mathbf{b}\left(\frac{V_p}{1 - V_p}\frac{B\varepsilon}{\mathbf{b}}\right)^{1/2}\left(\frac{1}{t}\right)^{1/2} \tag{17}$$

Figure 8 shows the increase of yield strength as a function of platelet diameter since the diameter is easier to measure experimentally. Two experimental points are included for comparison [25]. Generally, when the diameter of the platelet was increased, the effect of strengthening decreased drastically. Also, the experimental data fell near the theoretical curve when the diameter of the platelet is moderate. Yet, when the particle size is large, the experimental results do not agree as well as when the particle size is small.

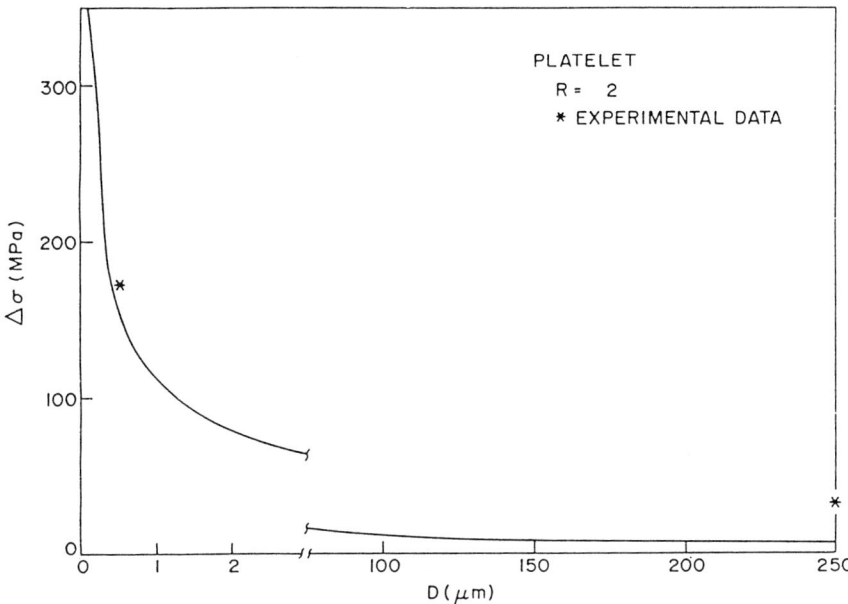

FIG. 8.   The calculated increase in yield strength of the composite over that of the matrix material on account of the dislocations due to the differential thermal contraction of aluminum and SiC, given by $\Delta\sigma = \sigma_{yc} - \sigma_{ym}$, where σ_{yc} is the yield stress of the matrix on the basis of Eq. (14) using $B = 8$, $A = 0.2$, $\mu = 26 \times 10^3$ MPa, $b = 2.86 \times 10^{-10}$ m, $R = D/d = 2$.

In the above discussion, the degree of strengthening is a function of a particle size; however, an explicit picture of how such a material property changes with morphology of the particle is also needed. Upon manipulating Eq. (11), the following equation can be obtained,

$$\rho = \frac{4V_p\varepsilon}{b(1 - V_p)}(R)^{2/3}\left(1 + \frac{2}{R}\right)\left(\frac{1}{v'}\right)^{1/3}, \tag{18}$$

where v' is the volume of the particle and $R = D/d$. It is not difficult to show that $R < 1$ represents whisker morphology, $R > 1$ is a representation of platelet morphology, and $R = 1$ is an equiaxial particle that is an approximation of a spherical reinforcement. Upon subsequent substitution into Eq. (16), it becomes

$$\Delta\sigma = 2\beta\mu\left(\frac{bV_p\varepsilon}{1 - V_p}\right)^{1/2}(R)^{1/3}\left(1 + \frac{2}{R}\right)^{1/2}\left(\frac{1}{v'}\right)^{1/6}. \tag{19}$$

Equivolume plots of ρ and $\Delta\sigma$ as functions of R are shown in Figs. (9) and (10), respectively, where the volume of the individual particle is assumed

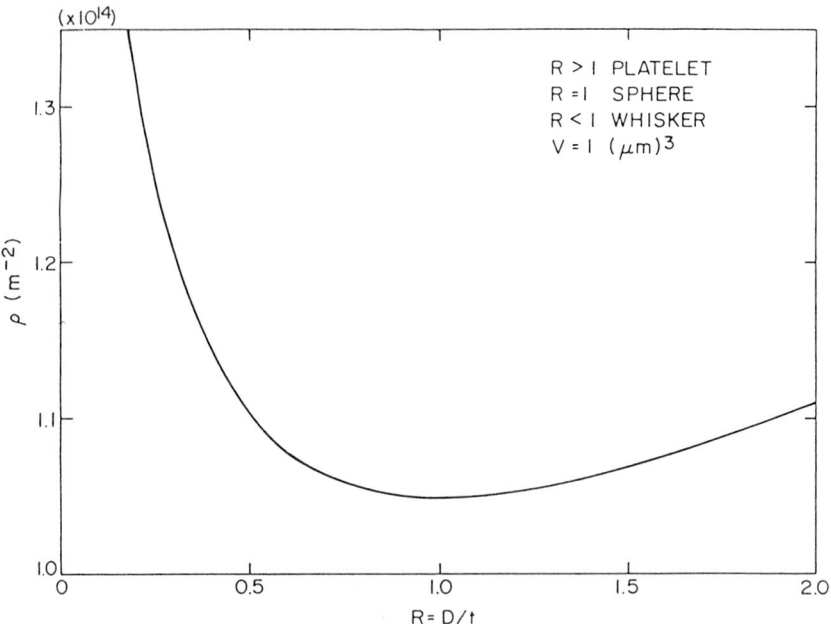

FIG. 9. Calculated dislocation density (ρ) versus aspect ratio (R); when $R < 1$, it represents platelet; when $R > 1$, it represents whisker; and when $R = 1$, it represents cube (sphere).

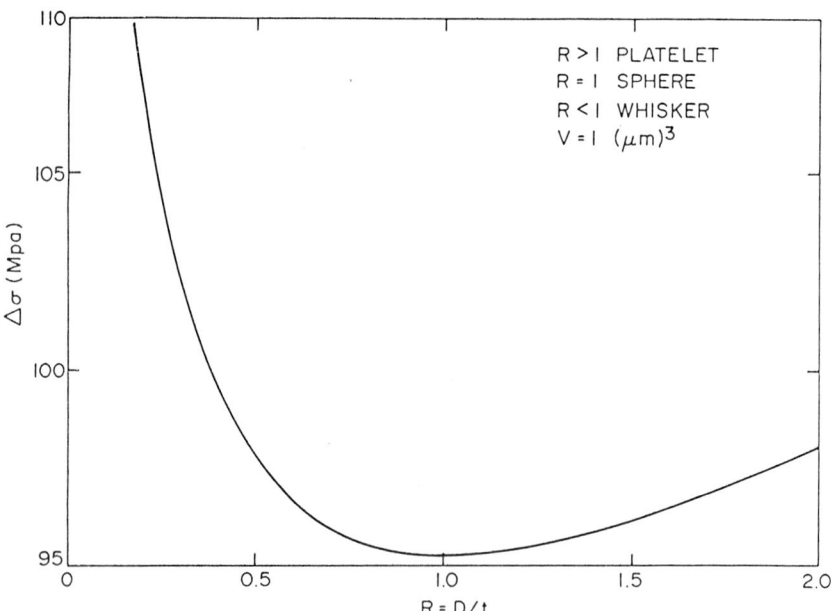

FIG. 10. The increase of yield strength as a function of the particle aspect ratio when the volume of the particle is constant and where $v' = 1\mu m^3$ is predicted from Eq. (19).

to be 1 μm^3, and have a constant V_p. Equation (19) predicts that a spherical composite is not recommended as far as increase of strengthening is concerned. For the same particle volume and volume fraction, the tensile strength of the platelet composite increases ~ 20 to 25% over that of the spherical composite as the aspect ratio increases from one to five.

Since experimentally producing equivolume particles of different morphologies is not a trivial task, the equation was checked with experimental data [25] by using reinforcements with different particle volumes, as shown in Table II. It is worthwhile to mention that although the increase of strengthening in the spherical composite is larger than that in the whisker sample, it does not contradict the result shown in Fig. 10, since the particle parameters are different.

The idea that high dislocation density in the matrix is due to a large difference in coefficient of thermal expansion and can account for the increase of strength of composites, as proposed by Arsenault and Fisher [7], has been proven to a first order by a simple prismatic-punching theoretical model. It is that the experimental data [25] in Fig. 8 and Table II agree quite well with the theoretical prediction when the particle size is moderate and are several factors off when the particle size is large. The prismatic-punching model is the most efficient model. Therefore, for larger particle sizes, the prismatic-punching model underestimates the density of dislocations generated. Future work must consider the fact that the dislocation density within the matrix is not uniform and that the density is highest near the reinforcement particle and decreases to the midpoint between particles.

It has been shown by FEM analysis [26] that plastic deformation begins at the boundary of the plastic zone created by thermal mismatch stress (ΔCTE). This effect is readily evident with a lower volume fraction, i.e., 10%

TABLE II

INCREASE OF STRENGTHENING OF DIFFERENT COMPOSITES[a]

(1100 Al MATRIX)

Particle parameter	Increase in strength $\Delta\sigma$ (MPa)	
	Theoretical prediction MPa	Experimental data MPa
$R = 1$ $V' = 0.125\ \mu$m^3	130.4	166
$R = 0.25$ $V' = 0.5\ \mu$m^3	124.2	152

[a] Additional experimental data is shown in Fig. 8. V' is the volume of the particulate.

or less. At a higher volume fraction, there is no well-defined boundary, but upon external loading, deformation initially begins in the least plastically deformed region. The significance of this result is that the proportional limit of the composite (when one is definable) is approximately the yield stress of the unreinforced matrix. Therefore, initially, there is a rapid rate of work hardening as the low dislocation regions are "exhausted"; and when the macro-yield stress range has been obtained, there are localized regions of more uniform dislocation density. It is apparent that these uniform regions control the macro-yield stress (0.2% plastic strain).

Although the mechanism based on dislocation generation produced results in reasonable agreement with the experiment (e.g., Table II), it is necessary to account for the contributions of all the strengthening mechanisms listed in the Introduction. There has been only one investigation [27] in which all

TABLE III
A LISTING OF THE STRENGTHENING COMPONENTS OF THE WHISKER
SiC 1100 Al MATRIX

$$\Delta\sigma^T_{yc} = \Delta\sigma_{disl} + \Delta\sigma_{sg} \pm \Delta\sigma_{res} + \Delta\sigma_{tex} + \Delta\sigma_{comp} + \Delta\sigma_{dep}$$

The calculated change in strength due to predicted increase in dislocation density;

$$\Delta\sigma_{disl} = 130.4 \text{ MPa}.$$

The empirical predicted increase in strength due to reduced subgrain size;

$$\Delta\sigma_{sg} = 55 \text{ MPa}.$$

The calculated change in strength due to the tensile thermal residual stress;

$$\Delta\sigma_{res} = 34.5 \text{ MPa}.$$

The predicted change in strength due to measured differences in texture between composite and matrix;

$$\Delta\sigma_{tex} = 0.$$

The calculated change in strength due to classical composite-strengthening theories;

$$\Delta\sigma_{comp} = 0.$$

The calculated change in strength due to the predictions of dispersion-strengthening theories;

$$\Delta\sigma_{dep} = 0.$$

Predicted

$$\Delta\sigma^T_{yc} = 130.4 + 55 - 34.5 = 150.9 \text{ MPa}.$$

Experimental

$$\Delta\sigma^T_{yc} = 144.9 \text{ MPa}.$$

of these factors have been considered to predict a composite yield stress. Table III and Table IV present comparisons of predicted composite yield stresses with the experimental result for SiC$_w$ and SiC$_s$, respectively. The experimental data and theoretical predictions are in reasonable agreement, confirming the validity of this approach.

B. Subgrain Strengthening

At present, a detailed mechanism of how a change in subgrain size results in a change in the strength is not known. However, an empirical relationship has been developed by McQueen and Jonas [28] from which the predicted increase in strength due to the reduced subgrain size is $\sim \frac{1}{2}$ the increase in strength due to the increased dislocation density.

TABLE IV

A LISTING OF THE STRENGTHENING COMPONENTS OF THE SPHERICAL
SiC 1100 Al MATRIX

$$\Delta\sigma_{yc} = \Delta\sigma_{disl} + \Delta\sigma_{sg} \pm \Delta\sigma_{res} + \Delta\sigma_{tex} + \Delta\sigma_{comp} + \Delta\sigma_{dep}.$$

The calculated change in strength due to predicted increase in dislocation density;

$$\Delta\sigma_{disl} = 124.2 \text{ MPa}.$$

The empirical predicted increase in strength due to reduced subgrain size;

$$\Delta\sigma_{sg} = 55 \text{ MPa}.$$

The calculated change in strength due to the tensile thermal residual stress;

$$\Delta\sigma_{res} = 0.$$

The predicted change in strength due to measured differences in texture between composite and matrix;

$$\Delta\sigma_{tex} = 0.$$

The calculated change in strength due to classical composite-strengthening theories;

$$\Delta\sigma_{comp} = 0.$$

The calculated change in strength due to the predictions of dispersion-strengthening theories;

$$\Delta\sigma_{dep} = 0.$$

Predicted

$$\Delta\sigma_{yc}^T = 124.2 + 55 = 179.9 \text{ MPa}.$$

Experimental

$$\Delta\sigma_{yc}^T = 172.5 \text{ MPa}.$$

References

1. J. D. Eshelby, *Proc. Roy. Soc. Lond.*, **A241**, 376 (1957).
2. M. Taya and R. J. Arsenault, *Scripta Metall.*, **21**, 349 (1987).
3. Y. Flom and R. J. Arsenault, *J. Metals*, **38**, 31 (1986).
4. R. J. Arsenault and M. Raya, *Acta Metall.*, **35**, 651 (1987).
5. R. J. Arsenault and S. B. Wu, *Mater. Sci. Eng.*, **96**, 77–88 (1987).
6. R. J. Arsenault, "Composite Structures" (I. H. Marshall, ed.), p. 70, Elsevier, London, 1987.
7. R. J. Arsenault and R. M. Fisher, *Scripta Metall.*, **17**, 67 (1983).
8. M. Vogelsang, R. Fisher, and R. J. Arsenault, *Metall. Trans.*, **17A**, 379 (1986).
9. M. F. Ashby and L. Johnson, *Phil. Mag.*, **20**, 1009 (1969).
10. K. K. Chawla and M. Metzger, *J. Mater. Sci.*, **7**, 34 (1972).
11. G. C. Weatherly, *Mater. Sci. J.*, **2**, 237 (1968).
12. A. M. F. Ashby, S. H. Gelles, and L. E. Tanner, *Phil. Mag.*, **19**, 757 (1969).
13. J. C. Williams and G. Garmong, *Metall. Trans.*, **6A**, 1699 (1975).
14. Y. Flom and R. J. Arsenault, *Mater. Sci. Eng.*, **77**, 191 (1986).
15. C. A. Hoffman, *J. Eng. Mater. Tech.*, **95**, 55 (1973).
16. G. Garmong, *Metall. Trans.*, **5**, 2183 (1974).
17. G. J. Dvorak, M. S. M. Rao, and J. Q. Tarn, *J. Comp. Mater.*, **7**, 194 (1973).
18. R. L. Mehan, "Metal Matrix Composites," p. 43. ASTM STP, Philadelphia, 1968.
19. V. C. Nardone and K. M. Prewo, *Scripta Metall.*, **20**, 43 (1986).
20. R. J. Arsenault and N. Shi, *Mater. Sci. Eng.*, **81**, 175 (1986).
21. N. Shi and R. J. Arsenault, unpublished results.
22. Mohan Misra, private communication. Martin Marrietta, Denver, 1987.
23. M. Taya and T. Mori, *Acta Metall.*, **35**, 155 (1987).
24. N. Hansen, *Acta Metall.*, **25**, 863 (1977).
25. Metallurgical Materials Laboratory, University of Maryland, College Park, unpublished research, 1985.
26. N. Shi and R. J. Arsenault, unpublished results.
27. R. J. Arsenault, *J. Comp. Res. Tech.*, ASTM, **10**, 140 (1988).
28. H. J. McQueen and J. J. Jonas, "Treatise on Materials Science and Technology" (R. J. Arsenault, ed.), Vol. 6, p. 394. Academic Press, New York, 1975.

Strengthening Behavior of In Situ Composites

T. H. COURTNEY

Department of Materials Science
University of Virginia
Charlottesville, Virginia

I. Introduction

In situ composites constitute a subclass of metal matrix composites having distinct advantages over many metal matrix composites on several accounts. One is cost. For example, the manufacturing of particulate metal matrix composites frequently involves costly, powder metallurgical techniques, and metal matrix composites utilizing long fibers are even more expensive to manufacture. Their processing can often be characterized as the "cottage industry" kind. For example, they are frequently made by laying down individual fibers between metal surfaces, followed by a consolidation process such as diffusion bonding. The high cost of such fabrication is exacerbated by the expense of the fibers. Economic considerations such as these usually restrict the conventional use of metal matrix composites to high performance applications, for which cost is a decidedly secondary concern.

101

In situ composites offer the potential for improved production economy, since a composite morphology is produced directly in such materials during their manufacture. In addition, the strengths of some in situ composites can be impressive. The idea of using in situ composites is not new; directionally solidified eutectic composites[1] were first investigated for their potential as high temperature structural materials in the 1960s [1–15]. A schematic of the solidification process is given in Fig. 1a (variants of the technique have been used to produce composites via directional decomposition of monotectics and eutectoids). The liquid is slowly solidified by withdrawing it from the furnace (growth rates are typically on the order of cm per hr) in the presence of a high-temperature gradient (usually on the order of 10^4 K/m or higher). Several factors have mitigated against the use of directionally solidified composites. Among them are the relatively low growth rates required to maintain plane front solidification, restrictions on alloy compositions, and other important considerations such as the different thermal expansion coefficients of the composite constituents, which can lead to significant thermal fatigue effects [16–21].

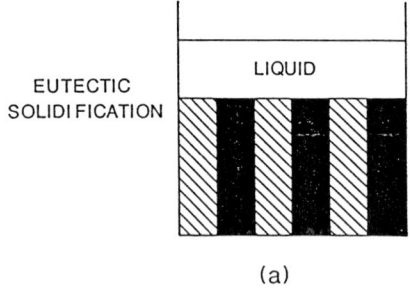

(a)

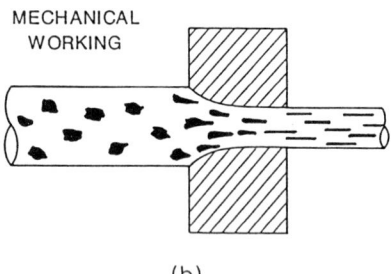

(b)

FIG. 1. Different techniques for producing in situ composites. (a) Directional solidification produces an aligned two-phase structure under appropriate solidification conditions. (b) Extensive cold deformation of a two-phase solid can yield a microstructure aligned along the deformation direction.

Strengthening in directionally solidified composites is provided by a reinforcing phase, e.g., Cr or TaC. Composite strengths usually obey a volume fraction rule; i.e., the flow behavior is a volumetric average of the flow behavior of the individual phases of the composite. Refinement of microstructural scale is associated with increased strengths. Incremental strengths generally correlate with interphase spacing to the $-1/2$ power [*11, 12, 22, 23*]; i.e., yield strengths are described by a Hall–Petch relationship [*24, 25*]. There are limitations to using microstructural refinement for increasing strengths of these materials. In particular, finer spacings necessitate increased solidification rates. However, beyond a critical (usually relatively low) solidification rate, a planar liquid–solid interface is no longer maintained, and the resulting solidification structure is cellular or dendritic, rather than compositelike. Reduced strengths are concomitant with such structural transitions.

Another way of fabricating in situ composites is illustrated in Fig. 1b. Such processing is the basis for producing heavily deformed in situ composites (HDISC), and the properties and structures of these materials are the focus of this chapter. In the process, a conventional primary fabrication scheme (e.g., casting) is used to produce a two-phase material. Following this, the material is subjected to extensive deformation via one or more of several processes (swaging, drawing, rod rolling, and sheet rolling have all been used to manufacture HDISCs). As indicated in Fig. 1b, the deformation produces a phase morphology akin to that found in conventional composites. However, this is not the reason this kind of processing is useful. Rather, it is that two-phase materials work-harden more rapidly than do single-phase ones. Indeed, impressive strengths can be generated in two-phase malleable materials subjected to extensive cold working. Patented steel wire is the prototypical HDISC [*26–28*]. This cold-worked pearlitic steel can manifest tensile strengths on the order of 4200 MN/m^2 ($\approx 600,000$ psi $= 0.02$ E, where E is the tensile modulus). Since the remarkable strengths that can be generated by cold working pearlite have been known for so long, it is somewhat surprising that only recently have efforts been made to investigate the strengths developed by cold working other two-phase metals. The results of these studies are described in the following sections. However, before we do this, we note that this phenomenon—i.e., the development of high strengths by extensive cold working of two-phase solids—has not yet been extensively commercialized. There is a good reason for this. Appreciable excess strengths (i.e., considerably above those expected on the basis of the work-hardening of the individual phases) are found only after extensive deformation. A reasonable rule of thumb is that modest increases in strength are found if a two-phase material is cold-worked to a true deformation strain of about 3 (RA = 95%), whereas truly impressive strengths are generated only by cold working to true strains of 7 (= 99.9% RA) or more. For example, if a 10-cm ingot is used as a starting workpiece, a true deformation strain of 7 yields a wire having a diameter of 0.30 cm. As a result, production of HDISCs is restricted to thin wires or foils at this stage of their development.

II. Processing

Details relative to HDISC processing are provided in the chapter by Spitzig in Volume 32 [29]. We summarize here only those salient features indicative of the flexibility of this manufacturing technique.

HDISC processing techniques have evolved from technology used to fabricate superconducting solenoids. Microstructural requirements for wires used in these devices include dispersing fine superconducting filaments in a high electrical and thermal conductivity matrix such as copper [30, 31]. Solenoids based on Nb–Ti alloys can be made by a "bundle-and-draw" technique. In this scheme, a cylinder of Nb–Ti is first inserted into a jacket of Cu. This bundle is then cold worked; e.g., by wire drawing. Following a certain deformation strain, the wire is cut into segments, rebundled, jacketed again in Cu, and the process is repeated as often as necessary to produce the size and spatial distribution of the Nb–Ti filaments required for solenoid operation. Somewhat similar techniques have been utilized for the manufacturing of solenoid wire based on the Al5 phase. This structure, for which Nb_3Sn serves as a prototype, is inherently brittle, and this precludes wire containing it being made in the same way as Nb–Ti solenoid wire. Instead, Nb wire is imbedded in a Cu or bronze matrix, and the "composite" is deformation processed to final size. The Nb is then reacted with Sn from either an external (if the matrix is Cu) or internal (if it is bronze) source.

Research on HDISCs (excepting the much earlier development of patented steel wire) was initiated in the 1970s. Seminal studies were carried out by Wasserman [32–35] and Bevk [36–40] and their coworkers. Most of the alloys first studied were initially processed by casting, but solid and liquid powder metallurgy techniques have also been used to produce starting material. As mentioned, a number of processes, including wire drawing, cold rolling, extrusion, and swaging have been used for the subsequent deformation processing. A large number of two-phase combinations have been made using these processing routes; a partial listing of these alloys and the techniques used to make them is given in Table I. It is of interest that some surprising combinations are amenable to producing HDISC. Copper–chromium alloys are one example [42, 49]. Chromium is among the most brittle of the *bcc* refractory metals; in polycrystalline form, it is not susceptible to extensive room-temperature plastic flow. Isolated dendritic single crystals of Cr are produced in Cu–Cr castings, provided the Cr volume fraction is not too great. During wire drawing, these dendrites deform commensurately with the Cu matrix and are elongated into thin filaments in the draw direction. If higher Cr content castings (such that the Cr volume fraction exceeds about 20%) are made, contiguity of the Cr phase is established. These alloys are incapable of being extensively cold worked. Fracture along Cr–Cr inter-particle contacts initiates composite fracture during working. Evidently the criterion for a two-phase material to be capable of being made into a HDISC is that the less ductile phase maintain strain compatibility across interphase or interparticle boundaries during processing. Thus, the less ductile phase must at least possess the requisite number of slip systems to do this. Chromium has the requisite number, and when it exists in isolated form, strain accommodation across Cu–Cr boundaries is maintained. When Cr–Cr contiguity is established, deformation of the composite is similar to that of polycrystalline Cr. In effect, Cu–Cr interphase boundaries are more "accommodating" than are Cr grain boundaries. Somewhat similar considerations apply during deformation processing of Ni–W composites. Directionally solidified eutectics of this system—containing approximately 6 vol-% of isolated W single crystals imbedded in a solid solution matrix—can be cold drawn to deformation

TABLE I

PARTIAL LISTING OF TWO-PHASE SYSTEMS
SUBJECTED TO LARGE DEFORMATION STRAINS

Systems	Fabrication techniques	References
Ag–30Cu	cast-drawn	[*32*]
Ag–(10–90)Fe	solid state sintered-drawn	[*33, 35*]
Ag–19Ni	solid state sintered-swaged-drawn	[*42*]
43Ni	liquid phase sintered-swaged-drawn	[*42*]
Cu–17Cr	cast-drawn	[*49*]
Cu–15Fe	cast-swaged-drawn	[*42*]
Cu–(52, 86)Fe	liquid phase sintered-swaged-drawn	[*63*]
Cu–10Mo	solid state sintered-swaged-drawn	[*49*]
Cu–(10,15,18)Nb	cast-drawn	[*36, 38*]
–(12,20)Nb	cast-rod rolled-drawn	[*44, 46*]
–20Nb	cast-rolled	[*47*]
Cu–20Ta	cast-rod rolled-drawn	[*46*]
Ni–6W	directionally solidified-swaged-drawn	[*50*]
–48W	liquid phase sintered-rolled	[*43*]

Key: Cu–20Nb indicates a composite containing 20 vol-% Nb, etc.

strains of at least 4.0 [*49, 50*].[2] Liquid-phase-sintered Ni–W structures—
containing 48 vol-% W in a highly connected form— cannot be cold-worked
directly following sintering [*43*]. But if such an alloy is hot-worked after
liquid-phase sintering, it can be cold-worked subsequently [*43*]. During hot
working, the W interparticle contacts are severed, but this is not accompanied by material failure. Instead, the matrix flows between the separated W particles, producing a structure of dispersed W single crystals. The
resulting microstructure is susceptible to cold working for much the same
reasons as are Cu–Cr alloys with a similar morphology.

Cold working of solid-state sintered Cu–Mo alloys [*49*] provides another
example relating to fabrication of two-phase metals. In this investigation,
the Mo powder used was polycrystalline. During the initial stages of cold
working, the Mo phase did not deform. After a certain critical process strain,
however, the internal stress generated as a result of matrix work hardening
and phase deformation incompatibility became sufficient to cause inter-
crystalline fracture of the Mo particles. The resulting single-crystal fragments
then plastically deformed to approximately the same degree as the matrix.
Copper–molybdenum alloys containing high Mo volume fractions could not
be deformation processed. Linkup between cracks formed via Mo inter-
crystalline fracture limited fabricability. This general description of "compatibility"—namely, that the less ductile phase be capable of strain accommodation for the volume fraction and phase morphology specific to the
composite—is verified further by results of preliminary experiments that
attempted to use intermetallics as HDISC constituents [*51*]. Even when the
intermetallic existed as isolated single crystals in a ductile matrix, the
two-phase material could not be moderately cold worked before material
failure. In this situation, the intermetallic does not have the requisite number
of slip systems to allow for strain accommodation across an interphase or
grain boundary despite its capability for single-crystal plasticity.

Having briefly described the potential of HDISC, their mechanical properties, which stamp them as such intriguing materials, are discussed in the
following section.

III. Properties

A. Strength

As noted, exceptional strengths can be developed by cold working two-
phase materials. Examples of the strengths generated by cold working some
simple two-phase combinations are provided in Fig. 2. Figure 2a illustrates

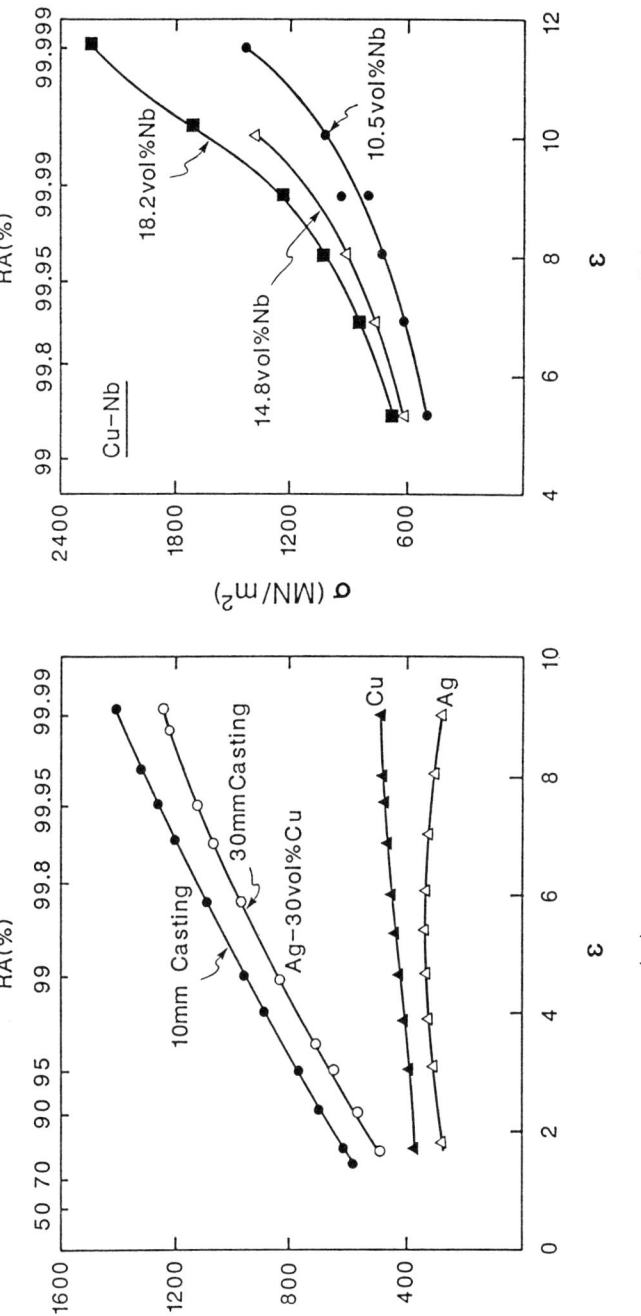

FIG. 2. Extensive deformation processing results in exceptional tensile strengths of two-phase solids. (a) The tensile strengths as a function of deformation strain for Ag–Cu eutectics (30 vol-% Cu). The strengths are high and depend on initial processing as well as on deformation strain. The 10-mm casting produced an initially finer microstructure than the 30-mm casting. This results in higher initial and as deformation processed strengths for the finer casting. The strengths of the single-phase composite constituents as they vary with process strain are also shown. Composite strengths are much in excess of these; this is clear evidence of the significant work hardening accompanying two-phase deformation. (Reprinted with permission from [32], Pergamon Press plc.) (b) The tensile strengths of several deformation-processed Cu-Nb composites as they vary with deformation strain. The strength increases with increases in Nb volume fraction. The work-hardening rate increases with processing strain. This is common to many HDISCs, particularly those containing a *bcc* transition metal. (Reprinted with permission, Fig. 3, Article by J. Bevk and K. R. Karasek, from *New Developments and Applications in Composites*, ed. by D. Kuhlman-Wilsdorf and W. C. Harrigan, Jr, 1979, p. 101.)

the variation of tensile strengths[3] of Ag–Cu rodlike eutectics [*32*] with prior wire-drawing strain. The work-hardening characteristics of the single-phase composite constituents are also shown in this figure. The different work-hardening behavior of the single- and two-phase materials is dramatic and becomes more pronounced with increased drawing strain. We also see that refinements in initial microstructural scale lead to higher strengths. For example, the alloy cast as the smaller ingot has a finer initial microstructure, and this translates to higher strengths following deformation.

Strengths of similarly processed Cu–Nb alloys are shown in Fig. 2b [*38*]. As with Cu–Ag alloys, the strengths of Cu–Nb mixtures are greater than expected on the basis of the work hardening of their constituents, and the strength differential becomes more pronounced with increasing process strain. Indeed, for this two-phase combination, the tensile strength increases approximately exponentially with deformation strain up to process strains on the order of 11 (= 99.998% RA).

A general idea of the strengths of HDISCs is provided in Fig. 3. In this figure, tensile strengths (normalized by the tensile modulus)[4] measured in a number of studies are plotted versus process strain. As a general rule, strengths of *bcc–fcc* mixtures are greater than those of *fcc–fcc* ones, although this might be attributed to the former combinations being processed to higher strains. Strengths approaching the theoretical strengths are observed at high deformation strains in some *bcc–fcc* combinations. While combinations involving two *fcc* metals are not as strong, they also manifest strengths well above those expected on the basis of the strain hardening of their constituents.[5]

There is considerable "scatter" in Fig. 3. Some of this can be removed by considering effects of microstructural scale, as will be discussed subsequently. Here, though, we note that refinements in initial microstructural scale translate into higher deformation-processed strengths. An example has been provided in Fig. 2a, and some of the variations in strength from system to system in Fig. 3 are a result of this factor.

Most studies on HDISCs have used wire drawing as the deformation process. However, as noted in Table I, rolling, extrusion, swaging, and various combinations thereof have also been used to produce HDISCs. Studies comparing the strengths generated by different deformation-processing schemes have been scant, but those that have been undertaken indicate that strength differentials resulting from different deformation-processing schemes are relatively small [*47*]. Thus, in further discussions, we will not differentiate between various deformation processes used to make HDISC. We note that the area is a fruitful one for further study, not only from a practical viewpoint but also because it might shed light on strengthening mechanisms.

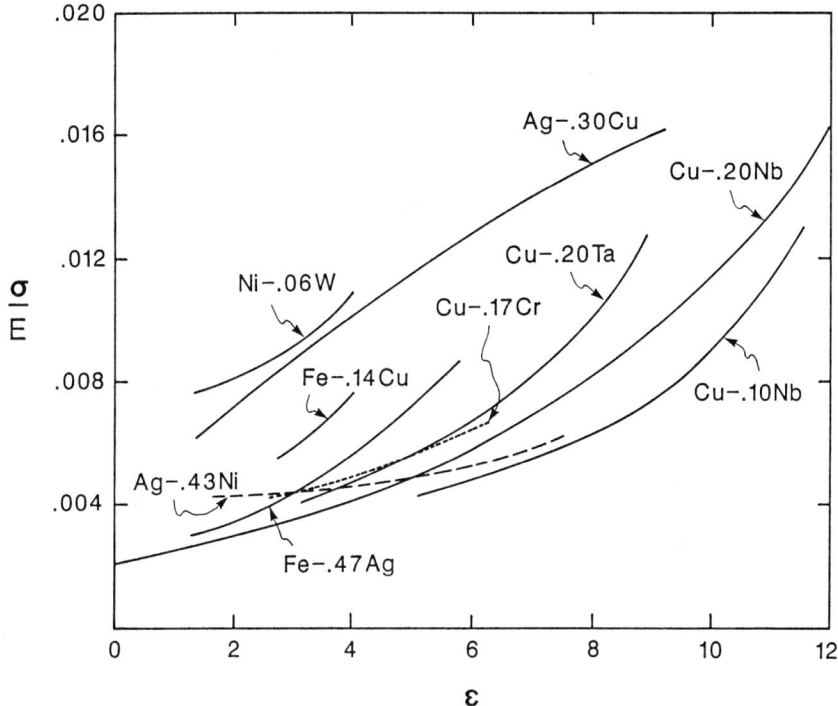

FIG. 3. Tensile strength, normalized in terms of the tensile modulus, for a number of deformation-processed two-phase materials. The strengths of the two-phase solids always exceed those expected on the basis of the work-hardening characteristics of their single-phase constituents. The extent to which they do, though, is system-specific and also depends on initial processing conditions. For example, finer initial microstructures result in greater as-deformation-processed strengths. Maximum values of strength are on the order of 0.02 times the tensile modulus; this is also the approximate strength of patented steel wire, the prototypical HDISC. Data for: Ag–.30Cu from [*32*], Ag–.43Ni and Cu–.17Cr from [*42*], Cu–.10Nb from (38), Cu–.20Nb from [*44*], Cu–.20Ta from [*46*], Fe–.47Ag from [*33*], Fe–.14Cu from [*42*], Ni–.06W from [*50*].

The data presented clearly indicate how extensive cold working can be used to generate high-strength two-phase materials. But what about their ductilities and toughnesses? These are considered in the following section.

B.   *Ductility and Toughness*

Data on ductilities of HDISC are limited. And fracture toughnesses have not been measured as a result of the small sizes of most HDISC. Studies that have measured tensile elongation of HDISCs indicate that these materials are not very ductile by this measure [44–47, 50]. Some results from the work of Spitzig *et al.* [44] and Trybus and Spitzig [47] are given in Fig. 4; they are typical of other, more limited, investigations. As indicated in Fig. 4, tensile fracture strains, as measured by tensile elongation, are modest for HDISC. However, this strain is not relevant for assesesing ductility and toughness of high-strength materials; essentially, all high-strength materials manifest limited tensile elongations even though some of them can be quite tough. A better measure of toughness is tensile reduction in area, and this can be appreciable (*cf.* Table II) for some HDISCs. For HDISC that fail in a cup-and-cone fashion, percent RA is usually respectable and often impressive given their strengths. When subjected to large deformation strains, tensile fracture modes are often shearlike [45], i.e., they are characteristic of failure preceded by flow localization. Tensile RAs are more limited for this fracture mode.

It is unfortunate that fracture toughnesses have not been, or can not be, measured for HDISCs. An approximate correlation of tensile behavior with toughness has been used to estimate fracture toughnesses of Ni–W HDISC [50]. The calculation used correlations between fracture toughness and tensile

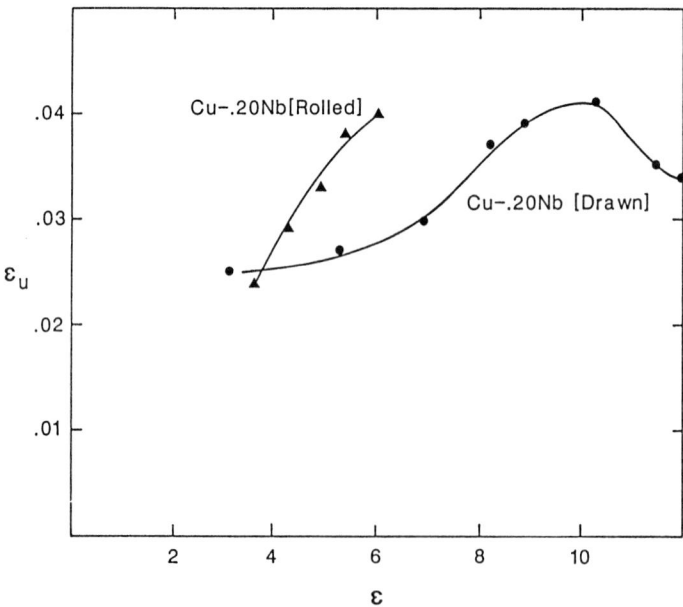

FIG. 4. Tensile elongation as a function of prior deformation strain for Cu–.20 Nb HDISCs processed by rolling and drawing. Tensile ductility is limited for HDISCs when it is assessed via tensile elongation. Tensile reduction in areas, though, can be appreciable for HDISCs. Data for drawn alloys from [44]; for rolled alloys from [47].

TABLE II
Tensile Reduction in Areas of Some
Copper-Based HDISCs[a]

System	Prior deformation strain	Tensile RA (%)
Cu–20Nb	3.1	80
	4.5	76,79
	5.5	76
	6.9	61
	7.8	44
	8.9	33
	10.1	34
	11.0	34
	11.5	24
	12.0	16
Cu–20Ta	3.1	85
	5.5	77
	6.9	48
	7.8	33
	8.0	22
	8.9	24

[a] RAs deduced from fracture strain data presented in [46]. The data presented here are from alloys with initial dendrite sizes on the order of 4 μm. Reference [46] also presents tensile fracture strains for alloys characterized by larger initial dendrite sizes. Tensile ductilities of these materials are somewhat less than they are for the alloys with a finer initial dendrite size.

properties applicable to conventional materials [54, 55]. These approximate estimates indicate that this material is quite tough. For example, it was estimated [50] that the toughness of a Ni–W alloy having a tensile strength of 2075 MN/m^2 (= 300,000 psi) is equivalent to that of a maraging steel having a strength of 1930 MN/m^2 = 280,000 psi). Similar calculations can be carried out for some previously discussed Cu-base composites. These are elaborated on in Table III. The calculated toughnesses shown there are not that impressive; they are comparable to the toughnesses of high-strength aluminum alloys but are less than those of steels and/or Ti alloys at equivalent strength levels. However, as indicated in Table III, toughnesses were calculated by using the uniform tensile strain as a measure of material work hardening. This leads to an underestimate of toughness, as described previously. If one were to use the (very) low strain flow behavior of HDISCs to determine n, the toughnesses calculated would be increased severalfold and would then begin to become impressive.

TABLE III

ESTIMATED FRACTURE TOUGHNESS FOR SOME Cu-BASED HDISC[a]

Cu–20Ta				
$E = 126$ GN/m^2				
Prior deformation strain	n	ε_f	σ_y(MN/m^2)	K_{Ic}(MN/m$^{3/2}$)
3.1	.014	1.9	600	22
6.9	.011	.65	1090	14
8.9	.0095	.29	1600	9
Cu–20Nb				
$E = 112$ GN/m^2				
3.1	.015	1.6	560	20
6.9	.014	1.0	970	19
8.9	.0165	.41	1350	17

[a] Data from [46] for same alloys whose ductilities are given in Table II. K_{Ic} values calculated from [54, 55]

$$K_{Ic} \text{ (English units)} = n[2E\sigma_y\varepsilon_f/3]^{1/2},$$

where n is taken as the uniform tensile strain and σ_y is approximated by the composite tensile strength.

Due to the sparseness of available data, our discussion of ductility and toughness has been limited. Yet it seems reasonable to infer that HDISC toughnesses are comparable—perhaps superior—to those of monolithic materials having equivalent strengths.

IV. Strengthening Mechanisms and Structure

A. *Strengthening Mechanisms*

What is responsible for the high strengths of HDISCs? Their strengths are undoubtedly related to structural and/or substructural differences between these materials and their cold-worked single-phase counterparts. Three different, but in some ways related, mechanisms have been postulated to explain HDISC strengths.

Bevk and coworkers suggested that HDISC strengths derive from exceptionally high dislocation densities generated in the composite matrix phase. Evidence in support of this view was provided by (limited) electron microscopic observations [38] and resistivity measurements [40] (which suggested an anomalously high dislocation scattering component). Their microscopic observations also indicated that at very large deformation strains, the Nb phase in their HDISC took on a dislocation-free, whiskerlike character. Support for the high dislocation density strengthening mechanism also comes from the work of Wasserman *et al.* [32], who observed significant dislocation densities (approximately 10^{16}–10^{18}/m^2) in heavily deformed (99% RA) Ag–Cu alloys.

Funkenbusch and Courtney [42] and Funkenbusch *et al.* [52] extended the basic dislocation strengthening idea of Bevk and coworkers. They asserted that enhanced strengths of HDISC result from the inherently greater strain incompatibility in two-phase materials as compared with single-phase ones. The physics of their model are couched in terms of geometrically necessary dislocations, a concept put forth by Ashby [56, 57]. Generation of this kind of dislocation provides for material contiguity across grain or interphase boundaries. Inherent strain differences (strain gradients) across grain boundaries in single-phase materials, for example, are accommodated by geometrical dislocations. In single-phase materials, strain incompatibilities across grain boundaries arise from differences in crystallographic orientation. In two-phase materials, the incompatibility results from both crystal orientation differences and the inherently different flow behavior of the two phases. The result is that greater numbers of geometrical dislocations are generated during two-phase deformation.

The geometrical dislocation concept is fundamentally sound. For example, overaged precipitation-hardened alloys containing nondeformable precipitates work harden more rapidly than do underaged ones containing precipitates that shear [58]. The strain gradient between a nondeforming particle and the surrounding matrix is accommodated by geometrical dislocations (an Orowan loop is a simple example of a geometrical dislocation). The increased work hardening of alloys containing nondeforming particles is a direct result of the generation of the additional geometrical dislocations.

Ashby showed that the geometrical dislocation density (ρ) varies with strain (ε) and a microstructural parameter (d) according to

$$\rho \approx \varepsilon/d. \tag{1}$$

In Eq. (1), d is the grain diameter for a polycrystal; for a two-phase alloy containing nondeforming particles, it is the mean surface-to-surface slip plane separation of the particles.

Funkenbusch and coworkers reasoned that the characteristic distance for geometrical dislocation generation in a HDISC is the mean interphase spacing. They also hypothesized that dislocations formed geometrically are subject to the same kinds of, and rates for, annihilation and generation processes as are ordinary (statistical) dislocations. Several models have been put forth to account for the variation in statistical dislocation density as it depends on these processes. The one used by Funkenbusch and coworkers is due to Kocks [59]. In his expression, the change in statistical dislocation density (ρ_s) with strain is given by[6]

$$d\rho_s/d\varepsilon = C_1\rho_s^{1/2} - C_2\rho_s. \tag{2}$$

In Eq. (2), the first term signifies that dislocation multiplication is proportional to the mean dislocation spacing, consistent with the idea that new dislocation length is created primarily by interactions between mobile dislocations and the existing dislocation structure. The second term represents dynamic recovery; the form of it is consistent with the assumption that the loss of dislocation line length per increment of strain is proportional to the total line length. The constants C_1 and C_2 can be determined by measuring material flow stress as a function of strain and by fitting the behavior to the following equation [42];

$$\sigma = \sigma_0 + \alpha M Gb(\rho_s)^{1/2}. \tag{3}$$

In Eq. (3), α is a constant of order unity, G is the shear modulus, b the material's Burgers vector, M the Taylor factor, and σ_0 is related to the intrinsic (lattice) resistance to dislocation motion.

Funkenbusch and coworkers used the geometrical dislocation concept in conjunction with Eq. (2) to express the variation of total dislocation density (ρ_T) with deformation strain in phases "A" and "B" of a HDISC as

$$d\rho_{TA,B}/d\varepsilon = C_{1A,B}(\rho_{TA,B})^{1/2} - C_{2A,B}\rho_{TA,B} + P_{A,B}K/V_{A,B}d. \tag{4}$$

In this equation, ρ_T is the *total* dislocation density (regardless of whether dislocations are generated statistically or geometrically), and $V_{A,B}$ are the volume fractions of A and B. All dislocations are assumed to multiply and annihilate as they do in the single-phase material; this is reflected in the first two terms on the right-hand side of Eq. (4). The last term on the right-hand side of this equation represents geometrical dislocation generation. It contains two unknown (but coupled) constants: P, a dislocation-partitioning coefficient (dislocations should partition preferentially to the softer phase), and K, a measure of the inherent strain incompatibility between the composite constituents. (For example, K = 0 corresponds to a single-phase material. For this case, composite strengths obey a conventional volume fraction rule.) It was then postulated that composite tensile strengths follow a modified volume fraction rule,

$$\sigma_c = V_A \sigma_A + V_B \sigma_B \qquad (5)$$

where the strengths of the individual phases are determined by their dislocation densities arrived at via Eq. (4).

Some basis physics of the model can be found in Eq. (4). In particular, since interphase spacing decreases with process strain, geometrical dislocation generation becomes increasingly important at large strains. In turn, this results in greater work hardening; this is in general agreement with observations. In addition, materials containing linear work-hardening metals (such as Fe, Nb, and some other *bcc* transition elements) should work harden more rapidly than two-phase materials comprised of saturation hardeners (e.g., *fcc–fcc* combinations).

To quantify HDISC strengths, Eq. (4) must be numerically integrated (for each phase) and then used in Eq. (5). To do this, the variation of interphase spacing with processing strain is needed. Funkenbusch et al. [49] used the relationship

$$d = d_0 \exp(\varepsilon/2), \qquad (6)$$

which was based on limited microstructural observations.[7] In addition, eight other parameters must be known to predict strength. These are C_1, C_2, and σ_0 (for the individual phases), and P (note $P_A + P_B = 1$) and K. The coefficients for the individual phases are determined from their flow behavior, as described above. A value of P is arbitrarily selected[8] (P = 0.5 was taken most often by Funkenbusch and coworkers), and then K is determined from *one* strength datum for the cold-worked two-phase material. That is, if the flow behavior of the composite constituents, and one strength datum for a HDISC at some arbitrary deformation strain and phase volume fraction, are known, the model allows strengths to be predicted for other combinations of strain and phase volume fraction.

The Funkenbusch model is remarkably successful in terms of "predicting" HDISC strengths. This is shown in Figs. 5a, b, and c, which indicate its usefulness in predicting strengths and strain-hardening behavior of Ag–Fe and Cu–Nb composites [52]. Funkenbusch and coworkers noted that their formulation might not apply at very large deformation strains where substructure might not evolve in the same way it does at low strains (e.g., Eq. (3)). However, it is clear that the model is still useful at large strain deformation.

Spitzig *et al.* [44] and Spitzig and Krotz [46] have advanced a strengthening model based on Hall–Petch behavior. In this scheme, the major phase of the composite is strengthened by barriers to dislocation flow provided by the interphase boundaries. The effects of varying phase volume fraction and

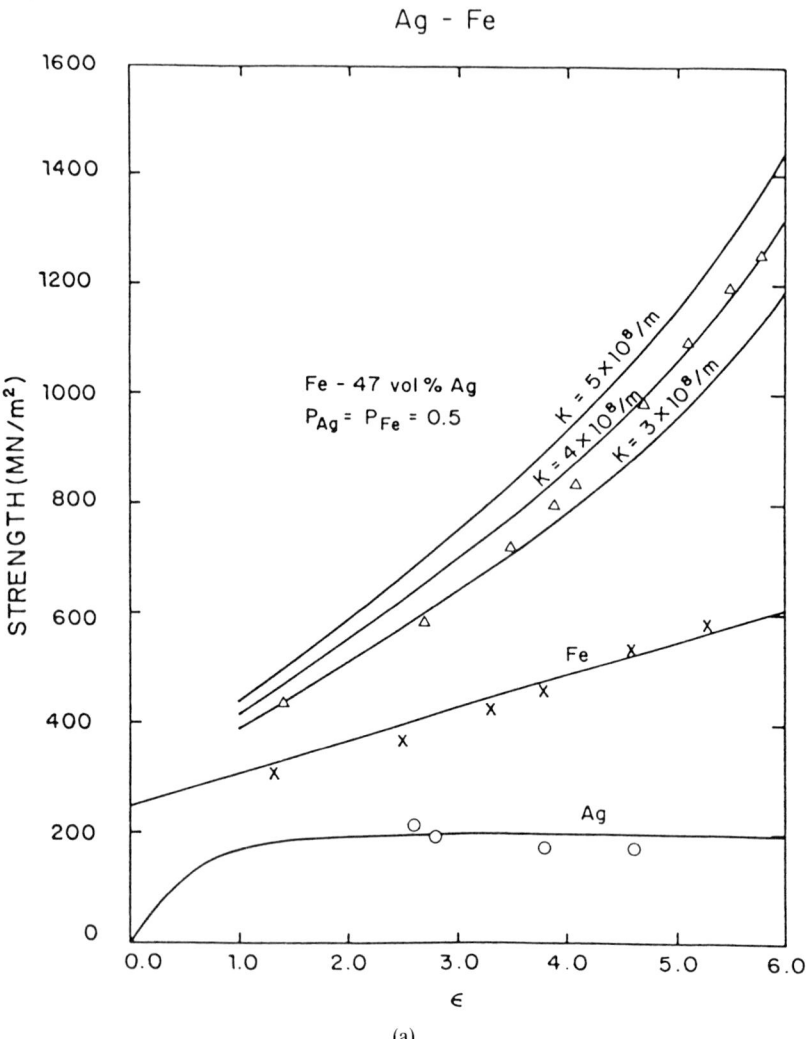

(a)

FIG. 5.   The Funkenbusch model is useful for mimicking the work-hardening behavior of HDISCs. In (a) the triangles represent the strengths of Fe–.47Ag HDISCs. The solid curves represent the model predictions for several values of K; a value of K of about $3.5 \times 10^8/m$ fits the experimental results. The crosses and circles represent the work hardening of the single-phase composite constituents. The solid lines drawn through the data represent the strengths as predicted by the Kocks model. (b) The model also mimics well the work-hardening behavior of Cu–Nb HDISCs, provided K is taken as approximately 1.0–$1.5 \times 10^8/m$. The line marked $K = 0$ represents the strengths predicted on the basis of the work-hardening characteristics of the composite constituents. (c) The model is also useful for predicting the volume fraction dependence on strength, as shown here for Fe–.47Ag HDISCs. From [52].

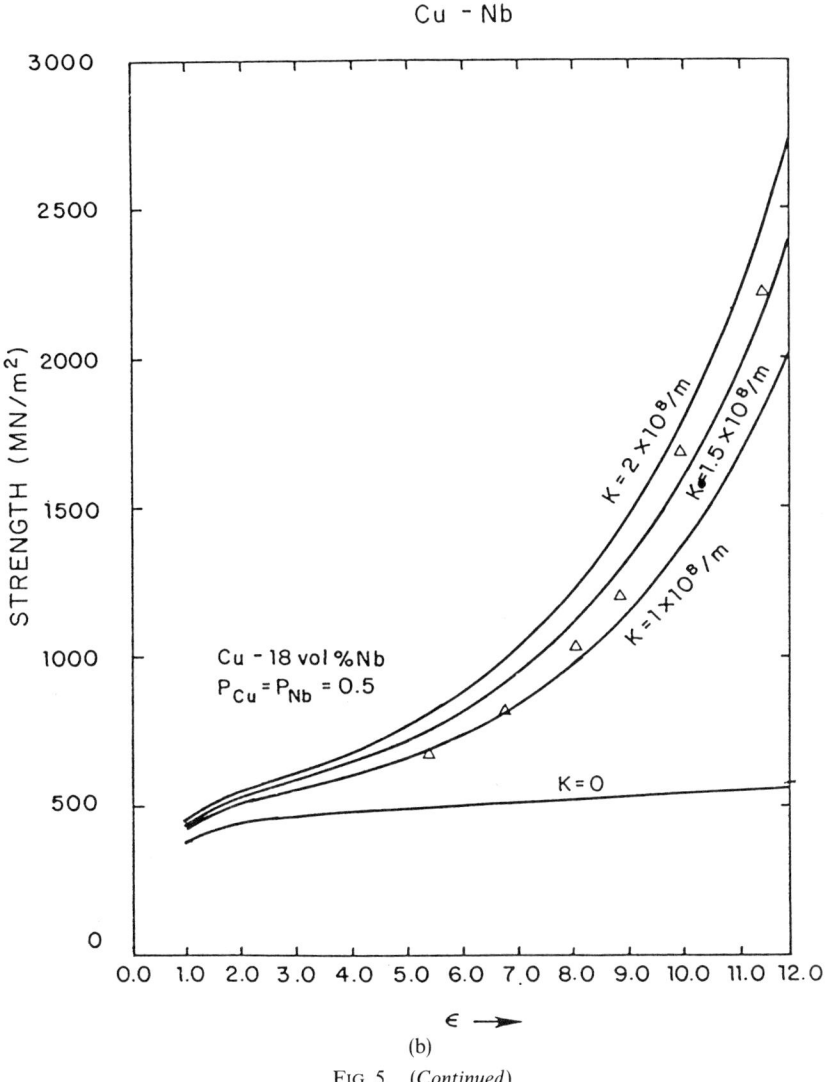

FIG. 5. (*Continued*)

composite constituents are accounted for by their influence on matrix slip
length (the distance between the interphase boundaries) and modulus (which
affects the value of the Hall–Petch coefficient). Experimental results correlate
well with this description of strengthening [*44, 46, 49*].[9]

Spitzig and coworkers considered their model successful on the basis of
such correlations. Moreover, they interpreted results of transmission electron
microscopic studies [*44, 47, 48*] conducted on composite constituents (Cu

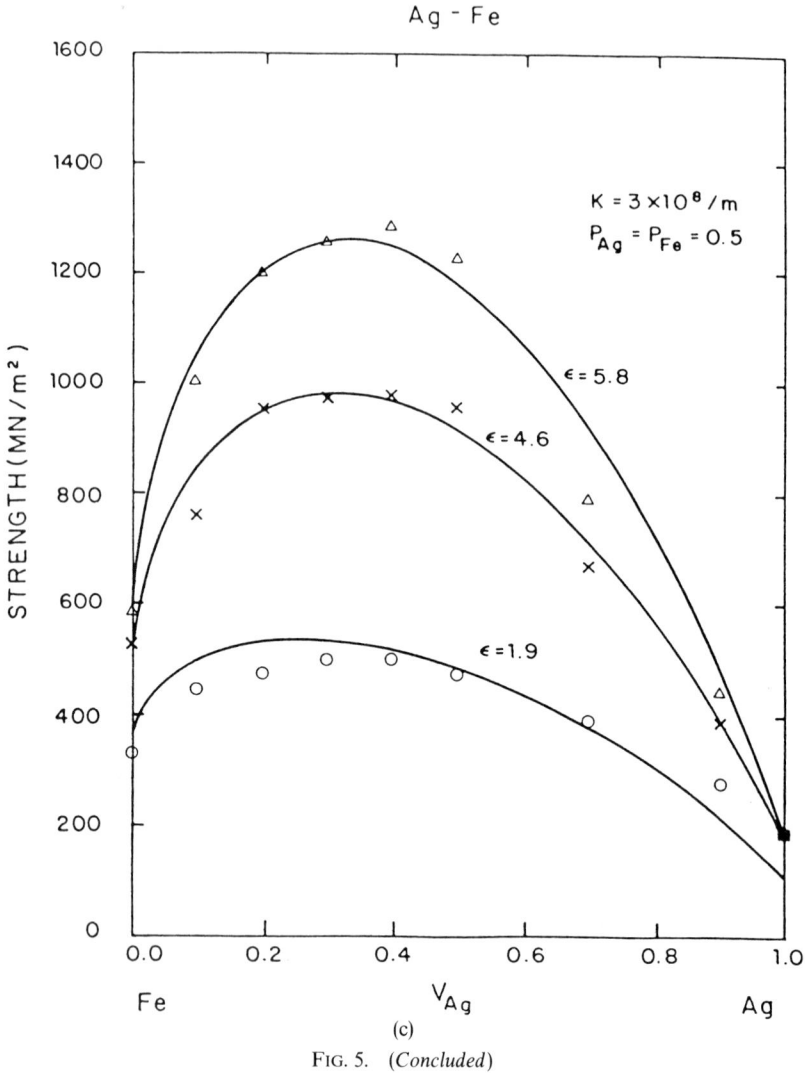

FIG. 5. (*Concluded*)

and, to a lesser extent, Nb) as refuting the Funkenbusch model. The microscopic studies indicate that the Cu matrix in Cu–Nb HDISC undergoes significant recovery at large deformation strains, where its structure is characterized by fine grains or subgrains (although, at relatively low deformation strains, a dislocation cell structure is developed). According to their reading [44] of the Funkenbusch model, Cu dislocation densities on the order of 10^{17}–10^{18}/m^2 are required to account for observed composite strengths.

However, the Funkenbusch *et al.* model predicts that most of the composite incremental strength derives from the linear work hardening *bcc* phase (i.e., this phase is presumed not to dynamically recover). The Funkenbusch model does predict higher Cu dislocation densities in the composite than in single-phase form. But these densities are nowhere near as large [60] as those stated by Spitzig and coworkers.

So, what is or are the strengthening mechanism(s) for HDISC? Their strengths correlate with interphase spacing to the $-\frac{1}{2}$ power, but this behavior is predicted by both the Funkenbusch and Spitzig models. In order to differentiate between them, other information is required. Some of this comes from structural analysis and some from results of studies that alter substructure without changing interphase spacing. These factors are considered in the following section.

B. Structural and Other Considerations

This section begins with its conclusions; evidence for them is presented throughout the section. We believe that the high strengths of HDISC derive basically from the greater strain incompatibility of two-phase materials in comparison with single-phase ones. This leads to the generation of additional, geometrical dislocations in the two-phase material. The dislocations and/or the substructure that evolves from them is responsible for increased strength. Although this view may appear to be an unqualified endorsement of the Funkenbusch description, it is not. In particular, this model is deficient in terms of describing substructural evolution at high strains in HDISC.

We first re-examine the effect of interphase spacing on strength. According to the Spitzig description, this spacing is the overriding parameter in determining strength. If this is so, strengths of finely dispersed two-phase structures would be independent of the deformation to which they have been subjected. That is, the strengths of HDISC would depend only on microstructural scale. Everett [61] has recently compared strengths of HDISC to those of fine composites produced by vapor-phase deposition. The deposition process produces microstructures having interphase spacings comparable to those of HDISC, but the deposited layers are not subject to extensive plastic deformation. Some of Everett's results are shown in Fig. 6, which compare the strengths of HDISC and deposited materials. The comparisons clearly indicate that—at equivalent spacings—strengths of HDISC are higher, sometimes appreciably so. The only obvious difference between the two types of materials is substructure. Thus, interphase spacing alone does not control the strengths of fine composites, including HDISC.

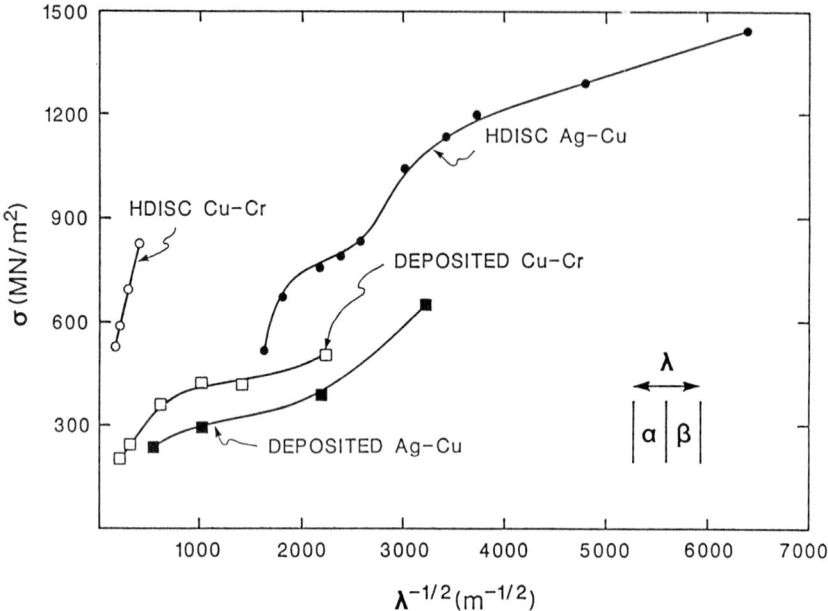

FIG. 6. The strengths of HDISCs are compared with those of layered composites of the same materials. The abcissa is the inverse square root of the interphase spacing. The deposited layer composites are not subjected to the extensive plastic strains that HDISC are. The higher strengths of the latter, therefore, reflect the effect of cold work on HDISC strengths. (Reprinted with permission from [*61*], Pergamon Press plc.)

Additional evidence relative to the effect of substructure on strength can be obtained by subjecting HDISC to elevated temperatures. Unambiguous results are obtained if the temperatures are high enough to effect recovery/recrystallization, yet low enough to prevent microstructural coarsening. Some results of studies of this type on Cu–Fe HDISC [*60, 62*] are given in Fig. 7. In these investigations, room-temperature strengths were measured following exposure to temperatures sufficient to fully recover/recrystallize Cu, but only to modestly recover the Fe phase. As indicated in Fig. 7, appreciable softening is found after exposure at 673 K, a temperature at which no microstructural coarsening was observed.[10] Since interphase spacings are not altered by such a heat treatment, the strength decreases must result from Cu recovery/recrystallization, i.e., from changes in Cu substructure. We also note that (single-phase) Fe precipitation hardens during this heat treatment. The aging time required to effect maximum hardening decreases with increasing prior deformation. This is additional, albeit in-

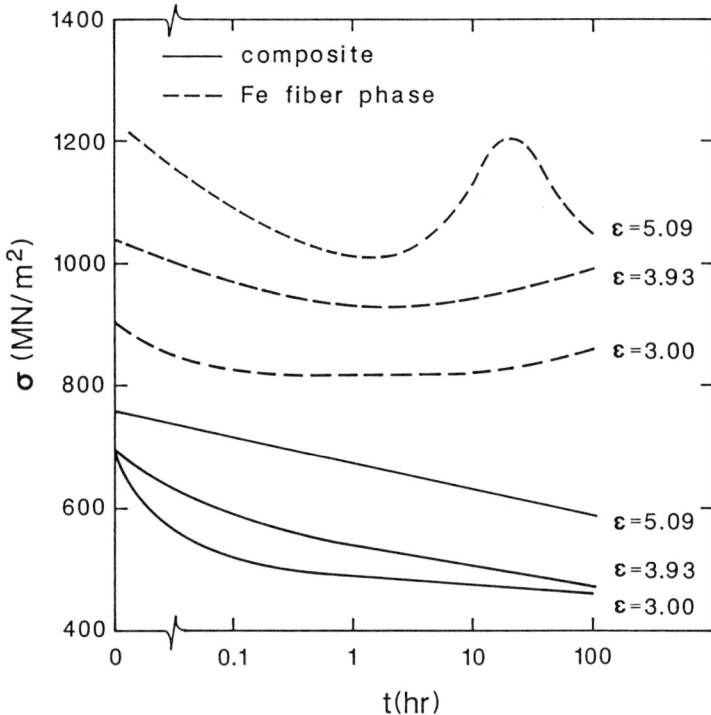

FIG. 7. Room-temperature strengths (solid lines) of Cu–.15Fe composites as a function of annealing time at 673 K for three different processing strains. Since the annealing temperature does not result in coarsening of the composite, the (sometimes) substantial decreases in strength are a result of substructural changes, primarily in the Cu. The dotted lines represent the strengths of single-phase Fe standards, containing 1% Cu in solid solution. The annealing temperature results in some Fe recovery, followed by age hardening due to Cu precipitation. The precipitation is accelerated by prior deformation strain. Data from [*60, 62*].

direct, evidence of the further development of dislocation-generated substructure with increasing deformation strain.

Krotz *et al.* [*45*] have conducted similar, more extensive, studies on Cu–Nb and Cu–Ta HDISC. In this work, composites were exposed to temperature and then tensile tested at room temperature (similar to the technique just described). In addition they were tensile tested at temperature, with and without prior exposure at the test temperature. Fiber coarsening did not occur at temperatures up to 300°C, and was minimal at temperatures on the order of 450°C (except perhaps in composites subjected to the largest deformation strains and which, therefore, manifested the finest interphase spacings). As shown in Fig. 8, temperature exposure leads to modest

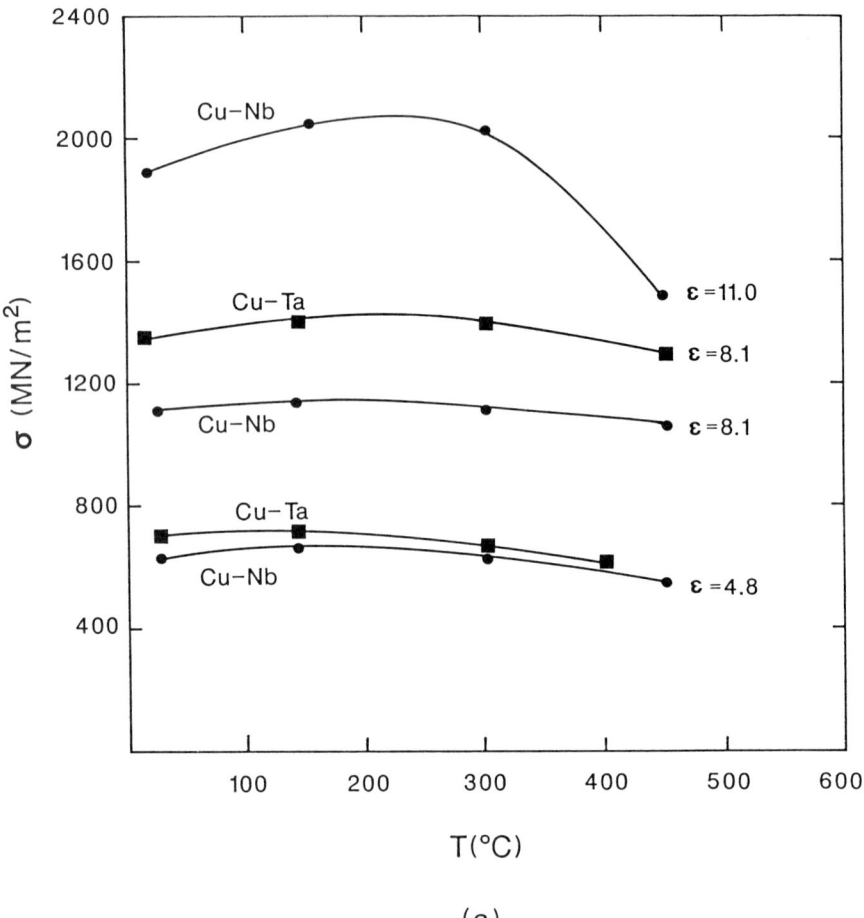

(a)

FIG. 8. (a) Room-temperature tensile strengths of some Cu-based HDISC as they vary with a 24-hr. annealing heat treatment. For temperatures of less than 450°C, the structure coarsens only slightly. Strength changes are minimal, for the most part, indicating only slight substructural rearrangements. However, extensive age hardening is observed in the Cu–Nb alloy drawn to the highest strain. This indicates that the substructure of the Nb continues to evolve with deformation strain (*cf.* Fig. 7). (b) The elevated temperature strengths of Cu–.20Nb HDISCs. The solid lines represesnt testing at temperature. The dotted lines represent tests in which the material was held at temperature for 24 hours prior to tensile testing. Since minimal coarsening was observed for temperatures below 450°C, the differences in strength must result from substructural changes. The greater the prior drawing strain, the greater are the differences in strength resulting from the different test procedures. Data from [*45*].

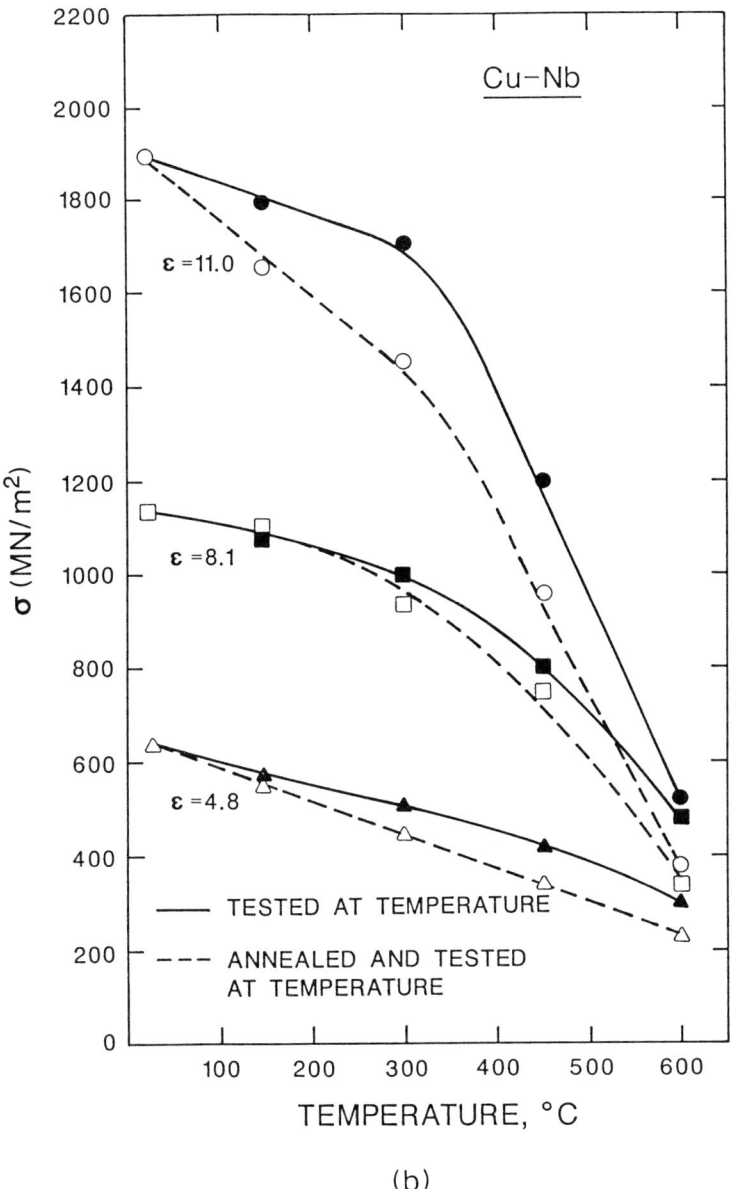

FIG. 8. (*Concluded*)

decreases in room-temperature strength. But elevated temperature exposure of materials prior to testing results in significant strength decreases. Moreover, strength decreases increase with prior deformation strain. These strength changes are clearly the result of substructural alteration. Finally, we note that age hardening of Nb and Ta, similar to that in Fe, has a pronounced effect on composite strength only at large composite strains. This is further evidence of continued evolution of the *bcc* substructure with strain. The decreases in composite strength illustrated in Figs. 7 and 8 result primarily from recovery processes taking place in the Cu. According to the description by Funkenbusch and coworkers, more substantial softening would be expected following recovery/recrystallization of Fe, Nb, or Ta, the presumed primary strengthening phases in their description. Unfortunately, temperatures at which such processes would be expected to occur are also the ones at which microstructural coarsening takes place.

The substructure of HDISCs depends on deformation strain in a way differently than it does for comparably deformed single-phase materials. However, it clearly does not evolve in the manner implicit to the Funkenbusch model, which presumes it would be totally manifested in increased dislocation densities. Unfortunately, there have been few quantitative measurements relative to substructural development in heavily deformed two-phase materials. Figure 9 [*28*] shows results of cell/subgrain size measurements on ferrite in pure Ferrovac Fe in comparison with those in deformed pearlite; it is clear that considerable substructural refinement is associated with deformation of ferrite in the two-phase material. Spitzig and coworkers have conducted electron microscopic observations of Cu substructural size in deformed single-phase material and in Cu–Nb HDISC. Some of their studies [*48*] indicated that the Cu cell size developed in HDISC is approximately the same as in comparably deformed single-phase Cu. On the other hand, the most extensive ones of their studies [*47*] indicate that cell size is refined in the two-phase material. These cell sizes are also shown in Fig. 9, where they can be compared to cell sizes measured in single-phase Cu. The only apparent cause for a reduced cell size is enhanced dislocation generation in the two-phase material. Thus, while it is clear that the dislocation morphology in some HDISC phases is not consistent with that implicit to the Funkenbusch model, the substructural arrangements found at high deformation strains in HDISC are consistent with increased dislocation generation in the two-phase solids. The Funkenbusch description, based as it is on dislocation evolvement at low strains, does not correctly anticipate structural evolution at high strains. In fairness, though, it should be noted that Funkenbusch and coworkers took pains to address this problem, explicitly noting that the constitutive equations used in their model might not be appropriate at large deformation strains.

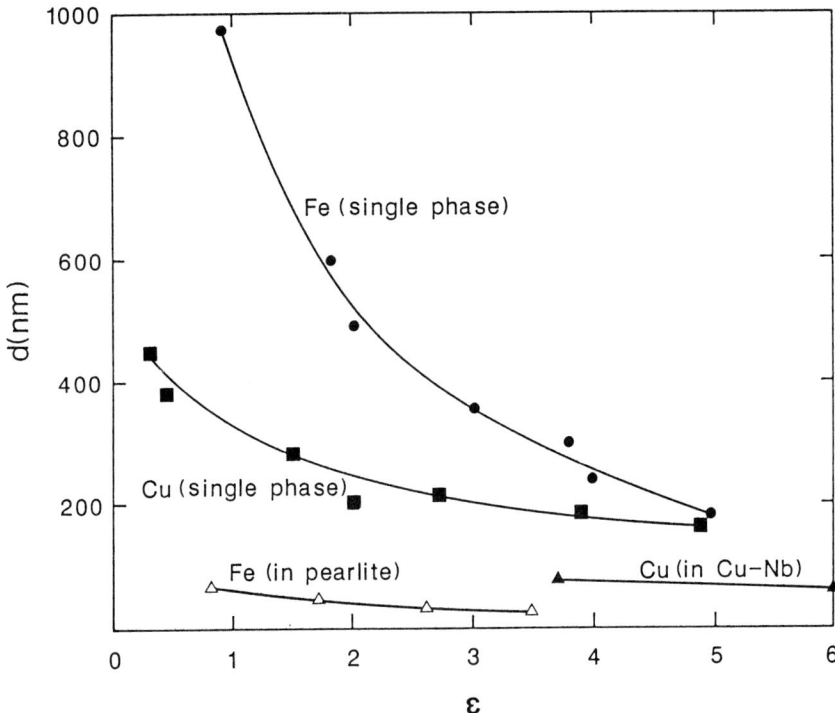

Fɪɢ. 9.   Cell sizes of Cu and Fe in single-phase form and in HDISC, as this parameter depends on deformation strain. The cell size in the single-phase material is greater than in the HDISCs (pearlite for the Fe, and Cu–Nb for the Cu). This apparently is indicative of redundant strain (i.e., additional dislocation generation) in the two-phase materials. Data for single-phase Cu, single-phase Fe, and pearlite from [28]; for Cu–Nb from [47].

There is another compelling reason that argues against a simple Hall–Petch mechanism determining the strengths of HDISC; the coefficients that are needed to "match" the strengths of HDISC are much higher than the Hall–Petch coefficients of their single-phase constituents [61]. Table IV compares the experimental values of the Hall–Petch coefficient for single-phase materials with those needed to account for the strengths of several HDISC. It seems unreasonable to expect values of this coefficient to be so much higher for the two-phase solid than it is for their constituents.

According to our interpretation of the data presented here, the following conclusions relative to HDISC strengthening can be made:

(1) The Funkenbusch model is significantly useful for predicting HDISC strengths.

TABLE IV
HALL–PETCH COEFFICIENTS FOR HDISCs AND FOR THEIR
SINGLE-PHASE CONSTITUENTS[a]

System	Reference	HDISC coefficients (MN/m$^{3/2}$)	Single-phase coefficients (MN/m$^{3/2}$)	
Ag–Cu	[32]	0.30	Ag:	0.07–0.17
			Cu:	0.11–0.17
Cu–Fe	[63]	3.10	Cu:	0.11–0.17
			Fe:	0.20–0.71
Cu–Nb	[44]	1.08	Cu:	0.11–0.17
			Nb:	0.04

[a] Adapted from [61]

(2) Substructure, however, does not evolve in a manner consistent with the original Funkenbusch formulation, which implies that additional dislocation generation in HDISCs is directly manifested in increased dislocation densities. Substructural measurements, though, do indicate that increased dislocation generation very likely takes place in cold-worked two-phase metals.

(3) Elevated temperature exposure of HDISCs leads to decreased strengths, which cannot be attributed to microstructural coarsening. The strength decrement appears to be caused by substructural alterations.

(4) The strengths of HDISCs can be linearly related to the interphase spacing to the $-\frac{1}{2}$ power. While this suggests a Hall–Petch mechanism, the values of the "Hall–Petch" coefficient in HDISCs are much greater than they are in single-phase materials. This argues against a simple interphase boundary barrier mechanism controlling HDISC strengths.

What is to be made of these somewhat inconsistent conclusions? Some possible explanations are provided in the following.

(1) Geometrical dislocations are generated during cold deformation of HDISCs. This is confirmed by the strength and substructural characteristics of HDISCs. Consequently, the physics underlying the Funkenbusch model appear valid, but their explanation of HDISC strengths being directly tied to dislocation density is not relevant for some HDISCs.

(2) Why is the Funkenbusch model phenomenologically so successful? In my judgement, it results from its "forgiving" nature. Clearly the

parameters C_1 and C_2, used in the model and reasonably suitable for materials at low deformation strains, are incapable of predicting the response of substructure at high deformation strains and/or dislocation densities. Nonetheless, the "error" is apparently "canceled" by fitting the data to the adjustable parameter K, which, in some fortuitous manner, takes into account that the excess dislocation density is less than predicted while simultaneously incorporating high-strain substructural development into it. That the constants C_1 and C_2 are useful in the model is reasonable, since these are material "rate constants," reflecting in some unspecified way the propensity of the material for work hardening and dynamic recovery regardless of the manner in which these phenomena are manifested.

V. Summary

Extensively cold-worked two-phase metals are strong solids. However, truly exceptional strengths are developed only by imposing very large deformation strains. At this stage of their development, therefore, use of HDISCs is restricted to fine wires and/or sheets.

A number of techniques can be used to produce HDISC. Starting materials can be made by solidification or powder processes, and the resulting material can be subsequently deformed by drawing, swaging, rolling, extrusion, or some combination of these processes. Interesting material combinations are capable of being processed to HDISC; examples include Cu–Cr and Ni–W. The material requirement is that the less malleable of the phases be capable of maintaining contiguity with the matrix in the microstructural arrangement within the specific composite.

The strengths of HDISCs are almost always described by a Hall–Petch relationship (one notable exception is the behavior of rolled Cu–Nb alloys [47]), in which the appropriate microstructural dimension is the interphase spacing. Even though this is the case, there is disagreement concerning the fundamental causes for HDISC strengths. The additional work hardening is thought to be due to additional geometrical dislocation generation according to Funkenbusch and coworkers. Their model has excellent predictive capabilities. Extensive microstructural studies of matrices of some HDISC do not reveal exceptional dislocation densities in them, although the substructural scale is sometimes refined in comparison with that of comparably deformed single-phase materials. Spitzig and coworkers assert that strengths of HDISC result from the barriers presented to matrix slip by the minor phase of the composite. This description also has shortcomings, as has been discussed in this chapter. Clearly, considerable additional work is needed to unambiguously define the mechanisms responsible for strengthening this interesting class of materials.

Acknowledgments

I would like to acknowledge the work of my former students Paul Funkenbusch, Jean Malzahn Kampe, and David Kubisch. Much of my understanding of this area is based on interactions with them. The Army Research Office supported their and my work in the field, and Professor J. K. Lee of Michigan Technological University collaborated with me in many aspects of it. Dr. William Spitzig kindly sent preprints of a number of papers relative to his and his colleagues' most recent studies on HDISC.

Endnotes

[1] Compositelike structures can be produced by directional solidification of materials not having a eutectic composition. However, the generic process is still often referred to as eutectic solidification in the lexicon of the composite community.

[2] This strain limit resulted from laboratory constraints on initial ingot size and on wire-drawing capabilities. It is believed that this alloy can be drawn to considerably greater strains in the absence of such limitations.

[3] Most studies have evaluated HDISC strength via the tensile, rather than by the more fundamental yield strength. This is associated with experimental difficulties in measuring yield strengths of very fine wires; in contrast, tensile strengths are reproducible and easily measured. Even when pains are taken to accurately measure yield strength, the low-strain work-hardening behavior of some HDISCs yields considerable scatter in offset strength [44]. Finally, because of their limited tensile ductilities, it is expected that HDISC tensile strengths mimic their yield strengths [52].

[4] The modulus used is the volumetric average of the constituent *polycrystalline* moduli. Deformation texture is often developed in HDISC [47, 48], so some "error" is introduced by using the polycrystalline moduli. Nonetheless, the resulting uncertainty does not negate the important conclusion that very high strengths are characteristic of some HDISC.

[5] Face-centered cubic metals are often described as saturation hardeners. That is, their flow stress attains a constant, limiting value at large deforma-

tion strains. Recent studies [53] indicate *fcc* metals work harden even at large deformation strains. This Stage-IV hardening, as it is called, is relatively small and is often not apparent in the flow curves of fcc materials (*cf.* Fig. 2a).
[6] The Kocks' model predicts saturation hardening for finite values of C_2. This model, like the Funkenbusch one, which uses it, is microstructurally "blind" in the sense that details regarding dislocation arrangements are not considered, and these arrangements are presumed not to change with strain (as is implicit by using strain-independent values for C_1 and C_2).
[7] More recent and more thorough studies [46] indicate that the coefficient of 0.5 for the strain in Eq. (6) is an overestimate. The recent work indicates that the coefficient is closer to 0.4 than to 0.5.
[8] The accuracy of the model is not critically dependent on the selection of P. The choice of P affects the resulting value of K required to produce a good fit of "theory" and experiment. That is, a "reasonable error" in P produces a compensating correction in K.
[9] The Funkenbusch model also predicts composite strength to vary with $d^{-1/2}$ (d = interphase spacing). This arises from the dependence of interphase spacing on deformation strain (*cf.* Eqs. (3), (4), and (6)).
[10] It is unfortunate that temperatures high enough to recrystallize Fe (and many other *bcc* transition metals) also induce composite coarsening. Thus, only a narrow low-temperature "window" is available for studies designed to delineate between substructural and interphase boundary strengthening.

References

1. H. Bibring, "Proc. Conf. on In Situ Composites," Vol. II, p. 1, Publication NMAB-308-II, National Academy of Sciences–National Academy of Engineering, Washington, D.C., 1973.
2. R. W. Kraft and D. L. Albright, *Trans. TMS–AIME*, **221**, 95 (1961).
3. E. R. Thompson and F. D. Lemkey, *Metall. Trans.*, **1**, 2799 (1970).
4. R. W. Hertzberg, F. D. Lemkey, and J. A. Ford, *Trans. TMS–AIME*, **233**, 342 (1965).
5. B. J. Bayles, J. A. Ford, and M. J. Salkind, *Trans. TMS–AIME*, **239**, 844 (1967).
6. W. H. Lawson and H. W. Kerr, *Metall. Trans.*, **2**, 2853 (1971).
7. A. R. T. de Silva and G. A. Chadwick, *Metal Sci. J.*, **3**, 168 (1969).
8. M. R. Bertorello and H. Biloni, *Metall. Trans.*, **3**, 73 (1972).
9. F. W. Crossman, A. S. Yue, and A. E. Vidoz, *Trans. TMS–AIME*, **245**, 397 (1969).
10. B. R. Butcher, G. C. Weatherly, and N. R. Petit, *Metal Sci. J.*, **3**, 7 (1969).
11. J. L. Walter and H. E. Cline, *Metall. Trans.*, **1**, 1221 (1970).
12. R. Kossowsky, W. C. Johnson, and B. J. Shaw, *Trans. TMS–AIME*, **245**, 1219 (1969).
13. T. H. Courtney, *Metall. Trans.*, **1**, 2965 (1970).
14. C. G. Rhodes and G. Garmong, *Metall. Trans.*, **3**, 1861 (1972).
15. M. J. Salkind and F. D. George, *Trans. TMS–AIME*, **242**, 1237 (1968).

16. D. A. Koss and S. M. Copley, *Metall. Trans.*, **2**, 1557 (1971).

17. E. M. Breinan, E. R. Thompson, and F. D. Lemkey, "Proc. Conf. on In Situ Composites," Vol. II, p. 201, Publication NMAB-308-II. National Academy of Sciences–National Academy of Engineering, Washington, D.C., 1973.

18. G. Garmong, "Proc. Conf. on In Situ Composites," (M. R. Jackson, J. L. Walter, F. D. Lemkey, and R. W. Hertzberg, eds.) Vol. II, p. 137. Xerox Individualized Publishing, Lexington, Mass., 1976.

19. G. Garmong, *Metall. Trans.*, **5**, 2183 (1974).

20. H. R. Gray and W. A. Sanders, "Conf. on In Situ Composites" (M. R. Jackson, J. L. Walter, F. D. Lemkey, and R. W. Hertzberg, eds.), Vol. II, p. 201. Xerox Individual Publishing, Lexington, Mass., 1976.

21. D. A. Woodford, "Conf. on In Situ Composites" (M. R. Jackson, J. L. Walter, F. D. Lemkey, and R. W. Hertzberg, eds.), Vol. II, p. 211. Xerox Individualized Publishing, Lexington, Mass., 1976.

22. H. E. Cline and D. F. Stein, *Trans. TMS–AIME*, **245**, 841 (1969).

23. B. J. Shaw, *Acta Metall.*, **15**, 1169 (1967).

24. E. O. Hall, *Proc. Roy. Soc. London*, **B64**, 747 (1951).

25. N. J. Petch, *J. Iron Steel Inst.*, **174**, 25 (1953).

26. J. D. Embury and R. M. Fisher, *Acta Metall.*, **14**, 147 (1966).

27. G. Langford, *Metall. Trans. A*, **8A**, 861 (1977).

28. J. D. Embury, "Strengthening Mechanisms in Crystals" (A. Kelly and R. B. Nicholson, eds.), p. 331. Wiley, New York, 1971.

29. W. A. Spitzig, "Metal Matrix Composites. Volume 1: Processing and Interfaces" *Treatise Mat. Sci. Tech.*, Vol. 32, (R. Everett and R. Arsenault, eds.), Academic Press, Boston, Mass., 1991.

30. S. Foner, E. J. McNiff, Jr., B. B. Schwartz, and R. Roberge, *Appl. Phys. Lett.*, **31**, 853 (1977).

31. J. D. Verhoeven, D. K. Finnemore, E. D. Gibson, J. E. Ostenson, and L. F. Goodrich, *Appl. Phys. Lett.*, **33**, 101 (1978).

32. G. Frommeyer and G. Wasserman, "Microstructures and Mechanical Properties of IN SITU Produced Silver–Copper Composite Wires," *Acta Metall.*, **23**, 1353 (1975).

33. H. P. Wahl and G. Wassermann, *Z. Metall.*, **61**, 326 (1970).

34. G. Wassermann, *Z. Metall.*, **64**, 844 (1973).

35. C. Leissner and G. Wassermann, *Metall.*, **23**, 414 (1969).

36. J. Bevk, J. P. Harbison, and J. D. Bell, *J. Appl. Phys.*, **49**, 6031 (1978).

37. K. R. Karasek and J. Bevk, *Scripta Metall.*, **14**, 431 (1980).

38. J. Bevk and K. R. Karasek, "New Developments and Applications in Composites" (D. Kuhlmann-Wilsdorf and W. C. Harrigan, Jr., eds.), p. 101, TMS–AIME, Warrendale, Pennsylvania, 1979.

39. J. Bevk, W. A. Sunder, G. Dubon, and E. Cohen, "In Situ Composites" (F. D. Lemkey, H. E. Cline, and M. McLean, eds.), Vol. IV, p. 121. Elsevier, Amsterdam, 1982.

40. K. R. Karasek and J. Bevk, *J. Appl. Phys.*, **52**, 1370 (1981).

41. W. A. Spitzig and P. D. Krotz, *Scripta Metall.*, **21**, 1143 (1987).

42. P. D. Funkenbusch and T. H. Courtney, *Acta Metall.*, **33**, 913 (1985).

43. Y. Leng, T. H. Courtney, and J. C. Malzahn Kampe, *Mat. Sci. Eng.*, **94**, 209 (1987).

44. W. A. Spitzig, A. R. Pelton, and F. C. Laabs, *Acta Metall.*, **35**, 2427 (1987).

45. P. D. Krotz, W. A. Spitzig, and F. C. Laabs, *Mat. Sci. Eng.*, (1989).

46. W. A. Spitzig and P. D. Krotz, *Acta Metall.*, **36**, 1709 (1988).

47. C. L. Trybus and W. A. Spitzig, *Acta Metall.*, (1989).

48. A. R. Pelton, F. C. Laabs, W. A. Spitzig, and C. C. Cheng, *Ultramicroscopy*, **22**, 251 (1987).

49. P. D. Funkenbusch, T. H. Courtney, and D. G. Kubisch, *Scripta Metall.*, **18**, 1099 (1984).

50. D. G. Kubisch and T. H. Courtney, *Metall. Trans. A*, **17A**, 1165 (1986).

51. Y. Leng, M. S. Thesis. Michigan Technological University, Houghton, Michigan, 1986.

52. P. D. Funkenbusch, J. K. Lee, and T. H. Courtney, *Metall. Trans. A*, **18A**, 1249 (1987).

53. J. Gil Sevillano, P. van Houtte, and E. Aernoudt, *Prog. Mat. Sci.*, **25**, 69 (1980).

54. G. T. Hahn and A. R. Rosenfield, *ASTM*, **STP432**, 5 (1968).

55. V. F. Zackay, E. R. Parker, J. W. Morris, Jr., and G. Thomas, *Mat. Sci. Eng.*, **16**, 201 (1974).

56. M. F. Ashby, *Phil. Mag.*, **21**, 399 (1970).

57. M. F. Ashby, "Strengthening Methods in Crystals" (A. Kelly and R. B. Nicholson, eds.), p. 137. Wiley, New York, 1971.

58. A. Kelly and R. B. Nicholson, *Prog. Mat. Sci.*, **10**, 151 (1963).

59. U. F. Kocks, *J. Eng. Mater. Tech.*, **98**, 76 (1976).

60. J. C. Malzahn Kampe, Ph.D. Thesis. Michigan Technological University, Houghton, Michigan, 1987.

61. R. K. Everett, *Scripta Metall.*, "Strengthening Mechanisms in Reformation Processed Composite Materials," **22**, 1227 (1988).

62. J. C. Malzahn Kampe and T. H. Courtney, *Scripta Metall.*, **20**, 285 (1986).

63. T. A. Nielsen, M. S. Thesis. Michigan Technological University, Houghton, Michigan, 1982.

4

npressive Properties of
etal Matrix Composites

R. J. ARSENAULT

Metallurgical Materials Laboratory
University of Maryland
College Park, Maryland

I.   Introduction

A consideration of the low-temperature tensile and compressive strengths of metal matrix composites (MMCs) could be divided into two main groups, i.e., tensile and compressive, or continuous (CMMCs) and discontinuous (DMMCs) metal matrix composites. (The high-temperature strength of composites will be discussed in the following chapter by Taya.) A division based on tensile and compressive properties would result in two very unequal parts, since there have been few studies of the compressive properties of MMCs. The division will be in terms of DMMCs and CMMCs. However, it is recognized that there is much more data for CMMCs as compared with DMMCs.

133

II. Discontinuous Metal Matrix Composites

This discussion will be almost exclusively in terms of tensile properties, with some discussion of compressive properties and of the mechanisms of strengthening where appropriate.

There are numerous potential reinforcements for DMMCs; however, only a few, Al_2O_3, mullite, TiB_2, B_4C, and SiC, have been considered in Al alloy matrices. From this small group, only SiC DMMCs have been studied extensively. In the case of the others, there are very few available publications. Therefore, the initial discussion of DMMCs will be confined to SiC/Al followed by a discussion of other reinforcements.

A. *Silicon Carbide/Aluminum*

Silicon carbide (SiC), either as particulate (also nodules) or whiskers additions to Al and Al alloys, has been the DMMC most widely studied. The single investigation that considered the most extensive range of particle morphologies and alloy matrices is the one by McDanels [1]. The reason for considering the results from a single investigation is due to the fact that there are generally differences in the data obtained by different investigators. The major causes of these differences are as follows:

- Lack of reproducibility from batch to batch. However, this problem has been reduced as the producing companies have gained experience, which has allowed them to produce composites that have a very homogeneous distribution of SiC in the matrix.
- Alloy additions that were added to the wrought alloys for grain refinement form intermetallic compounds during the production of the composites. Upon removal of these alloying additions, the intermetallic compounds disappeared.
- In terms of ductility, the amount of hot working has a significant effect.

I am sure there are numerous other factors that account for the differences in the data reported for the same composite, i.e., SiC size, size and volume fraction, and the same alloy matrix and heat treatment. Also, this discussion will be limited to SiC/Al DMMCs produced by the powder metallurgy route.

1. MODULUS OF ELASTICITY

The modulus of elasticity of a composite is strongly dependent upon the mode of measurement. In general, if the modulus is measured dynamically, it is higher than that determined from the initial slope of the stress-versus-strain curve. Also, it has been shown by Arsenault and Wu [2] that if the sample is tested in tension, the apparent modulus is greater than if tested in compression (Fig. 1). The reason for this difference is due to the presence of thermal residual stresses. The presence of the tensile thermal residual stress also accounts for the difference in modulus measured dynamically and statistically. Due to the ΔCTE (coefficient of thermal expansion) effect, the sample is already plastically deformed in tension, and when the initial tensile load is applied, the sample begins to deform plastically. The net result is that total strain consists of an elastic and plastic component. Therefore, a modulus determined from the initial slope of the stress–strain curve will be smaller than that determined dynamically.

McDanels [1] investigated composites that had three different SiC morphologies, whiskers ($l/d \approx 2\text{–}3$ $d \approx 0.5$ μm), nodules (flattened spheres 1–5 μm in diameter), and particulates (2–7 μm in diameter) in platelet form.

The modulus of elasticity of 6061 Al matrix composites increases with increasing reinforcement content, as shown in Fig. 2. The reinforcement content is the dominant factor in increasing the modulus of elasticity. For

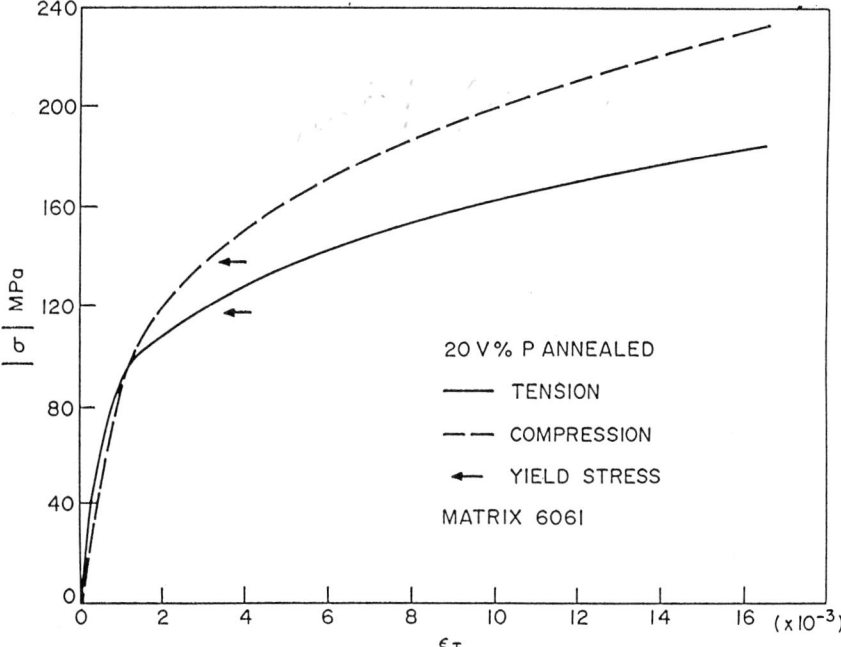

FIG. 1. Typical plots of the absolute stress versus the total strain for the annealed 20 vol-% SiC$_p$ composite (matrix, aluminum alloy 6061): ——, sampled-tested in tension; –––, another sample tested in compression; ←, yield stress. Reprinted with permission from Elsevier, S. A. Sequoia, R. J. Arsenault, and S. B. Wu, *Mater. Sci. Eng.*, **96**, 77 (1987).

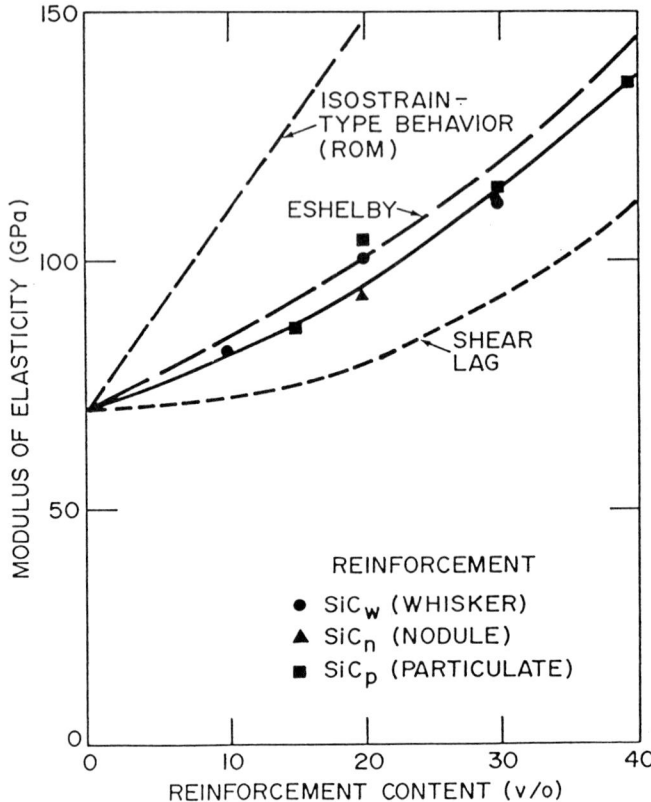

FIG. 2.   Effect of reinforcement content on modulus of elasticity of discontinuous SiC/6061 Al composites. The experimental data are from [*1*].

a given reinforcement content, the modulus tends to be isotropic with nearly equal values obtained from tests in the longitudinal and transverse directions. Moreover, the modulus appears to be insensitive to the type of reinforcement used [*1*]. The modulus is also independent of the matrix alloy, but heat treatment may have a slight effect upon it if measured as the initial slope of the σ–ε curve. The modulus of the composite in the T6-temper is slightly lower than the modulus measured on composites in the as-received F-temper. This reduction of 3 to 4% was not consistent among all matrix alloys tested by McDanels [*1*] and may be due to scatter in the data. This difference is probably due to the effects the heat treatment has on the initial stages of the stress versus strain curve. The observed modulus is much less than that predicted by iso-strain rule of mixtures (ROM). However, the predictions of Taya and Arsenault [*3*] based on the Eshelby model result in reasonable agreement with the experimental data, whereas the shear lag model [*3*] underestimates the modulus of the composite.

2. Yield and Ultimate Tensile Strengths

If we consider the σ_{yc} or ultimate tensile strength (UTS) of DMMCs, there are significant differences. Figure 3 summarizes the stress–strain behavior of 20 vol-% SiC_w/Al with different matrix alloys, with the other parameters held constant. Composites with 5083 Al matrix failed in a brittle manner. Heat treatment affects the transition into plastic flow. The F-temper matrix strains elastically before passing into a normal decreasing-slope plastic flow. The HT-6 samples exhibit a slightly greater amount of elastic strain, i.e., a higher proportional limit.

The increase in the UTS of DMMCs as compared with the UTS of the matrix is shown in Fig. 4. It appears from this data that the increase in strength decreases as the UTS of the matrix increases. Arsenault [*4*] proposed

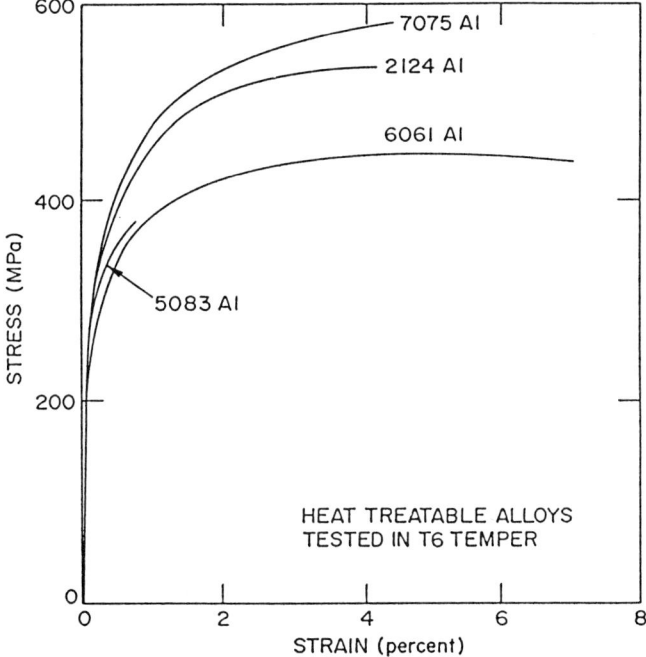

FIG. 3. Effect of Al matrix alloy on stress–strain behavior of composites with 20 vol-% SiC_w reinforcement (tested in direction parallel to final rolling directions. Reprinted with permission from The Minerals, Metals & Materials Society & ASM International, D. L. McDaniels, *Metall. Trans.* **16A**, 1105 (1985).

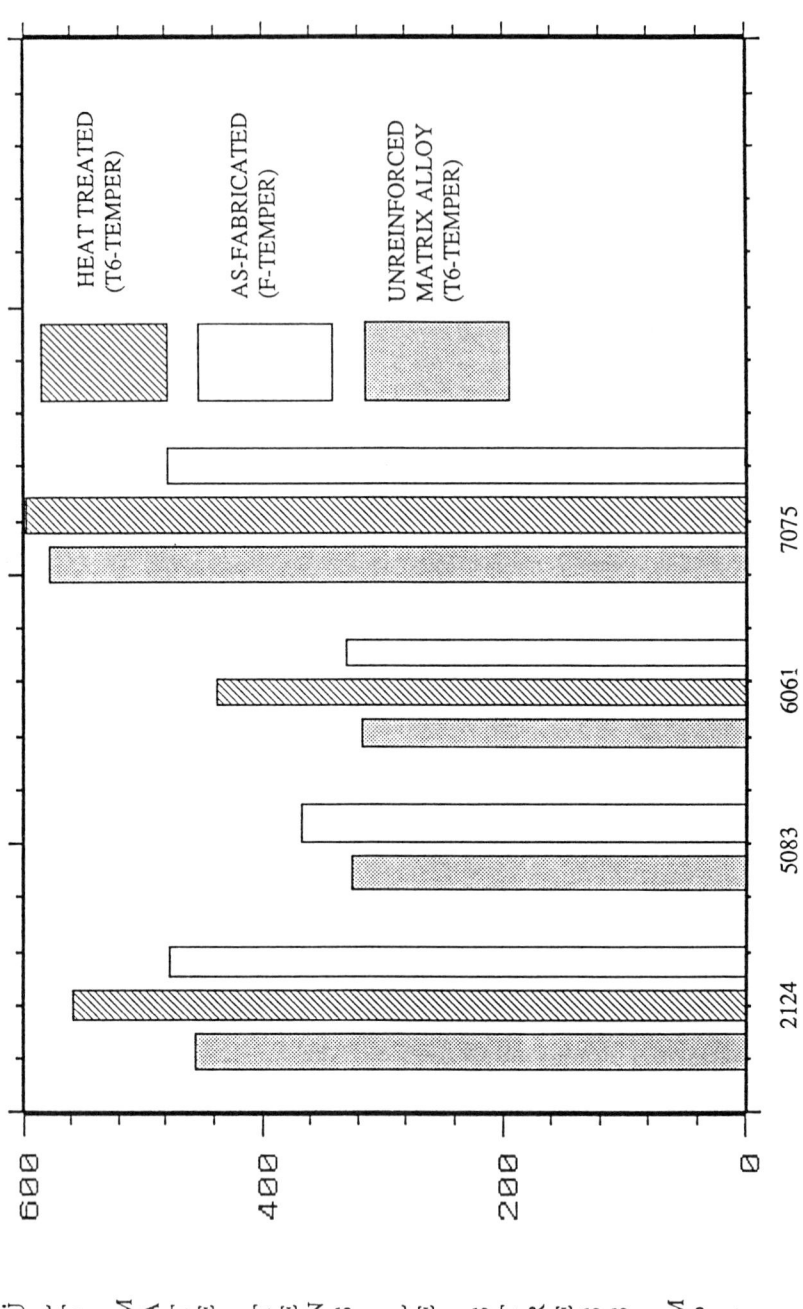

FIG. 4.   Effect of matrix alloy and heat treatment on UTS of 20 vol-% SiC_w/Al composites. Reprinted with permission from The Minerals, Metals & Materials Society & ASM International, D. L. McDaniels, *Metall. Trans.* **16A**, 1105 (1985). The properties of the unreinforced matrix are taken from [5].

that this was due to the fact that a higher σ_{ym} would result in the generation of fewer dislocations due to ΔCTE and a higher tensile thermal residual stress. The combination of these two factors would result in a smaller increase in strength of the composite. However, some other data generated by Taya and Arsenault [3] indicate that $\sigma_{yc} - \sigma_{ym}$ is relatively independent of σ_{ym} of the matrix.

Figures 5a and b present a summary of the data generated by McDanels [1] for various volume fractions of reinforcement morphologies and for various alloys. The comparisons are of the HT-6 condition except for the 5083 alloy, which is in the F-temper condition. There is a general trend of an increase in strength with an increase in volume fraction, but the data are not uniformly consistent. If we consider a specific case (Fig. 1., page 81), then there is consistent increase in strength with an increase in volume fraction.

As measured by strain-to-failure, the ductility of discontinuous SiC/Al composites is again a complex interaction of parameters. The primary factors are reinforcement content and orientation matrix alloy. Increases in failure strain are attributed to two main factors [1]. First, fabricators are constantly striving for cleaner, more uniform Al alloy powders, better powder/reinforcement mixing techniques that can help control powder size distribution, interparticle spacing, and homogeneity of structure. When comparing data for discontinuous SiC/Al composites, the fabrication data of the composites becomes important because it defines the state of the art. In turn, this helps determine the state of evolution of strength and ductility behavior.

The second factor affecting improvement of SiC/Al ductility is the degree of working from billet to final product. The higher degree of reduction by mechanical working helps to increase composite ductility by (1) reducing matrix porosity while generating fresh surfaces to improve Al powder binding; (2) breaking up inclusions and more effectively stringering them; and (3) making the dispersion of reinforced particles finer, and increasing uniformity.

However, in general, it can be stated that, as to be expected, the ductility decreases to a low level upon increasing the volume fraction [1]

3. COMPRESSIVE YIELD STRESSES

As stated before, there is a dearth of experimental data related to the compressive properties of DMMCs. From this limited data, a few general comments can be made: First, the $\sigma_{yc}^C > \sigma_{yc}^T$; secondly, the compressive ductility is greater than the tensile ductility.

The only systematic study of the difference in the compressive versus tensile properties is the one by Arsenault and Wu [6]. The data obtained for aluminum alloy 6061 as a function of the volume fraction of SiC whiskers

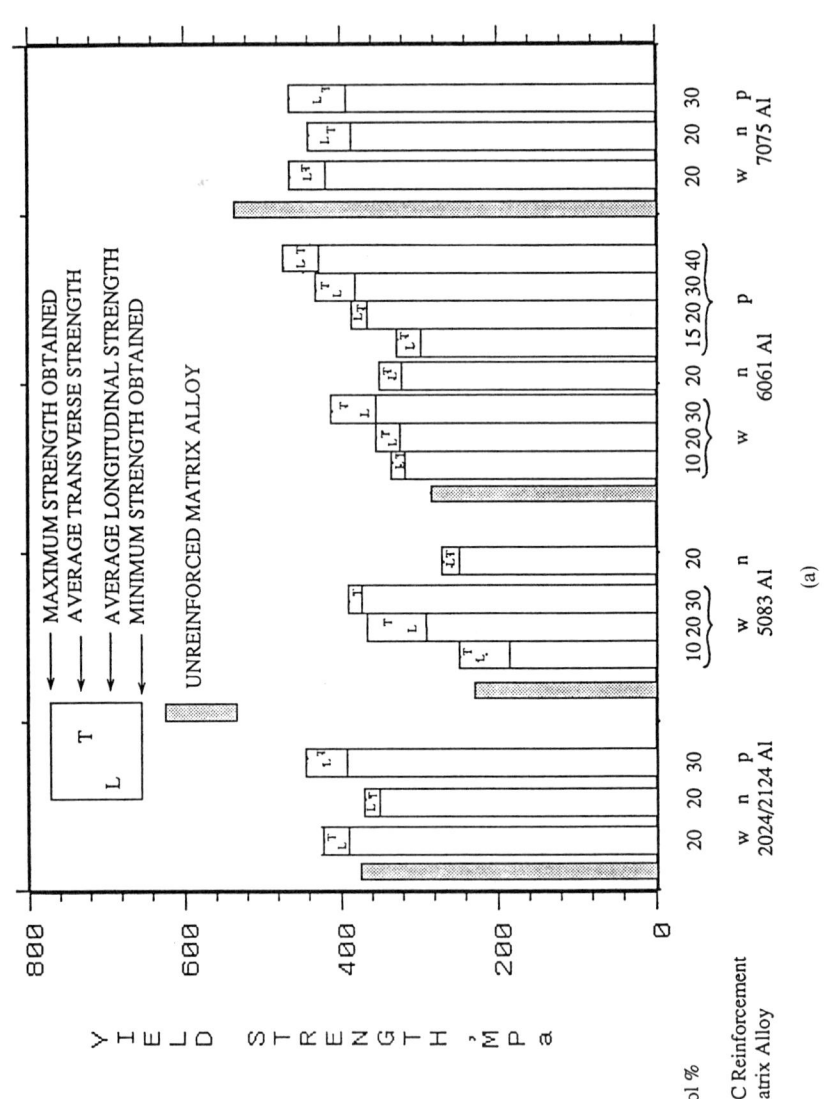

(a)

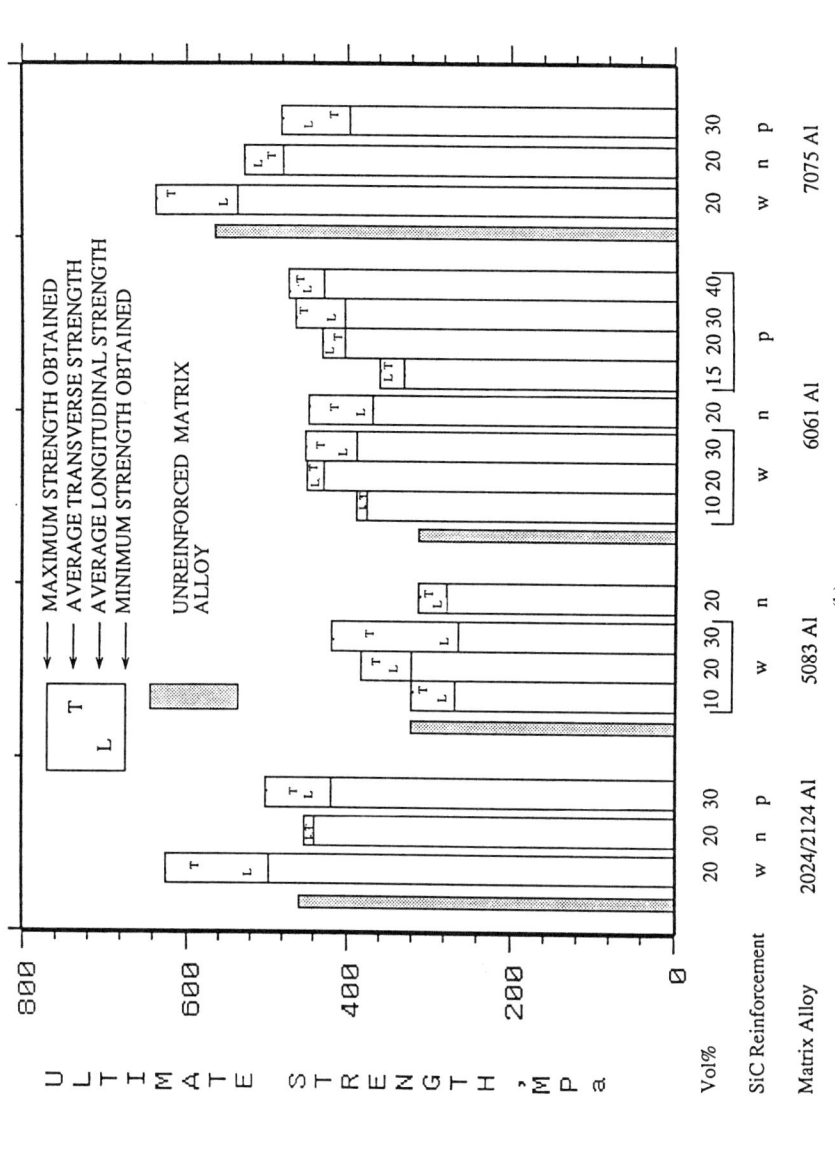

FIG. 5. Effect of SiC reinforcement type and content on tensile properties of discontinuous SiC/Al composites. (a) 0.2% offset yield strength; (b) UTS. Reprinted with permission from The Minerals, Metals & Materials Society & ASM International, D. L. McDaniels, *Metall. Trans.* **16A**, 1105 (1985). The properties of the unreinforced matrix are taken from [5].

as well as wrought commercial aluminum alloy 6061 (WR) and 99.99% Al (HP) are plotted in Fig. 6. For HP, there was no difference between the compressive and tensile yield stress, as expected. It was assumed that there would be no difference between wrought aluminum alloy 6061 and the 0 vol-% SiC$_w$/aluminum alloy 6061, since there is no difference in the yield stresses. However, for the heat-treated wrought aluminum alloy 6061, there was a difference between the compressive and tensile yield stress, whereas in the 0 vol-% SiC$_w$/Al alloy 6061, there was no difference between the compressive and tensile yield stress.

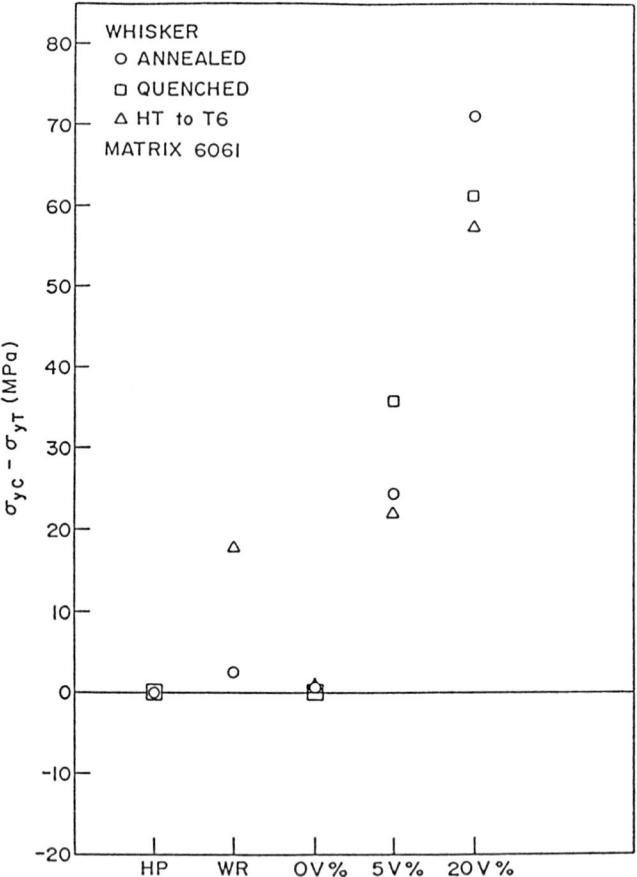

FIG. 6. A plot of the difference between the compressive yield stress and the tensile yield stress for various volume fractions of whisker composite materials (matrix, aluminum alloy 6061): ○, annealed; □, quenched; △, heat-treated to T6 condition. Also plotted is the difference in yield stress for high-purity aluminum and wrought aluminum alloy 6061.

The effect of increasing the volume fraction of SiC whiskers was to increase the magnitude of the difference between σ_{yc}^C and σ_{yc}^T. For all three conditions, i.e., annealed, quenched, and heat treated to T6, as the volume fraction of SiC whiskers increases, the magnitude of $\Delta\sigma$ ($\Delta\sigma = \sigma_{yc}^C - \sigma_{yc}^T$) increases. However, the relative increase is not consistent. In other words, for the 5 vol-% SiC$_w$ composite, the quenched condition results in a larger $\Delta\sigma$, but for the 20 vol-% SiC$_w$ composite, the annealed condition results in the largest $\Delta\sigma$.

For samples that were quenched or heat treated, there was also an increase in the magnitude of $\Delta\sigma$ with an increasing volume fraction of SiC whiskers. The data obtained from samples that were annealed, quenched, and heat treated did not exhibit any consistent pattern, e.g., the quenched samples did not consistently produce a larger value of $\Delta\sigma$.

Particulate SiC in an aluminum alloy 6061 matrix had the same general trends as those of the whisker composite samples. That is, as the volume fraction of particulate increases, so does $\Delta\sigma$, but there are differences as plotted in Fig. 7. The data of HP and WR are the same as those plotted in Fig. 6 and are plotted here for the purpose of comparison. The data from the particulate composites do not follow any uniform trend. There is not a uniform increase in $\Delta\sigma$ with increasing volume fraction of SiC particulates, e.g., there is very little difference between the 0 and 5 vol-% SiC$_p$ composites. For the quenched 20 vol-% SiC$_p$ composite, $\Delta\sigma$ is much larger than $\Delta\sigma$ for the annealed 20 vol-% SiC$_p$ composite, whereas $\Delta\sigma$ for the annealed 20 vol-% SiC$_w$ composite was slightly larger than $\Delta\sigma$ for the quenched 20 vol-% SiC$_w$ composite. When the 20 vol-% SiC$_p$ composite was subjected to a liquid nitrogen quench after a water quench, a much smaller difference between σ_{yc}^C and σ_{yc}^T resulted. The water quench only resulted in a $\sigma_{yc}^C - \sigma_{yc}^T$ value of 70 MPa, whereas the liquid-nitrogen quench after a water quench resulted in a $\sigma_{yc}^C - \sigma_{yc}^T$ value of 8 MPa. If the difference between σ_{yc}^C and σ_{yc}^T was initially small, e.g., for a heat-treated 5 vol-% SiC$_p$ composite, then the effect of the subsequent liquid-nitrogen quench was minimal, as expected.

For the 1100 aluminum alloy matrix, the volume fraction was held constant at 20 vol-%, and the morphology of the SiC was changed from whisker to spherical SiC 0.5 μm in diameter. From a comparison of the data for the 20 vol-% SiC$_w$ composites in Table I with that in Fig. 6, it is obvious that the difference between σ_{yc}^C and σ_{yt}^T is smaller than that of the 6061 aluminum alloy matrix composite by a factor of 2. In Table I, it is also possible to compare the effect of different reinforcement morphologies. The theoretical prediction is that $\Delta\sigma$ should be zero for the spherical SiC case and have a large positive value for the whisker case. The experimental data (Table I) indicate that $\Delta\sigma$ is negative for the spherical SiC morphology, i.e., $\sigma_{yc}^T > \sigma_{yc}^C$.

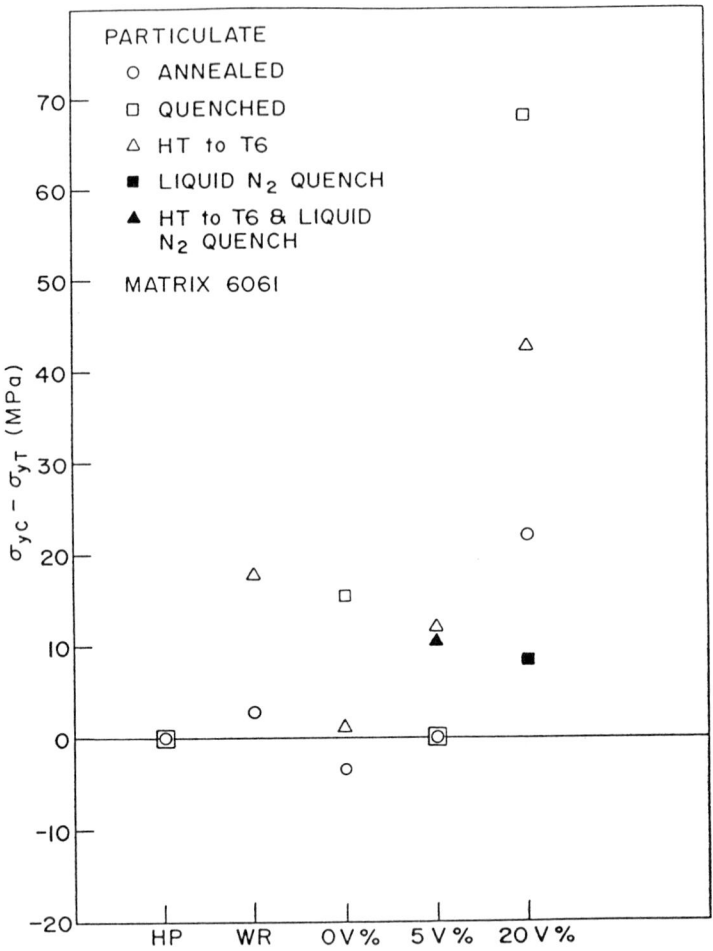

FIG. 7. A plot of the difference in yield stress for various volume fractions of particulate composite materials (matrix, aluminum alloy 6061): ○, annealed; □, quenched; △, heat-treated to T6 condition; ■, liquid-nitrogen-quenched; ▲, heat-treated to T6 condition and liquid-nitrogen-quenched. Also plotted is the difference in yield stress for high-purity aluminum and wrought aluminum alloy 6061.

B. Other Discontinuous Reinforcements

As stated in the beginning of the discussion of DMMCs, a few other systems have been investigated.

These have included, Al_2O_3/Al, TiB_2/Al, and mullite/Al. The addition of the reinforcement in the case of Al_2O_3/Al [7] and mullite/Al [8] results in no

TABLE I
Difference $\Delta\sigma = (\sigma^{C}_{yc} - \sigma^{T}_{yc})$ in the Yield Stresses

Matrix	Condition	Volume fraction (vol-%)	$\Delta\sigma$ (MPa)
Al alloy 1100	annealed	0	3.80
Al alloy 1100	annealed	20 SiC_w	42.92
Al alloy 1100	annealed	20 SiC_s	−15.46

appreciable increase in strength. In other words, the composite had the same strength as the unreinforced matrix. A probable reason for this lack of strengthening by the short fibers of Al_2O_3 and mullite is the lack of a good bond.

The strength of TiB_2/Al [9] is greater than the corresponding Al matrix alloy, but incremental strengthening, due to the same volume fraction of reinforcement, is less for the case of addition of SiC as the reinforcement. The probable reason for this difference is the smaller ΔCTE for the TiB_2/Al systems as compared with SiC/Al, i.e., a factor of 2 versus a factor of 4 to 10.

III. Continuous-Filament Metal Matrix Composites

As stated before, most of the experimental investigations have been concerned with tensile σ^{T}_{yc} and UTS. There are very few investigations associated with determining the compressive σ^{C}_{yc}. However, in CMMCs, the difference between the longitudinal (in a direction parallel to the filament) and transverse properties (i.e., tested perpendicular to the filament axis) have been investigated to some extent. The discussion of CMMCs will be divided into four parts: A. ROM, B. CMMCs that follow ROM predictions, C. CMMCs that do not follow the predictions of the ROM, D. a comparison of the longitudinal and transverse properties, and E. compression versus tensile strengths.

A. Rule of Mixtures

A discussion of elastic properties (modulus) and strengths of CMMCs in general begins with a comparison of the ROM-type predictions and the experimental data. This rule involves the following assumptions: (1) filaments and matrix are strained by equal amounts, (2) the phases act as homogeneous materials and are well-bonded, and (3) Poisson ratio differences between fiber and matrix are ignored.

Before discussing some pertinent experimental data, a brief overview of the classical continuum models of the ROM, as applied to continuous-filament composites, will be given. This treatment is given with permission by Chawla [10].

1. PREDICTION OF ELASTIC CONSTANTS

Consider a unidirectional composite such as the one shown in Figs. 8a and b. Assume that plane sections of this composite remain plane after deformation. Let us apply a force P_c in the fiber direction. Now, if the two

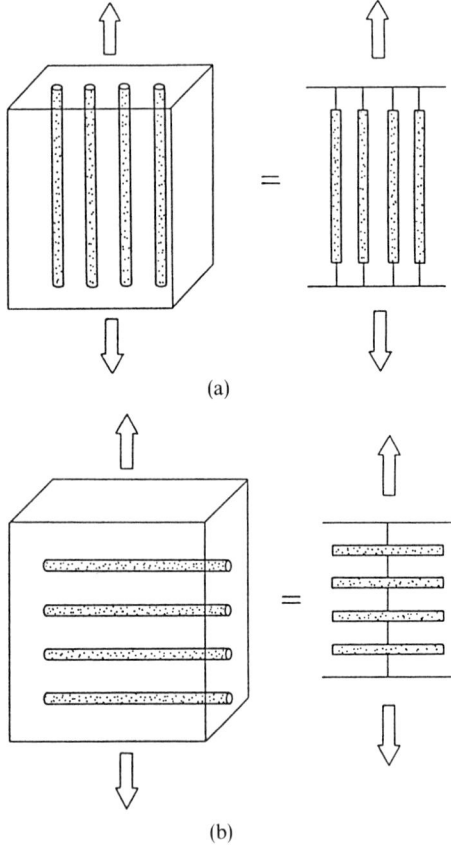

(a)

(b)

FIG. 8.   Unidirectional composite: (a) isostrain or action in parallel, (b) isostress or action in series.

components adhere perfectly and if they have the same Poisson ratio, then each component will undergo the same longitudinal elongation Δl. Thus, we can write for the strain in each component,

$$\varepsilon_f = \varepsilon_m = \varepsilon_{cl} = \frac{\Delta l}{l}, \tag{1}$$

where ε_f is the strain in the fiber, ε_m is the strain in the matrix and ε_{cl} is the strain in the composite in the longitudinal direction. This is called the *isostrain* or *action in parallel* situation and was first treated by Voigt [11]. If both fiber and matrix are elastic, we can relate the stress σ in the two components to the strain ε_l by Young's modulus E. Thus,

$$\sigma_f = E_f \varepsilon_{cl} \quad \text{and} \quad \sigma_m = E_m \varepsilon_{cl}.$$

Let A_c be the cross-sectional area of the composite, A_m, that of the matrix, and A_f, that of all the fibers. Then, from the equilibrium of forces in the fiber direction, we can write

$$P_c = P_f + P_m$$

or

$$\sigma_{cl} A_c = \sigma_f A_f + \sigma_m A_m. \tag{2}$$

From Eqs. (1) and (2), we get

$$\sigma_{cl} A_c = (E_f A_f + E_m A_m)\varepsilon_{cl}$$

or

$$E_{cl} = \frac{\sigma_{cl}}{\varepsilon_{cl}} = E_f \frac{A_f}{A_c} + E_m \frac{A_m}{A_c}.$$

Now, for a given composite length, $A_f/A_c = V_f$ and $A_m/A_c = V_m$. Then the above expression can be simplified to

$$E_{cl} = E_f V_f + E_m V_m = E_{11}. \tag{3}$$

Equation (3) is called the ROM for Young's modulus in the fiber direction. A similar expression can be obtained for the composite longitudinal strength from Eq. (2), namely,

$$\sigma_{cl} = \sigma_f V_f + \sigma_m V_m, \tag{4}$$

For properties in the *transverse* direction, we can represent the simple unidirectional composite by what is called the *action-in-series* or *isostress* situation; see Fig. 8b. In this case, we group the fibers together as a continuous phase normal to the stress. Thus, we have equal stresses in the two components, and the model is equivalent to that treated by Reuss [12]. For loading transverse to the fiber direction, we have

$$\sigma_{ct} = \sigma_f = \sigma_m,$$

while the total displacement of the composite in the thickness direction, t_c, is the sum of displacements of the components; that is,

$$\Delta t_c = \Delta t_m + \Delta t_f.$$

Dividing throughout by t_c, the composite gauge length in the transverse direction, we obtain

$$\frac{\Delta t_c}{t_c} = \frac{\Delta t_m}{t_c} + \frac{\Delta t_f}{t_c}.$$

Now $\Delta t_c/t_c = \varepsilon_{ct}$, the composite strain in the transverse direction, while Δt_m and Δt_f equal the strains in the matrix and fiber times their respective gauge lengths; that is, $\Delta t_m = \varepsilon_m t_m$ and $\Delta t_f = \varepsilon_f t_f$. Then,

$$\varepsilon_{ct} = \frac{\Delta t_c}{t_c} = \frac{\Delta t_m}{t_c}\frac{t_m}{t_c} + \frac{\Delta t_f}{t_c}\frac{t_m}{t_c}$$

or

$$\varepsilon_{ct} = \varepsilon_m\frac{t_m}{t_c} + \varepsilon_f\frac{t_f}{t_c}. \tag{5}$$

For a given composite cross-sectional area under the applied load, the volume fractions of fiber and matrix can be written

$$V_m = \frac{t_m}{t_c} \quad \text{and} \quad V_f = \frac{t_f}{t_c}.$$

This simplifies Eq. (5) to

$$\varepsilon_{ct} = \varepsilon_m V_m + \varepsilon_f V_f. \tag{6}$$

Considering both components to be in the elastic regime and remembering that $\sigma_{ct} = \sigma_f = \sigma_m$, in this case, we can write Eq. (6) as

$$\frac{\sigma_{ct}}{E_{ct}} = \frac{\sigma_{ct}}{E_m}V_m + \frac{\sigma_{ct}}{E_f}V_f$$

or

$$\frac{1}{E_{ct}} = \frac{V_m}{E_m} + \frac{V_f}{E_f} = \frac{1}{E_{22}}. \tag{7}$$

Relationships given by Eqs. (3), (4), (6), and (7) are commonly referred to as

ROMs. Figure 9 shows the plots of Eqs. (1) and (3). The reader should appreciate that these relationships and their variants are but rules of thumb obtained from a simple strength-of-materials approach. More comprehensive micromechanical models, based on the theory of elasticity, can be and should be used to obtain the elastic constants of fibrous composites. In the following, we describe, albeit very briefly, some of these. A critique of these models has been made by Chamis and Sendeckyj [13].

2. Micromechanical Approach

In the most general case, an anisotropic body has 21 independent elastic constants [14]. An isotropic body, on the other hand, has only two independent elastic constants. In such a body, when a tensile stress is applied in the z-direction, a tensile strain ε results in that direction. In addition to this, because of the Poisson-ratio effect, two equal compressive strains ($\varepsilon_x = \varepsilon_y$) result in the x- and y-directions. In a generally anistropic body, the two transverse strain components are not equal. In fact, in such a body, tensile loading can result in tensile as well as shear strains. The large number of independent elastic constants (21 in the most general case, i.e., no symmetry elements) represents the complexity of the situation. Any symmetry elements present will reduce the number of independent elastic constants [14].

A composite containing uniaxially aligned fibers will have a plane of symmetry perpendicular to the fiber direction (i.e., material on one side of the plane will be the mirror image of the material on the other side). Such a material will have 13 independent elastic constants. Additional symmetry elements, depending on the fiber arrangement, can be present. Figure 10

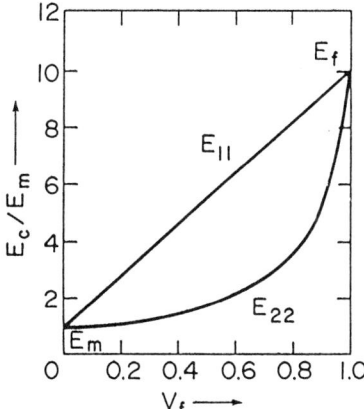

FIG. 9. Variation of longitudinal modulus (E_{11}) and transverse modulus (E_{22}) with fiber volume fraction (V_f).

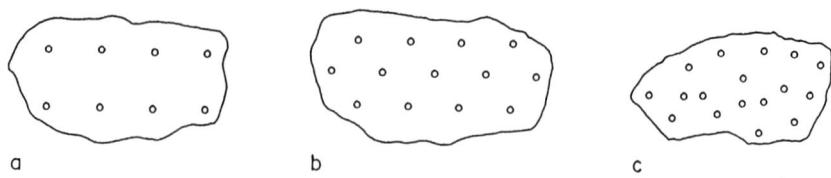

FIG. 10.   Various fiber arrays in a matrix: (a) square, (b) hexagonal, and (c) random.

shows square, hexagonal, and random fiber arrays in a matrix. A square array of fibers, for example, will have symmetry planes parallel to the fibers as well as perpendicular to them. Such a material is an orthotropic material (three mutually perpendicular planes of symmetry) and possesses nine independent elastic constants [14]. Hexagonal and random arrays of aligned fibers are transversely isotropic and have five independent elastic constants. These five constants as well as the stress–strain relationships, as derived by Hashin and Rosen [15] and Rosen [16], are given in Table II. There are two Poisson ratios; one gives the transverse strain caused by an axially applied stress and the other gives the axial strain caused by a transversely applied stress. The two are not independent but are related. Thus, the number of independent elastic constants for a transversely isotropic composite is five. Note that the total number of independent elastic constants in Table II is five (count the number of C's).

TABLE II
ELASTIC MODULI OF A TRANSVERSELY ISOTROPIC FIBROUS COMPOSITE

$$E = C_{11} - \frac{2C_{12}^2}{C_{22} + C_{23}} \qquad\qquad K_{23} = \tfrac{1}{2}(C_{22} + C_{23})$$

$$G = G_{12} = G_{13} = C_{44} \qquad\qquad G_{23} = \tfrac{1}{2}(C_{22} - C_{23})$$

$$v = v_{13} = v_{31} = \frac{1}{2}\left(\frac{C_{11} - E}{K_{23}}\right)^{1/2} \qquad v_{23} = \frac{K_{23} - \phi G_{23}}{K_{23} + \phi G_{23}}$$

$$E_2 = E_3 = \frac{4G_{23}K_{23}}{E_{23} + \phi\Gamma_{23}} \qquad\qquad \phi = 1 + \frac{4K_{23}v^2}{E}$$

Stress–Strain Relationships

$$\varepsilon_{11} = (1/E_1)[\sigma_{11} - v(\sigma_{22} + \sigma_{33})] \qquad \varepsilon_{22} = \varepsilon_{33} = (1/E_2)(\sigma_{22} - v\sigma_{33}) - (v/E)\sigma_{11}$$

$$\gamma_{12} = \gamma_{13} = (1/G)\sigma_{12} \qquad\qquad \gamma_{23} = \frac{2(1 + v_{23})}{E_2}\sigma_{23}$$

Source: Adapted with permission from Springer-Verlag, New York, K. K. Chawla, "Composite Materials," 1987.

Chamis [*17*] has summarized the elastic constants for a transversely isotropic composite in terms of the elastic constants of the two components. He gives the relationships for a thin sheet or ply, but they are more general and valid for any transversely isotropic composite such as the one shown in Fig. 11. Since the plane 2–3 is isotropic in Fig. 11, the properties in directions 2 and 3 are identical. Chamis treats the matrix as an isotropic material, whereas the fiber is treated as an anisotropic material. Thus, E_m and v_m are the two constants required for the matrix, whereas five constants (E_{f1}, E_{f2}, G_{f12}, G_{f23}, and v_{f12}) are required for the fiber. Table III gives these equations. The five independent constants are E_{11}, E_{22}, G_{12}, G_{13}, and v_{12}.

Frequently, composite structures are fabricated by stacking up thin sheets of unidirectional composites, called plies, in an appropriate orientation sequence dictated by the elasticity theory. It is of interest to know the properties of a ply, such as its elastic constants and strength characteristics. In particular, it is of value if we are able to predict the lamina characteristics starting from the individual component characteristics. Later in the macromechanical analysis, a ply is treated as a homogeneous but thin orthotropic material. Elastic constants in the thickness direction can be ignored in such a ply, leaving four independent elastic constants, namely E_{11}, E_{22}, v_{12}, and G_{12}, one less than the number for a thick but transversely isotropic material. This missing constant is G_{23}, the transverse shear modulus in the 2–3 plane normal to the fiber axis.

A brief description of the various micromechanical techniques used for predicting the elastic constants is given below. Chamis and Sendeckyj [*13*]

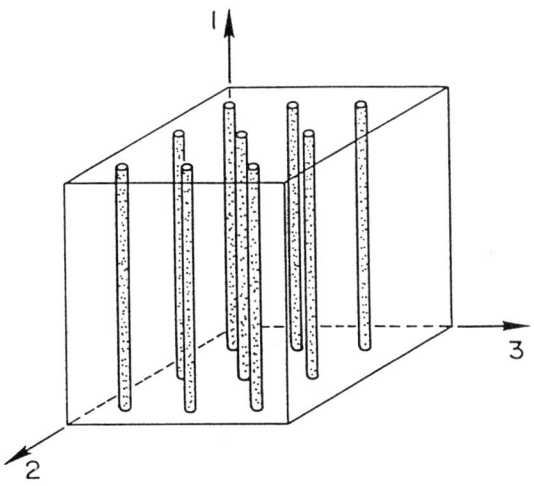

FIG. 11.   A transversely isotropic fiber composite: Plane transverse to fibers (2–3 plane) is isotropic.

TABLE III

ELASTIC CONSTANTS OF A TRANSVERSELY ISOTROPIC
COMPOSITE IN TERMS OF COMPONENT CONSTANTS
(MATRIX ISOTROPIC, FIBER ANISTROPIC)

Longitudinal modulus	$E_{11} = E_{f1}V_f + E_m V_m$
Transverse modulus	$E_{22} = E_{33} = \dfrac{E_m}{1 - (V_f)^{1/2}(1 - E_m/E_{f2})}$
Shear modulus	$G_{12} = G_{13} = \dfrac{G_m}{1 - (V_f)^{1/2}(1 - G_m/G_{f12})}$
Shear modulus	$G_{23} = \dfrac{G_m}{1 - (V_f)^{1/2}(1 - G_m/G_{f23})}$
Poisson ratio	$v_{12} = v_{13} = v_{f12}V_f + v_m V_m$
Poisson ratio	$v_{23} = \dfrac{E_{22}}{2G_{23}} - 1$

Source: Adapted with permission from Springer-Verlag, New York,
K. K. Chawla, "Composite Materials," 1987.

have presented a critique of these methods. Following this brief description, an account is given of a set of empirical equations called Halpin–Tsai equations that has been developed for predicting the elastic constants of a fiber composite.

In the so-called self-consistent field methods [13], approximations of phase geometries are made and a simple representation of the response field is obtained. The phase geometry is represented by one single fiber embedded in a matrix cylinder. This outer cylinder is embedded in an unbounded homogeneous material whose properties are taken to be equivalent to those of average composite properties. The matrix under a uniform load at infinity introduces a uniform strain field in the fiber. Elastic constants are obtained from this strain field. The results obtained are independent of fiber arrangements in the matrix and are reliable at low fiber volume fractions, V_f, reasonable at intermediate V_f, and unreliable at high V_f [18]. Exact methods deal with specific geometries, for example, fibers arranged in a hexagonal, square, or rectangular array in a matrix. The elasticity problem is then solved by a series development, a complex variable technique, or a numerical solution.

The variational or bounding methods focus on the upper and lower bounds of the elastic constants. They do not predict properties directly. If, however, the upper and lower bounds coincide, then the property is determined exactly. Frequently, the upper and lower bounds are well separated.

When these bounds are close enough, we can safely use them as indicators of the material behavior. It turns out that this is the case for longitudinal properties of a unidirectional lamina. Hill [18] derived bounds for the ply elastic constants that are analogous to those derived by Hashin and Rosen [15] and Rosen [16]. In particular, Hill put rigorous bounds on the longitudinal Young's modulus, E_{11}, in terms of the bulk modulus in plane strain (k_p), Poisson's ratio (v), and the shear modulus (G) of the two phases. No restrictions were made on the fiber form or packing geometry. The term k_p is the modulus for lateral dilatation with zero longitudinal strain and is given by

$$k_p = \frac{E}{2(1 - 2v)(1 + v)}.$$

The bounds on the longitudinal modulus, E_{11}, are

$$\frac{4V_f V_m (v_f - v_m)^2}{(V_f/k_{pm}) + (V_m/k_{pf}) + 1/G_m} \leq E_{11} - E_f V_f - E_m V_m$$

$$\leq \frac{4V_f V_m (v_f - v_m)^2}{(V_f/k_{pm}) + (V_m/k_{pf}) + 1/G_f}. \tag{8}$$

Equation (8) shows that the deviations from the ROM (Eq. (3)) are quite small ($<2\%$). We may verify this by substituting some values of practical composites such as carbon or boron fibers in an epoxy matrix or a metal matrix composite such as tungsten in a copper matrix. Note that the deviation from the ROM value comes from the $(v_m - v_f)^2$ factor. Because if $v_f = v_m$, we have E_{11} given exactly by the ROM.

Hill [19] also showed that for a unidirectionally aligned fiber composite

$$v_{12} \gtrless v_f V_f + v_m V_m \quad \text{accordingly as} \quad (v_f - v_m)(k_{pf} - k_{pm}) \gtrless 0. \tag{9}$$

Generally, $v_f < v_m$ and $E_f \gg E_m$. Then, v_{12} will be less than that predicted by the ROM ($= v_f V_f + v_m V_m$). It is easy to see that the bounds on v_{12} are not as close as the ones on E_{11}. This is because $v_f - v_m$ appears in the case of v_{12} (Eq. (9)), whereas $(v_f - v_m)^2$ appears in the case of E_{11} (Eq. (8)). In the case where $v_f - v_m$ is very small, the bounds are close enough to allow us to write

$$v_{12} \simeq v_f V_f + v_m V_m. \tag{10}$$

3. HALPIN–TSAI EQUATIONS

Halpin and Tsai [20], Halpin and Kardos [21], and Kardos [22] have empirically developed some generalized equations that readily give quite satisfactory results compared with the complicated micromechanical equations. These equations are quite accurate at low fiber volume fractions. They are also useful in determining the properties of composites that contain discontinuous fibers oriented in the loading directions. One writes a single equation of the form

$$\frac{p}{p_m} = \frac{1 + \xi \eta V_f}{1 - \eta V_f}, \tag{11}$$

$$\eta = \frac{p_f/p_m - 1}{p_f/p_m + \xi}, \tag{12}$$

where p represents composite moduli, for example, E_{11}, E_{22}, G_{12}, or G_{23}; p_m and p_f are the corresponding matrix and fiber moduli, respectively; V_f is the fiber volume fraction; and ξ is a measure of the reinforcement, which depends on the boundary conditions (fiber geometry, fiber distribution, and loading conditions). The term ξ is an empirical factor that is used to make Eq. (11) conform to the experimental data.

The function η in Eq. (12) is constructed in such a way that when $V_f = 0$, $p = p_m$ and when $V_f = 1$, $p = p_f$. Furthermore, the form of η is such that

$$\frac{1}{p} = \frac{V_m}{p_m} + \frac{V_f}{p_f} \quad \text{for} \quad \xi \to \infty$$

and

$$p = p_f V_f + p_m V_m \quad \text{for} \quad \xi \to \infty.$$

These two extremes (not necessarily tight) bound the composite properties. Thus, values of ξ between 0 and ∞ will give an expression for p between these extremes. Some typical values of ξ are given in Table IV. Thus, we can cast the Halpin–Tsai equations for the transverse modulus as

$$\frac{E_{22}}{E_m} = \frac{1 + \xi \eta V_f}{1 - \eta V_f}, \eta = \frac{E_f/E_m - 1}{E_f/E_m + \xi}. \tag{13}$$

TABLE IV
VALUES OF ξ FOR SOME UNIAXIAL
COMPOSITES

Modulus	ξ
E_{11}	$2(l/d)$
E_{22}	0.5
G_{12}	1.0
G_{21}	0.5
K	0

Source: Adapted with permission from Springer-Verlag, New York, K. K. Chawla, "Composite Materials," 1987.

Comparing these expressions with exact elasticity solutions, one can obtain the value of ξ. Whitney [23] suggests $\xi = 1$ or 2 for E_{22}, depending on whether a hexagonal or square array of fibers is used.

Nielsen [24] has modified the Halpin–Tsai equations to include the maximum packing fraction ϕ_{max} of the reinforcement. His equations are

$$\frac{p}{p_m} = \frac{1 + \xi \eta V_f}{1 - \eta \Psi V_f},$$

$$\eta = \frac{p_f/p_m - 1}{p_f/p_m + \xi}, \tag{14}$$

$$\Psi \simeq 1 + \left(\frac{1 - \phi_{max}}{\phi_{max}^2}\right) V_f,$$

where ϕ_{max} is the maximum packing factor. It allows one to take into account the maximum packing fraction. For a square array of fibers, $\phi_{max} = 0.785$, whereas for a hexagonal arrangement of fibers, $\phi_{max} = 0.907$. In general, ϕ_{max} is between these two extremes and near the random packing, $\phi_{max} = 0.82$.

4. TRANSVERSE STRESSES

When a fibrous composite consisting of components with different elastic moduli is uniaxially loaded, stresses in transverse directions arise because of the Poisson ratio differences between the matrix and fiber, that is, because the two components have different contraction tendencies. We follow here Kelly's [25] treatment of this important but, unfortunately, not well appreciated subject.

Consider a unit fiber-reinforced composite consisting of a single fiber (radius a) surrounded by its shell of matrix (outer radius b) as shown in Fig. 12. The composite as a whole is thought to be built of an assembly of such unit composites, a reasonably valid assumption at moderate fiber volume fractions. We apply an axial load (direction z) to the composite. Owing to the obvious cylindrical symmetry, we treat the problem as polar coordinates, r, θ, and z. It follows from the axial symmetry that the stress and strain are independent of angle and are functions only of r, which simplifies the problem. We can write Hooke's law for this situation as

$$\begin{bmatrix} \varepsilon_r & 0 & 0 \\ 0 & \varepsilon_\theta & 0 \\ 0 & 0 & \varepsilon_z \end{bmatrix} = \frac{1 + v}{E} \begin{bmatrix} \sigma_r & 0 & 0 \\ 0 & \sigma_\theta & 0 \\ 0 & 0 & \sigma_z \end{bmatrix} - \frac{v}{E}(\sigma_r + \sigma_\theta + \sigma_z)\begin{bmatrix} 1 & 0 & 0 \\ 0 & 1 & 0 \\ 0 & 0 & 1 \end{bmatrix}, \tag{15}$$

where ε is the strain, σ is the stress, v is the Poisson ratio, E is Young's modulus in the longitudinal direction and the subscripts r, θ, and z refer to

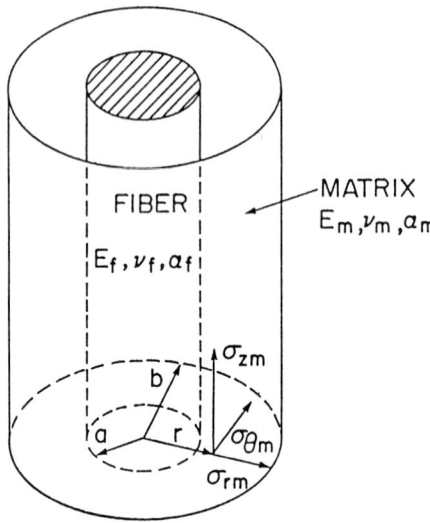

FIG. 12.   A single fiber surrounded by its matrix shell.

the radial, circumferential, and axial directions, respectively. The only equilibrium equation for this problem is

$$\frac{d\sigma_r}{dr} + \frac{\sigma_r - \sigma_\theta}{r} = 0. \tag{16}$$

Also, for the plane strain condition, we can write for the strain components, in terms of displacements,

$$\varepsilon_r = \frac{du_r}{dr}, \qquad \varepsilon_\theta = \frac{u_r}{r}, \qquad \varepsilon_z = \text{const}, \tag{17}$$

where u_r is the radial displacement.

From Eq. (15) we have, after some algebraic manipulation,

$$\frac{\sigma_\theta}{K} = (1 - v)\varepsilon_\theta + v(\varepsilon_r + \varepsilon_z),$$

$$\frac{\sigma_r}{K} = (1 - v)\varepsilon_r + v(\varepsilon_\theta + \varepsilon_z), \tag{18}$$

where

$$K = \frac{E}{(1 + v)(1 - 2v)}.$$

From Eqs. (17) and (18), we get

$$\frac{\sigma_\theta}{K} = v\frac{du_r}{dr} + (1 - v)\frac{u_r}{r} + v\varepsilon_z,$$

$$\frac{\sigma_r}{K} = (1 - v)\frac{du_r}{dr} + v\frac{u_r}{r} + v\varepsilon_z. \tag{19}$$

Substituting Eq. (19) into Eq. (16), we obtain the following differential equation in terms of the radial displacement u_r;

$$\frac{d^2u_r}{dr^2} + \frac{1}{r}\frac{du}{dr} - \frac{u_r}{r^2} = 0. \tag{20}$$

Equation (20) is a common differential equation that is used in elasticity problems with rotational symmetry [26], and its solution is

$$u_r = Cr + \frac{C'}{r}, \tag{21}$$

where C and C' are constants of integration to be determined by using boundary conditions. Now Eq. (21) is valid for displacements in both components, that is, fiber and matrix. Let us designate the central component by subscript 1 and the sleeve by subscript 2. Thus, we can write the displacements in the two components as

$$u_{r1} = C_{1r} + \frac{C_2}{r},$$

$$u_{r2} = C_{3r} + \frac{C_4}{r}. \tag{22}$$

The boundary conditions can be expressed as follows:

(1) At the free surface, the stress is zero, that is $\sigma_{r2} = 0$ at $r = b$.
(2) At the interface, the continuity condition requires that at $r = a$, $u_{r1} = u_{r2}$ and $\sigma_{r1} = \sigma_{r2}$.
(3) The radial displacement must vanish along the symmetry axis, that is, at $r = 0$, $u_{r1} = 0$.

The last boundary condition immediately gives $C_2 = 0$, because otherwise, u_{r1} will become infinite at $r = 0$. Applying the other boundary conditions to Eqs. (19) and (22), we obtain three equations with three unknowns. Knowing these integration constants, we obtain u and thus the stresses in the two components. It is convenient to develop an expression for radial pressure, p, at the interface. At the interface $r = q$, if we equate σ_{r2} to $-p$, then after some tedious manipulations, it can be shown that

$$p = \frac{2\varepsilon_z(v_2 - v_1)V_2}{V_1/k_{p2} + V_2/k_{p1} + 1/G_2},$$
(23)

where k_p is the plane strain bulk modulus equal to $E/2(1 + v)(1 - 2v)$. The expressions for the stresses in the components involving p are

Component 1:

$$\sigma_{r1} = \sigma_{\theta 1} = -p,$$
$$\sigma_{z1} = E_1\varepsilon_z - 2v_1 p.$$
(24)

Component 2:

$$\sigma_{r2} = p\left(\frac{a^2}{b^2 - a^2}\right)\left(1 - \frac{b^2}{r^2}\right),$$

$$\sigma_{\theta 2} = p\left(\frac{a^2}{b^2 - a^2}\right)\left(1 + \frac{b^2}{r^2}\right),$$
(25)

$$\sigma_{z2} = E_2\varepsilon_z + 2v_2 p\left(\frac{a^2}{b^2 - a^2}\right).$$

Note that p is positive when the central component 1 is under compression, that is when $v_1 < v_2$.

Figure 13 shows the stress distribution schematically in a fiber composite (1 = fiber, 2 = matrix). We can draw some inferences from Eqs. (24) and (25) and Fig. 13.

(1) Axial stress is uniform in components 1 and 2, although its magnitude is different in the two components and depends on the respective elastic constants.

(2) In the central component 1, σ_{r1} and $\sigma_{\theta 1}$ are equal in magnitude and sense. In the sleeve 2, σ_{r2} and $\sigma_{\theta 2}$ vary as $1 - b^2/r^2$ and $1 + b^2/r^2$, respectively.

(3) When the Poisson ratio difference $(v_2 - v_1)$ goes to zero, σ_r and σ_θ go to zero; that is, the rheological interaction will vanish.

(4) Because of the relatively small difference in the Poisson ratios of the components of a metallic composite, the transverse stresses that develop in the elastic regime will be relatively small.

Kelly and Lilholt [27] invoked the existence of these transverse stresses owing to the Poisson ratio difference to explain the observed high strength and high work-hardening rate of the copper matrix in the tungsten fiber/copper matrix composite compared with these same characteristics in un-

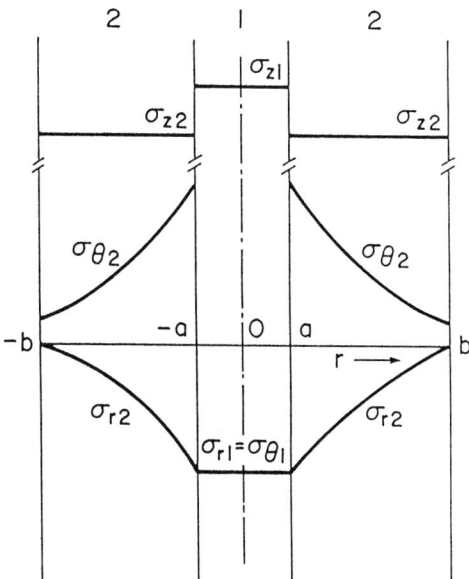

FIG. 13. Three-dimensional stress distribution (schematic) in unit composite shown in Fig. 12. Transverse stresses (σ_r and σ_θ) result from the differences in the Poisson ratios of fiber and matrix.

reinforced copper. The fact that the derived in situ copper stress–strain curve dropped after tungsten entered the plastic regime, thus eliminating the Poisson ratio difference, showed that the transverse stresses resulting from different contractile tendencies of the components do indeed alter the matrix state of stress in the composite.

B. Experimental Verification of the Rule of Mixtures

The measured tensile strengths of W wire and stainless-steel wire-reinforced Al usually match the predictions [28]. For B/Al composites, the prediction cannot be adequately tested because of the scatter in the strength of B fibers before incorporation into the Al. Other factors contributing to the discrepancy between predictions and experimental data result from the fact that the ROM ignores the effects of residual stresses, interfacial reactions between fiber and matrix, fiber surface finish and surface damage and fractures induced by fabrication, flaw density in the fibers, dislocations generated by ΔCTE, and dimensions of the phases.

The application of ROM predictions depends upon a determination of the strengths of the matrix and the filament. Determining the strength of the matrix is not difficult, but since most filaments are intrinsically brittle, there is significant variation in the strength of these filaments. The variation in strength of brittle fibers reflects their sensitivity to flaws. Moreover, because the probability of finding critical flaws increases with increasing length, these fibers exhibit a decrease in strength with increasing gage lengths. Nunes [29] carried out experiments to determine the gauge-length dependence and statistical strength of Al_2O_3 fibers. These results were used to predict the failure strength of uniaxially reinforced Al_2O_3/Al and Al_2O_3/Mg composites. Figure 14 shows the filament tensile failure strength as a function of corrected gauge length. Elastic tensile modulus values were determined statically and dynamically, giving 390 GPa, which is somewhat higher than the value of 379 GPa attributed to polycrystalline Al_2O_3. Exceptionally good agreement (within 2%) was obtained from a theoretical lower-bound stress prediction based on a statistical failure mode and published tensile strength data on

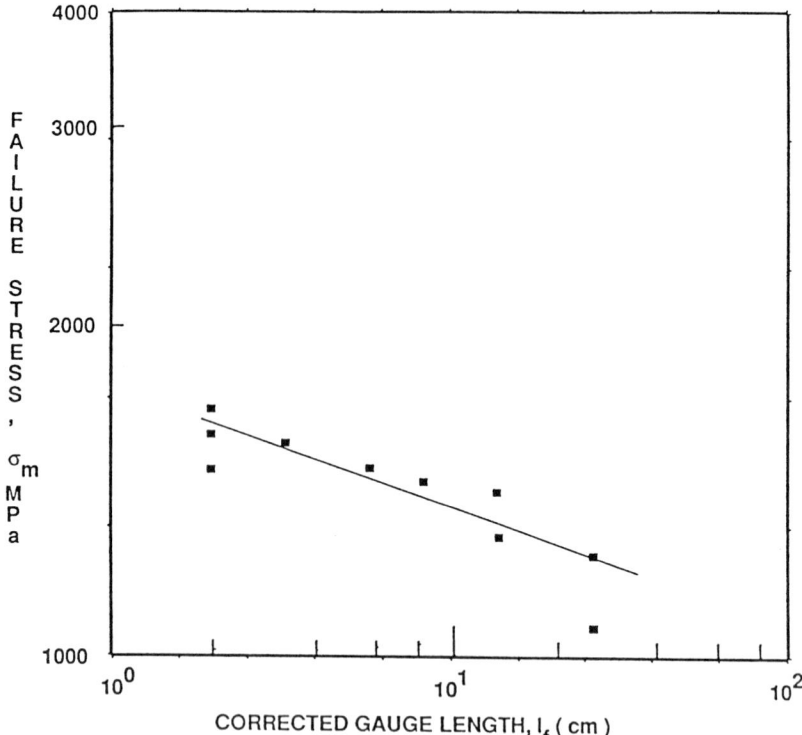

FIG. 14. Filament tensile failure stress versus corrected gauge length of polycrystalline Al_2O_3. Reprinted with permission from the Am. Soc. for Testing Materials, J. Nunes, *J. Comp. Tec. Res.*, **5**, 2, 53–60 (1983).

Al_2O_3/Al and Al_2O_3/Mg composite castings. No correlation was obtained when the same experimental data were compared with a ROM prediction based on mean failure stress, as shown in Fig. 15.

The classic CMMC W/Cu follows the ROM prediction without any modifications. This occurs because the W filaments make a very good bond to Cu, there are no reaction products, and the residual stresses are small due to the smaller ΔCTE effect.

C. Combinations of Filament and Reinforcement Where ROM Is Not Valid

Interfacial reactions are a significant cause of the lack of agreement between the ROM predictions and the experimental data. The interaction reaction products are generally extremely brittle, and as a result are crack initiation sites that result in the fracture of filament. Also, as a result of the

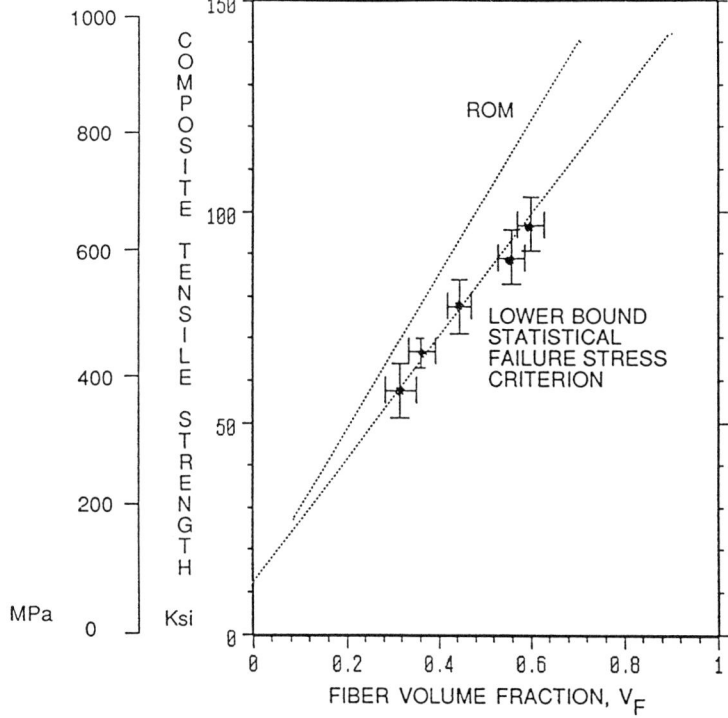

FIG. 15. Comparison of theoretical curves with published (Du Pont) experimental data for Al_2O_3/Al. Reprinted with permission from the Am. Soc. for Testing Materials, J. Nunes, *J. Comp. Tec. Res.*, **5**, 2, 53–60 (1983).

reaction, there is consumption of the filament, thereby reducing the cross-sectional area of the filament. Therefore, the filament cannot withstand as high a load as expected.

In the case of C/Al and C/Ni, the unidirectional tensile strengths are generally lower than the ROM predictions. This reduction is due to formation of Al_4C_3 at the interface [30, 31]. The effect of Al_4/C_3 on the longitudinal and transverse strength C/Al composites is shown in Fig. 16. A review by Baker [32] and a more recent review by Islam and Wallace [33] give an overview of the strength of C/Al CMMCs.

The other major cases where ROM predictions are not valid are B/Al and B/Ti and SiC/Ti; of these, B/Al with its low density and relatively high modulus has received the most attention for application. A very comprehensive review of the state of the art of B/Al was prepared in 1974 [34]. It was found that interfacial reactions had a very large effect on the mechanical properties, e.g., the tensile strength decreased from 3.6 to 1 GPa (12 vol-% B/Al) due to the increase in AlB_2 coating, which increased from a few Å to 300 Å [35]. Therefore, efforts to prevent the interfacial reaction have been attempted by coating the boron filaments with B_4C or SiC (defined as borsic).

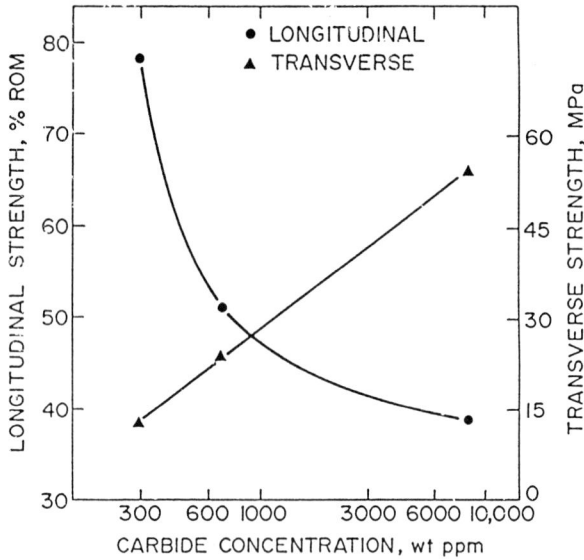

FIG. 16. Relationship between strength properties and carbide formation of the standard composites. Reprinted with permission, "The Effect of Processing on the Transverse Strength of Gr/Al Composites," by M. F. Amateau and D. L. Duhl, figure 16 of chapter 4 in *Failure Modes in Composite IV*, Proceedings of Symposium, TMS-AIME, edited by J. A. Cornie and F. W. Corssman, p. 336, 1978, The Metallurgical Society of AIME, 420 Commonwealth Drive, Warrendale, PA 15086.

TABLE V
COMPOSITE TENSILE ROOM TEMPERATURE PROPERTIES[a]

| | Ultimate tensile strength | | Modulus[b] longitudinal GPa |
System	Longitudinal MPa	Transverse MPa	
Ti–6Al–4v[c]	890	890	120
SiC/Ti–6Al–4V[b]	820	380	225
Borsic/Ti–6Al–4V[b]	895	365	205
B$_4$C/B/Ti–6Al–4V[b]	1,055	310	205
SCS–6/Ti–6Al–4V[b]	1,455	340	240

[a] Four-ply, undirectionally reinforced, 35 to 40 vol-%.
[b] After fabrication and low temperature, 596°C for 512 hours exposure.
[c] Mill annealed, 732°C for 2 hours air cool.
Adapted from Ref. [37].

The coating [36] does not appear to adversely affect the strength of the composite, but it does reduce the degradation of the strength after high-temperature exposure. In the case of SiC/Ti and B/Ti [37], there are combinations (Table V) where the composite of 25–40 vol-% is no stronger than the matrix. However, with a coating of B$_4$C, there is a strength increase, and also, in the case of an AVCO SiC special coated filament, there is an increase in strength.

D. A Comparison of Longitudinal and Transverse Properties

The general observation is that the transverse properties are much less than longitudinal properties as shown in Fig. 17. In Fig. 17, this is approximately a factor-of-6 reduction in the strength for longitudinal-to-transverse orientations [38]. However, there are other cases where the longitudinal strength is 20 times larger than the transverse strength [39]. Also, there have been investigations by Everett and Skowronek [30] where the processing was modified, and it was shown that as longitudinal strength was increased, there was a corresponding decrease in transverse strength (Fig. 18). One of the main reasons for the reduction of the transverse strength is a result of direct contact of the fibers. This direct contact acts as a weak link. An investigation by Towata and Yamada [40] of hybrid-reinforced composite, continuous SiC or C with SiC whiskers or particulate in an Al matrix exhibited superior transverse strength due to the separation of the continuous filaments by the SiC whiskers or particulate.

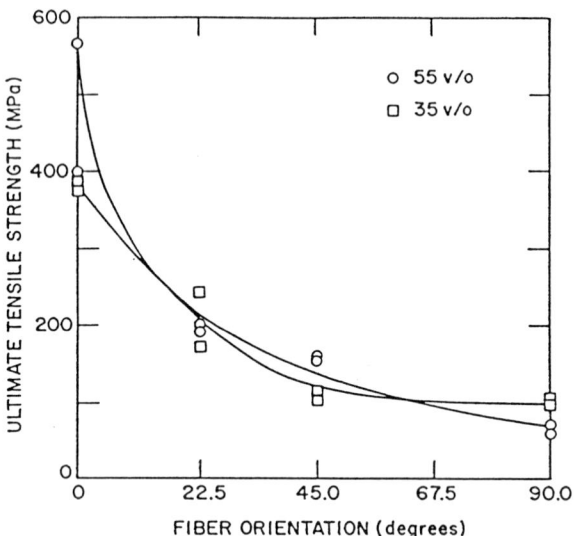

FIG. 17.   Effect of fiber orientation on UTS for both 35 and 55 vol-% Al_2O_3 fiber-reinforced CP Mg. Reprinted with permission from The Minerals, Metals & Materials Society & ASM Int'l, J. E. Hack, R. A. Page, and G. R. Leverant, *Metall. Trans.* **15A**, 1389–1396 (1984).

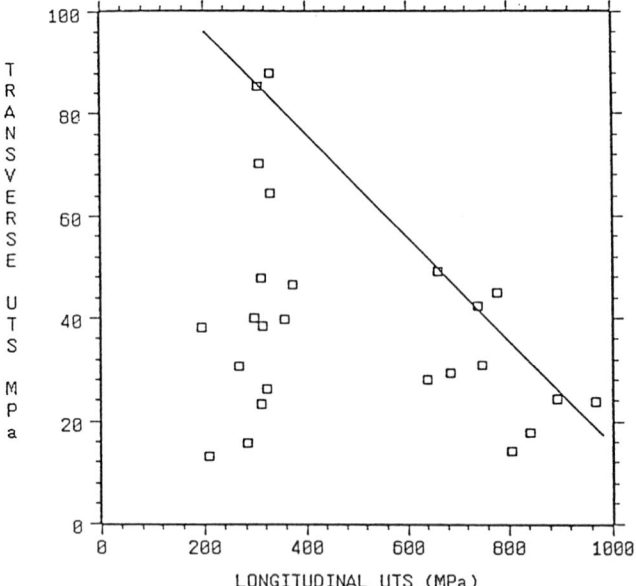

FIG. 18.   Transverse tensile strength versus longitudinal tensile strength plot for C/Al composites fabricated at various temperature, time, and pressure conditions. Reprinted with permission from the Cambridge University Press, from [30].

E. *Compression versus Tensile Strengths*

The compression strength of CMMCs is generally greater than the tensile strength. Figure 19 is such an example for the case of Al_2O_3/Al [*41*]. This is representative of a sample with brittle filament, and failure in compression generally begins with kinking of the filaments. The difference in σ_{yc}^C and σ_{yc}^T can be readily explained in terms of the tensile thermal residual stress [*42*].

If the filament is somewhat ductile and the ΔCTE is small, then the difference between σ_{yc}^C and σ_{yc}^T is small and the difference in the UTSs is smaller [*28*] (Fig. 20).

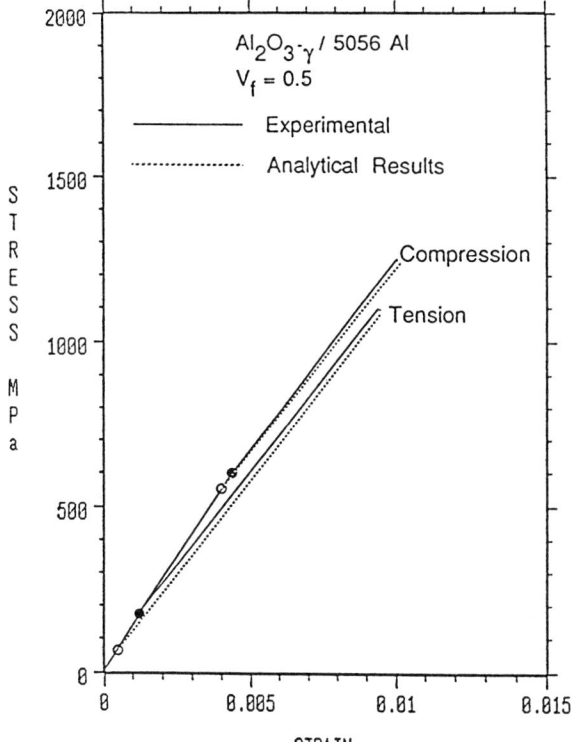

FIG. 19.   Tension and compression stress–strain curves of continuous Al_2O_3 fiber/5056 Al composite with $V_f = 0.5$. The experimental [*41*] and theoretical [*42*] results are denoted by solid and dashed curves, respectively. The solid and open circles denote the yield stress of the experimental and theoretical results, respectively.

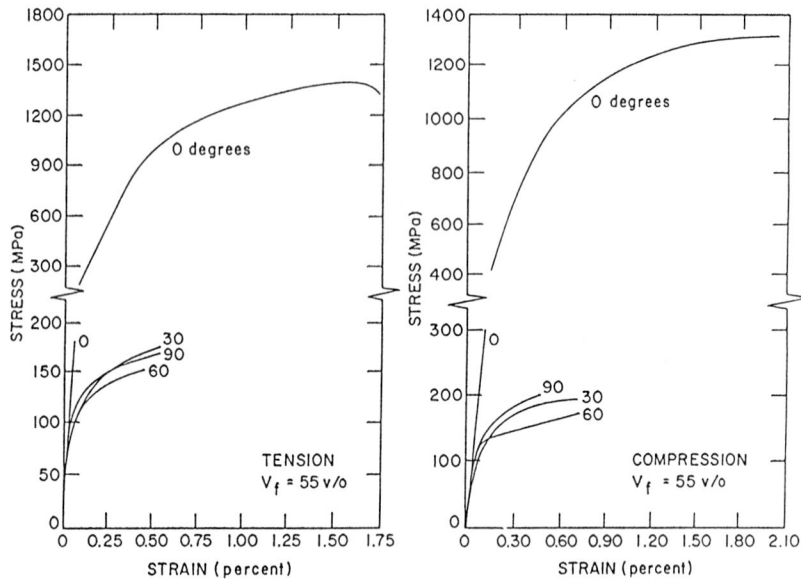

FIG. 20. Experimental stress–strain curves in tension and compression for unidirectional W/Al at various fiber orientations with respect to the loading axis. Reprinted with permission from [28].

References

1. D. L. McDanels, *Metall. Trans.*, **16A**, 1105 (1985).
2. R. J. Arsenault and S. B. Wu, *Mater. Sci. Eng.*, **96**, 77 (1987).
3. M. Taya and R. J. Arsenault, *Scripta Metall.*, **21**, 349 (1987).
4. R. J. Arsenault, *J. Comp. Tech. Res.*, ASTM, **10**, 140 (1988).
5. "Metals Handbook, 8th Ed" (T. Lyman, ed.), Vol. 1, ASM, Metals Park, Ohio, 1961.
6. R. J. Arsenault and S. B. Wu, *Mater. Sci. Eng.*, **96**, 77 (1987).
7. T. W. Clyne, *J. Mater. Sci.*, **20**, 85 (1985).
8. F. K. Chi, R. D. Maier, T. W. Kracek, and P. M. Boymel, in "ICCM-6/ECCM-2" (F. L. Mathews, N. C. R. Baskell, J. M. Hodgkinson and J. Morton, eds.), Vol. 2, p. 449. Elsevier, Amsterdam, 1988.
9. R. Aiken. Martin Marietta, private communication.
10. K. K. Chawla, "Composite Materials." Springer-Verlag, New York, 1987.
11. W. Voigt, "Lehrbuch der Kristallphysik." Teubner, Leipzig, 1910.
12. A. Reuss, *Z. Angew. Math. Mech.*, **9**, 49 (1929).
13. C. C. Chamis and G. P. Sendeckyj, *J. Comp. Mater.*, **2**, 332 (1968).
14. J. F. Nye, "Physical Properties of Crystals," p. 131. Oxford University Press, London, 1969.
15. Z. Hashin and B. W. Rosen, *J. Appl. Mech.*, **31**, 233 (1964).
16. B. W. Rosen, *Composites*, **4**, 16 (1973).
17. C. C. Chamis, NASA Tech. Memo 83320, 1983 [presented at 38th Annual Conf. of the Society of Plastics Industry (SPI), Houston, Texas, Feb., 1983].

18. R. Hill, *J. Mech. Phys. Solids*, **12**, 199 (1964).

19. R. Hill, *J. Mech. Phys. Solids*, **13**, 189 (1965).

20. J. C. Halpin and S. W. Tsai, "Environmental Factors Estimation in Composite Materials Design," AFML TR 67-423, 1967.

21. J. C. Halpin and J. L. Kardos, *Polym. Eng. Sci.*, **16**, 344 (1976).

22. J. L. Kardos, *CRC Crit. Rev. Solid State Sci.*, **3**, 419 (1971).

23. J. M. Whitney, *J. Structural Div. Am. Soc. Civil Eng.*, **113**, (Jan. 1973).

24. L. E. Nielsen, "Mechanical Properties of Polymers and Composites," Vol. 2. Marcel Dekker, New York, 1974.

25. A. Kelly, in "Chemical and Mechanical Behavior of Inorganic Materials," p. 523. Wiley-Interscience, New York, 1970.

26. A. E. H. Love, "A Treatise on the Mathematical Theory of Elasticity," 4th ed., p. 144. Dover, New York, 1952.

27. A. Kelly and H. Lilholt, *Phil. Mag.*, **20**, 175 (1971).

28. T. W. Chou, A. Kelly, and A. Okura, "Fiber Reinforced Metal-Matrix Composites," *Composites*, **16**, 187 (1985).

29. J. Nunes, *J. Comp. Tech. Res.*, **5**(2), 53–60 (1983).

30. R. Everett and C. J. Skowronek, "Failure Mechanisms in High Performance Materials" (J. G. Early, T. R. Shives, and J. H. Smith, eds.), p. 128. Cambridge University Press, Cambridge, England, 1984.

31. M. F. Amateau and D. L. Duhl, "The Effect of Processing on the Transverse Strength of Gr/Al Composites," in *Failure Modes in Composite IV, Proceedings of Symp. by TMS–AIME* (J. A. Cornie and F. W. Crossman, eds.), p. 336, ASM, Metals Park, Ohio, 1978.

32. A. A. Baker, *Mater. Sci. Eng.* **17**, 177 (1975).

33. M. U. Islam and W. Wallace, "Advanced Materials and Manufacturing Process," Vol. 3, p. 1, 1988.

34. K. G. Kreider and K. M. Prewo, "Boron-Reinforced Aluminum," in *Composite Materials* (K. G. Kreider, ed.), Vol. 4. Metal Matrix Composites. Academic Press, New York, 1974.

35. K. Hunda, Y. Shinohara and A. Okura, "Composites 86, Recent Advances in Japan and United States" (K. Kawata, S. Umekawa, and A. Kobayashi, eds.), p. 457. Japan Society for Composite Materials, 1986.

36. J. R. Hancock, "Fatigue of Metal-Matrix Composites," in *Composite Materials* (L. J. Broutman and R. H. Krock, eds.), pp. 371–414. Academic Press, New York, 1974.

37. P. R. Smith, F. H. Froes, and J. T. Cammett, "Correlation of Fracture Characteristics and Mechanical Properties of Titanium Matrix Composites," in *Fracture Modes in Composites VI* (J. A. Cornie and F. W. Crossman, eds.). Metall. Soc. of AIME, Warrendale, Pennsylvania, 1977.

38. J. E. Hack, R. A. Page, and G. R. Leverant, *Metall. Trans.*, **15A**, 1389–1396 (1984).

39. C. C. Chamis, "Laminated and Reinforced Metals," in *Encyclopedia of Chemical Technology* (Kirk-Othmer, Wile ed.), 3d ed. Vol. 13, p. 941. Wiley, New York, 1981.

40. S. Towata and O. Yamada, "Composites 86, Recent Advances in Japan and United States" (K. Kawata, S. O. Umekawa and A. Kobayashi, eds.), p. 497. Japan Society for Composite Materials, 1986.

41. Y. Abe, private communication. Sumito Chemical Co., Japan.

42. R. J. Arsenault and M. Taya, *Acta Metall.* **35**, 651 (1987).

5

Mechanical Behavior of Metal Matrix Composites under High Strain Rates and Impact Loadings

M. TAYA

Department of Mechanical Engineering
University of Washington
Seattle, Washington

I. Introduction

Among the environments in which metal matrix composites (MMCs) are to be used, high strain rate and impact loadings have been considered to a very limited extent. Assessment of the mechanical behavior of a MMC under high-strain-rate loadings is of vital importance if the MMC is designed to be used as engine components, which are likely to be subjected to sudden change in loading or to flying foreign objects. The mechanical response of the MMC component due to the variable loading rates can be understood adequately if the data on the high-strain-rate behavior of the MMC are known. These data can be obtained from the dynamic stress–strain curves at various strain rates. On the other hand, the mechanical behavior of a MMC impacted by a flying foreign object provides critical data by which the survivability of a jet engine under high velocity impact such as bird impact or impact by particles of unburnt debris from combustion chamber can be judged. These data are usually obtained by transverse impact testing or plate impact testing on a MMC specimen, and the residual mechanical properties of the impacted MMC are then measured. In the following, we will describe the two different modes of high-strain-rate loadings separately.

169

II. High-Strain-Rate Behavior

A. Direct Study

The high-strain-rate behavior of a material is best characterized by the stress–strain curves at various high strain rates. If the data on the stress–strain curves of a given metal matrix composite at various strain rates are available, then the three-dimensional constitutive equations of the MMC can be constructed. These are quite useful to design engineers, who can design a MMC engine component with some confidence. The stress–strain curves at high strain rates are usually obtained by split Hopkinson bar (SHB) tests in two modes: uniaxial tension/compression and shear. The SHB testing of SiC whisker/2124 Al composites has recently been conducted by Harding *et al.* [1] to generate the tensile strain–strain curves at strain rates up to 10^3/s. The SHB testing system used in this study is called Oxford type and its schematical view is shown in Fig. 1. The Oxford-type SHB testing system can generate tensile stress pulses directly in an input bar, thus eliminating the elastic precompressive stress pulse. In this respect, most of the existing SHB systems in the United States are based on the U.S. Air Force-type SHB, where the tensile specimen is more likely to incur a priori damage by the precompressive stress pulse. The tensile stress–strain curves of SiC_w/2124 Al composite at three different strain rates obtained by the Oxford SHB testing

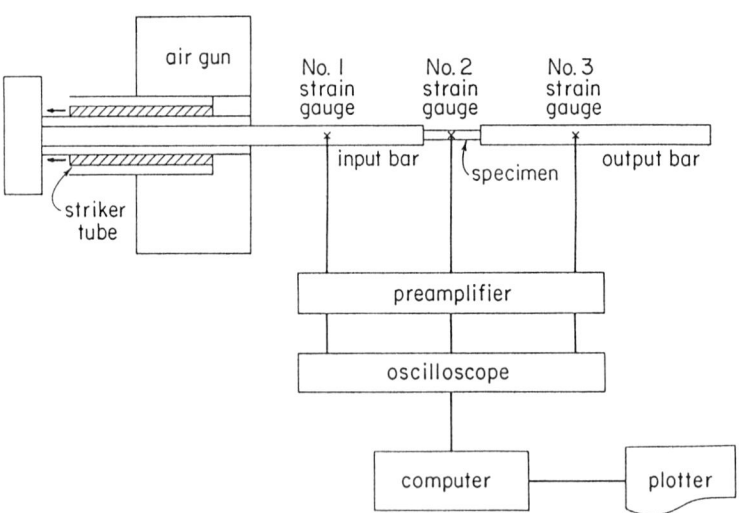

FIG. 1.   The split Hopkinson bar at the University of Oxford for tensile tests.

system are shown in Fig. 2, where (a) and (b) denote two cases of the volume fraction of whisker; $V_f = 15$ and 25%, respectively. The strain rate, $\dot{e}$ dependence of the apparent Young moduli, the flow stress (0.2% offset), and the failure strain of SiC–whisker/2124 aluminum–matrix composite are plotted as a function of $\log_{10} \dot{e}$ in Fig. 3a, b, and c, respectively. It is clear

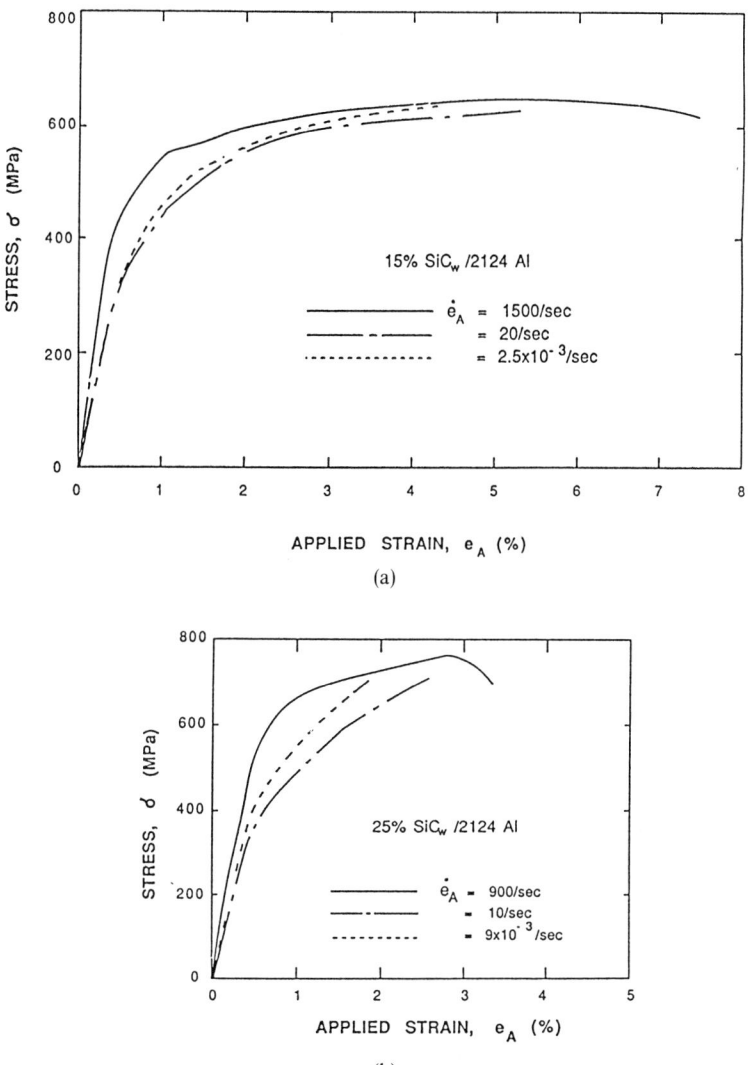

FIG. 2.   Effect of strain rate on tensile stress–strain curves of SiC$_w$/Al composite: (a) 15% V_f SiC whisker; (b) 25% V_f SiC whisker.

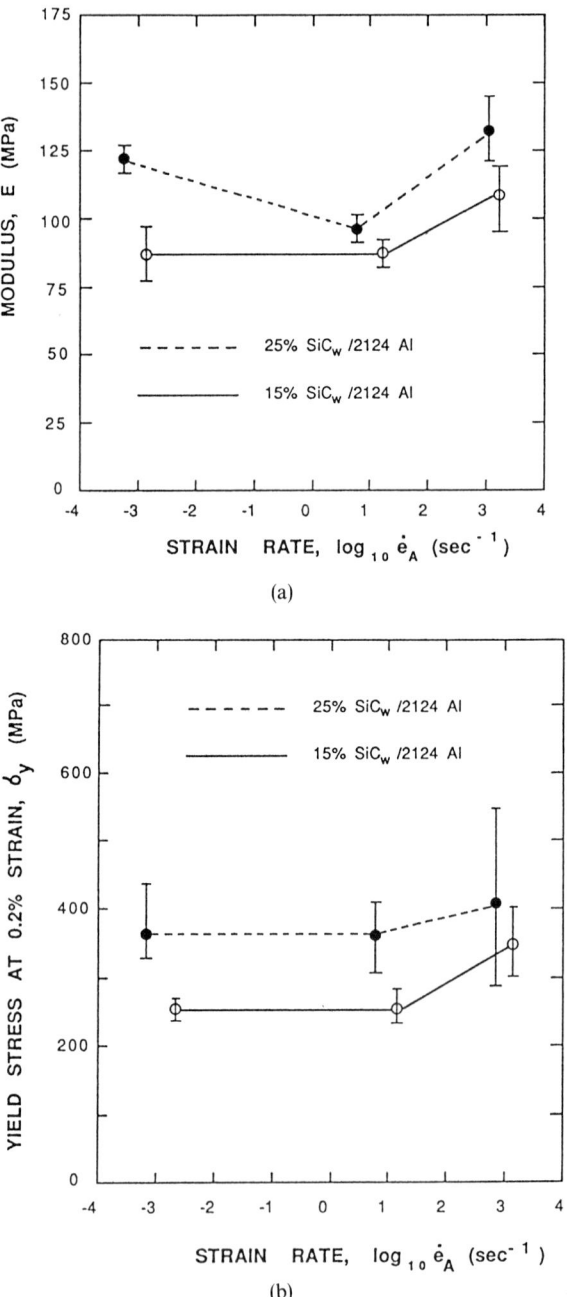

FIG. 3. Strain rate dependence of (a) Young's modulus, (b) yield stress, and (c) failure strain.

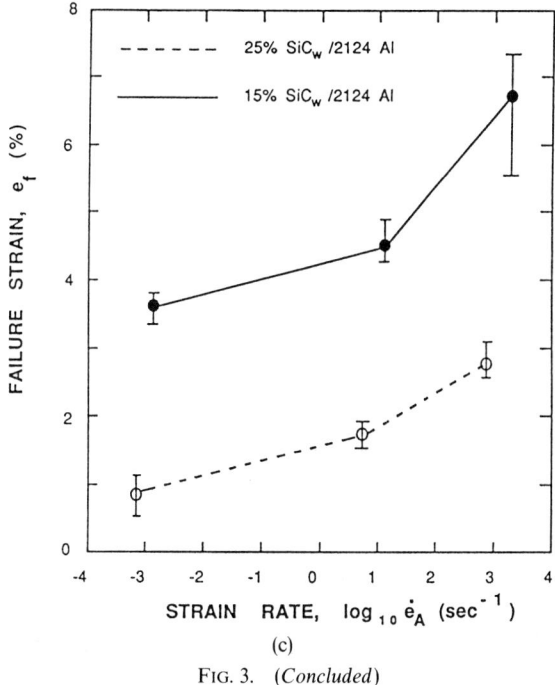

FIG. 3. (*Concluded*)

from Fig. 2 and Fig. 3a that the modulus of both types of composite increases with strain rate, although the apparent modulus of 25% V_f composite at intermediate strain rate dips below the quasistatic value. For both types of composite, the fracture strain was found to increase with strain rates as seen from Fig. 2 and Fig. 3c. The strain-rate dependence of fracture strain is quite a contrast to other materials where the fracture strain normally decreases with the increase in strain rate. In order to obtain some clues to explain this, Pickard *et al.* [2] made a transmission electron microscope (TEM) study on the quasistatically- ($\dot{\varepsilon} = 2.5 \times 10^{-3}/s$) and impact-loaded ($\dot{\varepsilon} = 1500/s$) 15% V_f composite specimens. Figure 4 shows typical morphology of dislocations on a plane perpendicular to the tensile axis reasonably close to the fracture surface where (a) and (b) denote the case of quasistatically, and high-strain-rate impact-loaded specimens, respectively. The average dislocation densities of these cases (a and b) are 3×10^{10} and 5×10^9 cm^2, respectively. It should be noted here that the dislocation density of high-strain-rate-loaded specimen is lower than that of quasistatically loaded specimen.

The shear stress–shear strain behavior of SiC$_w$/2124 Al composite has been studied experimentally by Marchand *et al.* [3], who used the SHB testing

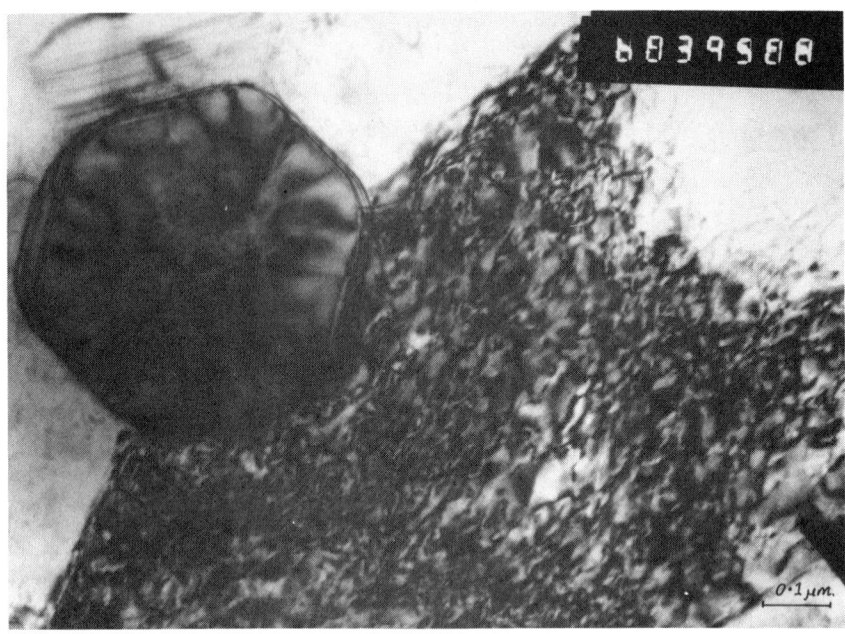

(a)

(b)

FIG. 4.   Typical microstructures of the 15% V_f SiC$_w$/Al composite: (a) quasistatically loaded, and (b) impact-loaded at $\dot{\varepsilon} = 1500/s$.

174

TABLE I
THE INFLUENCE OF WHISKER-REINFORCEMENT ON SHEAR STRENGTH

Specimen	Shear strain rate (s^{-1})	10^{-4}	900	1300	1600	3500
W	Ultimate strength in shear of aluminum alloy, τ_{um} (MPa)	315	295	300	310	305
Z1	Ultimate strength in shear of whisker-reinforced aluminum, τ_{uc} (MPa)	390	335	365	375	375
	$\Delta = 100\,(\tau_{uc} - \tau_{um})/\tau_{um}$	24	14	21	21	23
Z2	Ultimate strength in shear of whisker-reinforced aluminum, τ_{uc} (MPa)	382	—	—	381	388
	$\Delta = 100\,(\tau_{uc} - \tau_{um})/\tau_{um}$	21	—	—	23	27

system with torsional mode. The range of the shear strain rates tested is from 10^{-4} to 3.5×10^3/s. The results of the shear strength of 2124 Al (unreinforced matrix) and 13% V_f SiC$_w$/2124 Al composite are given in Table I, where two different composite specimens taken from different bars (Z1 and Z2) were used. The increase in the shear strength of the composite over the unreinforced matrix metal is about 20%. However, the strain-rate sensitivity of the composite is small, unlike the case of tensile mode.

The effect of strain rate on ductility (fracture strain) in shear was also investigated, and the results are given in Table II. It is clear from Table II that the ductility of the composite increases with strain rate although that of the unreinforced matrix metal does not. This strain rate dependance of the composite ductility in shear is very similar to that in uniaxial tension, see Fig. 3c.

They also conducted static and dynamic fracture initiation experiments, which provided the static stress intensity factor (K_{IC}) and dynamic stress intensity factor (K_{ID}), respectively, and the results are given in Table III. It can be concluded from the data presented in Table III that the higher strain

TABLE II
THE INFLUENCE OF THE WHISKER REINFORCEMENTS ON DUCTILITY IN SHEAR
(DUCTILITY IS APPROXIMATELY EQUAL IN BARS Z1 AND Z2)

Shear strain	10^{-4}		900		1300		1600		3500	
SiC vol-%	0	13.2	0	13.2	0	13.2	0	13.2	0	13.2
Ductility (%)	44	29	38	35	52	38	50	40	50	40

TABLE III

STATIC AND DYNAMIC VALUES OF THE CRITICAL STRESS INTENSITY FACTOR DURING
FRACTURE INITIATION AT ROOM TEMPERATURES

		Material	
Stress intensity factor (MPa $\sqrt{m}$)	Stress intensity rate (MPa $\sqrt{m/s}$)	2124-T6 Aluminum alloy matrix	2124-T6 Aluminum alloy with 13.2 vol-%
K_{IC}	$\dot{K}_{IC} = 1$	38 ± 2	21 ± 1
K_{ID}	$\dot{K}_{ID} = 2 \times 10^6$	41 ± 2	27 ± 1

rate increased the toughness of both unreinforced metal and composite, although the toughness of the composite is always lower than that of the unreinforced metal.

B. Indirect Study

When an engine component made of a MMC is subjected to unexpected high-velocity impact, its survivability and remaining properties are of main concern to a component design engineer. An assessment of the survivability of the MMC can be made by subjecting it to two types of impact loadings: plate impact loading and transverse impact loading. The first type is considered to be idealized impact testing, since its data are easier to analyze, whereas the second type provides more realistic impact loadings, but its data analysis is complicated. Despite the fact that limited data of the mechanical behavior of metal matrix composites under impact loadings are available to date, we will review these two cases of impact loadings.

A schematic view of plate impact testing is shown in Fig. 5, where a target MMC specimen flying with velocity (V_s) impacts a stationary specimen of the same MMC. Upon impact, a compressive wave travels, is reflected at

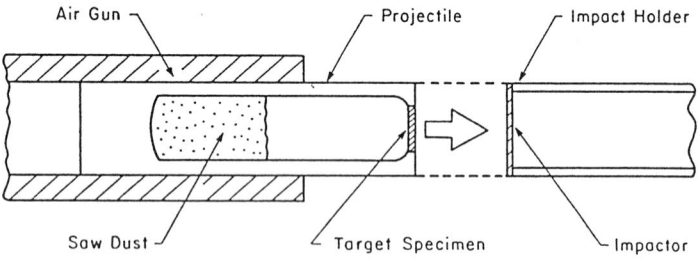

FIG. 5. Plate impact testing apparatus.

the free surface, and comes back as a tensile wave, causing a spall damage at some point of the target MMC specimen. Very limited studies have been made on the behavior of a metal matrix composite under plate impact loadings: SiO_2 particle/copper matrix composite [4] and boron fiber/6061 Al composite [5]. Taya et al. [4] conducted plate impact tests on a SiO_2/Cu composite. Figures 6a, b, and c show the extent of the spall damage in SiO_2/Cu composites that were plate impacted with a striking velocity of $V_s = 110$, 160, and 208 m/s, respectively. Figure 6 was used to measure the void density f across the plate thickness, and the results are plotted in Fig. 7, where the scattered band of the data at the peak value of f is shown by vertical lines. It follows from Figs. 6 and 7 that the higher the striking velocity V_s, the more extensively the voids are distributed. In the case of SiO_2/Cu composite, SiO_2 appeared to have provided the nucleation sites for voids, which thereafter grew under the increasing applied strain, although voids were nucleated at only a fraction of SiO_2 particles [4]. An attempt was then made to predict the peak value of f by the dynamic void growth model, which Yoon and Taya have recently developed [6]. A comparison between the experimental results (open symbols) of the peak values of f and the

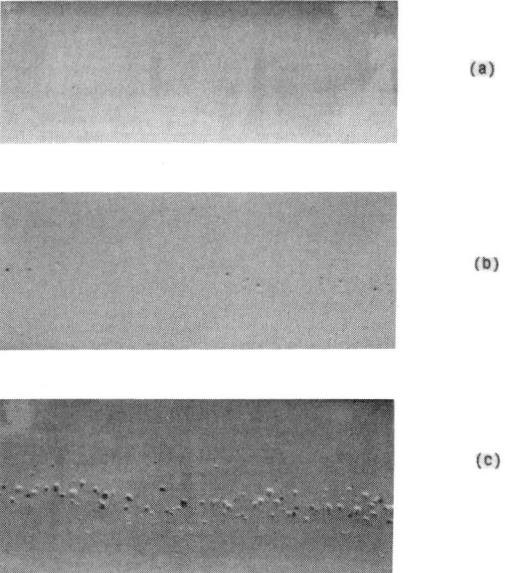

FIG. 6. SEM photographs showing voids across cross-section of SiO_2 = particulate/copper-matrix composite specimens, which were plate-impacted at striking velocities of (a) $V_s = 110$ m/s, (b) 160 m/s, and (c) 208 m/s.

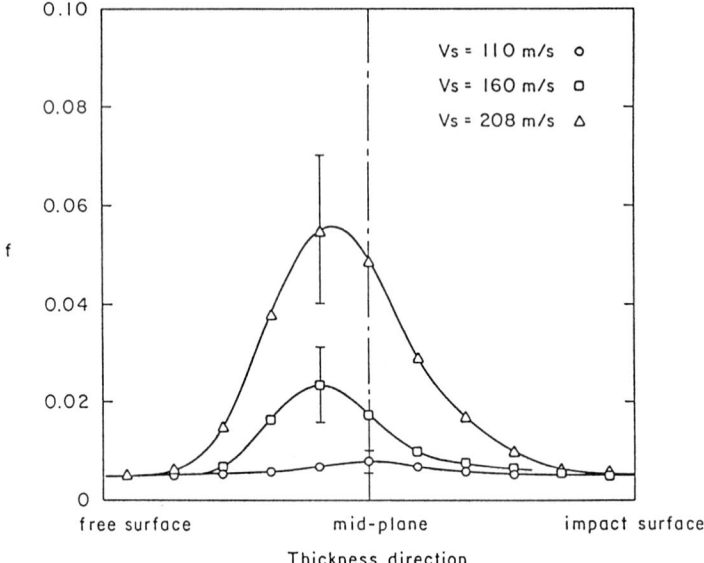

FIG. 7.   Distribution of the volume fractions of voids, f, across the plate thickness at various impact velocities (V_s). The error of the peak value is indicated by an arrow.

prediction by the analytical model of Yoon and Taya (solid and dashed lines) is shown in Fig. 8, where f_0 and n are the initial void volume fraction and a nonlinear parameter, respectively. The latter, n, appears in the one-dimensional constitutive equation of the metal matrix deforming at high strain rate [6], i.e.,

$$\sigma = \sigma_y + \sigma_0 \left(\frac{\dot{e}}{\dot{e}_0}\right)^{1/n} \tag{1}$$

where σ_y is the reference flow stress, σ_0, $\dot{e}_0$ are material constants, and $n = 1$ and ∞ correspond to Newtonian viscous material and perfect rigid plastic material, respectively. The residual mechanical properties of the plate-impacted MMC are expected to be degraded due to its porosity.

Schuster and Reed experimentally studied the extent and modes of the damage of two different B/6061 Al composites induced by plate impact with striking velocities up to 1700 m/s. In their experiment, the thickness of the flyer plate and the target B/6061 Al composite plate are 0.01 inch and $0.1 \sim 0.13$ inch, respectively; thus, the thickness ratio of the flyer to the target plate is about 0.1, quite different from the experiment by Taya and coworkers, where the ratio was 0.5. This thickness ratio will determine the location of spall damage across the plate thickness. Two B/6061 Al compo-

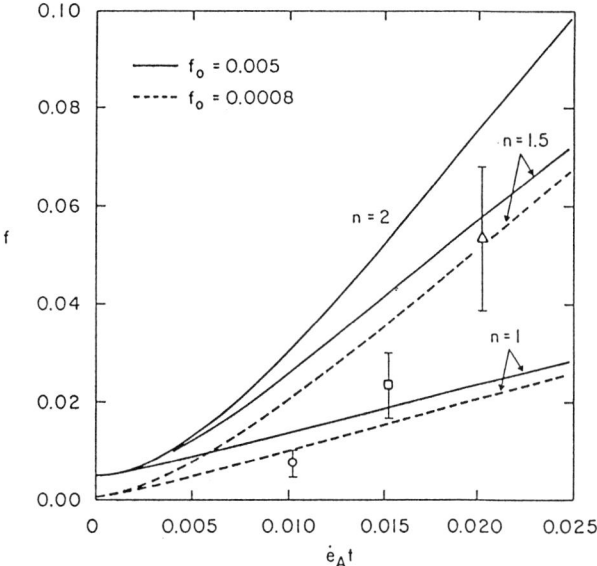

FIG. 8.   The volume fraction of voids (f) calculated by the dynamic void growth model. The results for $f_0 = 0.005$ and 0.0008 are plotted by solid and dashed curves, respectively, and the experimental results by open symbol with the error arrow.

sites used in the Schuster and Reed experiment were plasma-sprayed and brazed B/6061-713 Al matrix composite with $V_f = 58.5\%$ (plasma-sprayed/braced composite) and plasma-sprayed and diffusion-bonded B/6061–713 Al matrix composite with V_f 47.8% (plasma-sprayed/diffusion-bonded composite). In the former composite, the boron fibers are not always separated from each other, providing many contact points between fibers, whereas in the latter composite, fibers are generally well spaced, minimizing the fiber contacts. The plasma-sprayed/brazed composite exhibited both spall-type damage and fiber damage. Spall cracks appeared above a threshold striking velocity, (V_s)th $= 500$ m/s, as fissures along planes roughly parallel to and located about 1/5 plate thickness from the free interface. The width of the spall damage is narrow, indicating the narrowness of the tensile pulse refracted from the free surface, which is consistent with a very short initial compressive pulse (0.2 μs). It was suggested by Schuster and Reed that the fiber damage was caused by the initial compressive pulse, because most of the fibers damaged are located within the plate to the impact side. The damage in form of cracking appears to be caused in the region of fiber contacts and the matrix between fibers. In contrast, the plasma-sprayed and diffusion-bonded composite did not experience definite spall fracture at

$V_s \le 1700$ m/s, which is the maximum striking velocity possible by their experimental setup. However, fiber fractures were observed, which seem to increase with V_s.

Schuster and Reed also conducted plate impact tests on unreinforced 6061-T6 Al plate to obtain the data on spall damage: the threshold velocity (V_s)th and attenuation characteristics. A comparison between the composites and the unreinforced matrix indicated that the threshold velocity for spall damage of the composite is three times larger than that of the unreinforced matrix, showing the improvement in the spall damage resistance of the composite over the unreinforced matrix. This is attributed to the fact that the peak stress amplitude in the composite was reduced by geometric dispersion of the loading pulse due to the fiber arrangement. Hence, one could improve the spall damage resistance of a MMC by careful fiber arrangement in order to enhance the wave dispersion phenomena in the composite.

Transverse impact loading is considered to be more realistic than plate impact loading, and it gives rise to a three-dimensional indent or damage to a target material, which makes the analysis of the transient wave or stress state intractable. The damage in a MMC due to transverse impact loadings has been investigated by a limited number of researchers. These studies can be categorized into two types, depending upon impact (striking) velocity: high-velocity impact [7] and low-velocity impact [8–11].

Gray [7] studied the damage in B/6061 Al composite plate and Ti–6Al–4V (Ti alloy) plate impacted by projectile at various velocities up to 137 m/s and the residual fatigue life of the impacted B/6061 Al composite and Ti alloy plates. The thickness of both plates is about 1 mm. The threshold velocity above which the projectile penetrates and possess through the target plate, (V_s)th was approximately 137 m/s for the B/6061 Al composite and 243 m/s for Ti alloy. The summary of the impact damage of both B/6061 Al and Ti alloy is given in Table IV, where tested specimens are grouped depending on impact velocity (10 specimens for each group), and where group I and group VI are the unimpacted B/6061 Al and Ti alloy, respectively. All the specimens (or all the groups) were then subjected to bending fatigue tests, with the ratio of the minimum-to-maximum stress (R) being set 0.1. In these fatigue tests, the impacted side of the plate was designed to be in tension. The summary of the S–N curves of the B/6061 Al composites is shown in Fig. 9. The effect of increasing impact velocity on the residual fatigue strength is obvious from Fig. 9. Similar bending fatigue tests were conducted on Ti alloy specimens, and the result of S–N plots are given in Fig. 10, where the case of inverted tests (the impacted side was designed in compression) was also included. It can be concluded from Figs. 9 and 10 that B/6061 Al composite exhibited poorer fatigue strength than the unreinforced Ti alloy.

TABLE IV
BALLISTIC IMPACT TEST RESULTS

	B/Al-6061				
Group	Desired impact velocity, ft/s (m/s)	Average impact velocity, ft/s (m/s)	Rebound[a] velocity, ft/s (m/s)	Impact energy, ft·lb (J)	Average damage measurement, in (cm)
I	no impact	0	0	0	0
II	150(45.7)	154(46.9)	46(14.0)	0.259(0.351)	0.056(0.143)[b]
III	225(68.6)	230(70.1)	65(19.8)	0.584(0.791)	0.153(0.389)[b]
IV	300(91.4)	300(91.4)	61(18.6)	1.035(1.403)	0.270(0.686)[b]
V	450(137.2)	454(138.5)	34(10.4)	2.458(3.333)	0.450(1.143)[b]
	Ti–6Al–4V				
VI	no impact	0	0	0	0
VII	350(106.7)	349(106.4)	35(10.7)	1.446(1.960)	0.068(0.172)[c]
VIII	500(152.4)	502(153.0)	75(22.9)	2.955(4.006)	0.086(0.220)[c]
IX	650(198.1)	667(203.3)	114(34.7)	5.180(7.022)	0.105(0.267)[c]
X	800(243.8)	813(247.8)	96(29.3)	7.816(10.597)	0.126(0.321)[d]

[a] Rebound velocity was measured for only one test in each group.
[b] Transverse crack length.
[c] Dent diameter.
[d] Hole diameter.

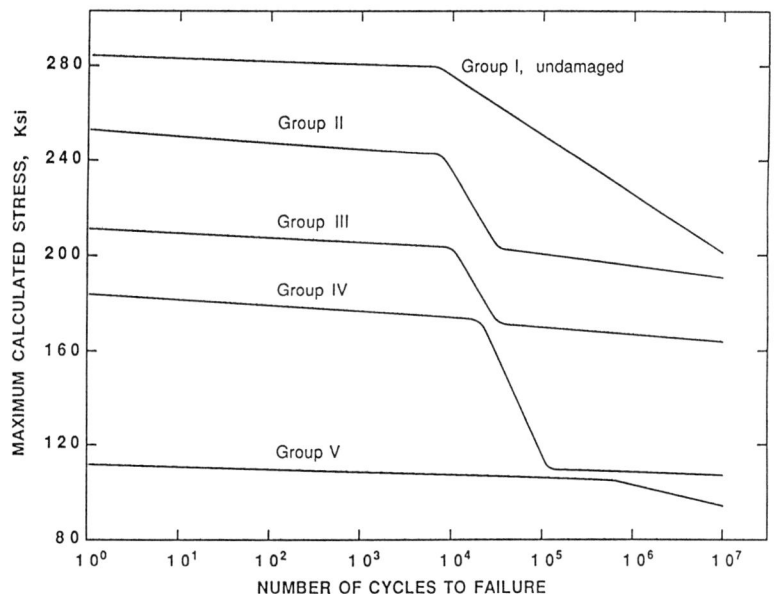

FIG. 9. *S–N* curves for groups I-V B/6061 Al composite.

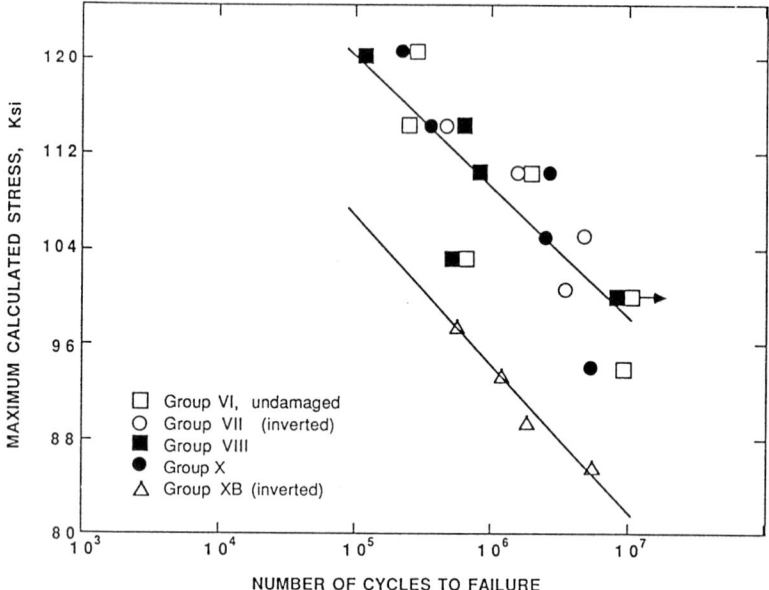

FIG. 10.   *S–N* curves for group VI–XB Ti–6Al–4V alloy.

C.   *Charpy Test*

The most popular method of evaluating the impact resistance of an MMC has been use of the Charpy test with a notched or unnotched specimen of rectangular cross-section, simply because the Charpy test is easy to conduct although the energy absorbed in the MMC specimen cannot be accurately estimated and the impact velocity used is normally low. Dardi and Kreider [8] studied the impact energy of BOSIC fiber/Al alloy composite, where the matrix aluminum alloys used were 6061-F grade, 6061-T6 grade, 2024-F grade, and 5052/56-I grade. The main parameters studied by Dardi and Kreider were the volume fraction of fiber (V_f) and fiber orientation. The results of the Charpy impact energy are shown as a function of V_f in Fig. 11, where three cases of fiber orientation are tested: longitudinally reinforced fiber composite with transverse notch (LT), transversely reinforced composite with transverse notch (TT), and that with longitudinal notch (TL). Figure 11 illustrates that the LT specimen possesses the best impact resistance among these three composites, and its impact energy increases with V_f, whereas the values of the impact energy of TT and TL composites are much smaller than that of LT composite, and they are insensitive to V_f. Ahmad and Barranco [9] conducted Charpy ∨ notch tests on W–2%ThO$_2$ fiber/

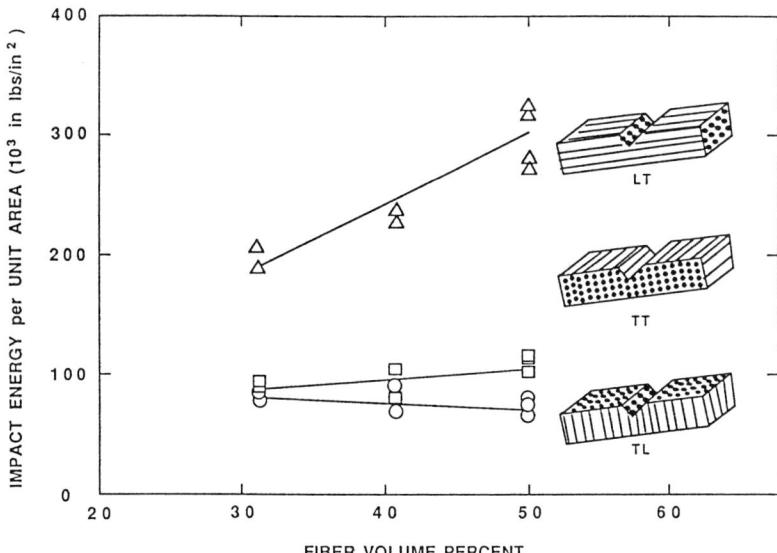

FIG. 11.   Impact energy per unit area for 4.2 mil BORSIC/6061–F Al composite as a function of fiber vol-%.

cobalt-based super alloy (W–ThO$_2$/cobalt alloy) composite with $V_f =$ 28.25% and the unreinforced cobalt alloy, which were tested at several different high temperatures of up to 835°C. The results of the Charpy impact energy of the W–ThO$_2$/ cobalt alloy composite and unreinforced cobalt alloy are plotted as a function of testing temperature in Fig. 12. It follows from Fig. 12 that the impact energy of the composite increases with testing temperature, whereas that of the unreinforced cobalt alloy is constant over the range of testing temperature. If the composite is held at a high temperature (1093°C) for a long time (200 hours), the reaction products at the matrix–fiber interface, which are usually brittle, are expected to form, resulting in the degradation in the mechanical properties of the composite. Both the composite and unreinforced matrix were thermally exposed at 1093°C for 200 hours and were then Charpy tested; the results are shown as filled circles in Fig. 12. This illustrates that the exposure to high temperature for a long time reduces the impact energy of the composite, whereas the same thermal environment did not change the impact energy of the unreinforced matrix.

Prewo [10] studied the impact energy of selected MMC systems by using the standard Charpy tester with modified specimen geometry. He observed two different fracture mechanisms that were responsible for the fracture of

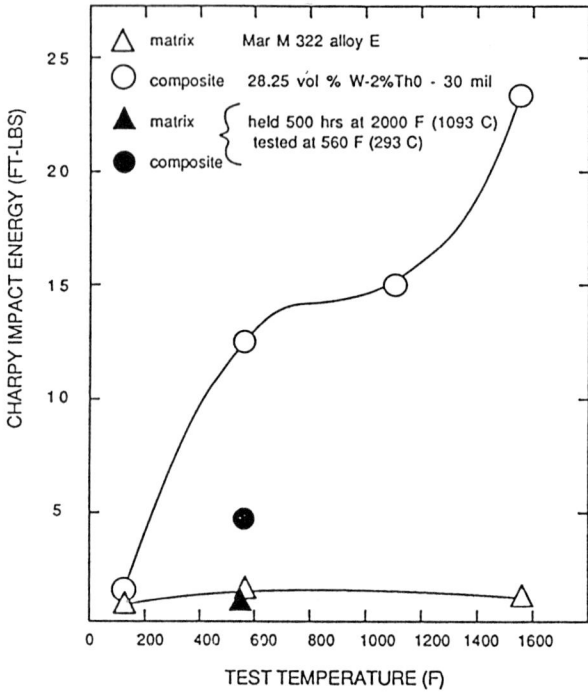

FIG. 12. Charpy impact energy for 28% V_f composite and matrix alloy Mar M322 as a function of test temperature.

the impacted MMC: the driven fracture maximum flexural (bend) stress (σ_{max}) and the maximum shear stress (τ_{max}). Simple beam theory provides the ratio of σ_{max} to τ_{max} given by

$$\frac{\sigma_{max}}{\tau_{max}} = \frac{2L}{h}, \qquad (2)$$

where L and h are the span and height of a Charpy-type impact specimen, and h for notched specimens does not include the notch depth. If the flexural and shear strengths of a given MMC are known, one can conclude that the larger the L/h ratio becomes, the larger the maximum flexural stress (σ_{max}) becomes, thus resulting in the flexural fracture, whereas for the smaller L/h ratios, the shear fracture is more likely to occur. Prewo also proposed a formula to evaluate the impact energy of various MMCs, i.e., the impact energy (I) is given by

$$I = c \frac{\sigma_f^2 \, d_f V_f}{12\sigma_{90}}, \qquad (3)$$

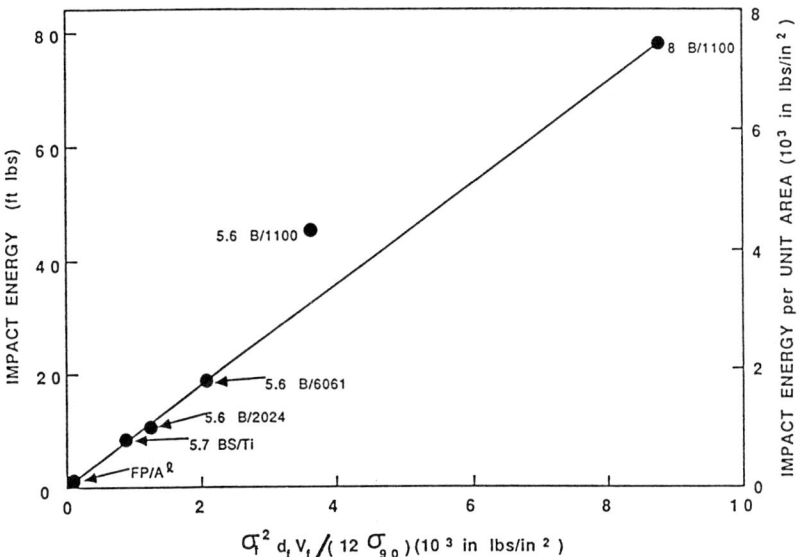

FIG. 13.   Impact energy dissipated by full-sized notched impact specimens as a function of composite material properties.

where c is a constant, σ_f, d_f, and V_f are the fiber strength, fiber diameter, and volume fraction of the fiber, and σ_{90} is the composite transverse tensile strength. When all the impact data of several MMCs are plotted as a function of $\sigma_f^2 d_f V_f / (12 \, \sigma_{90})$ in Fig. 13, most of the impact energy data appear to be on a straight line, justifying Eq. (3). It can be concluded from Fig. 13 that the MMCs with higher-to-lower impact resistance in that order are 8 B/1100 Al, 5.6 B/1100 Al, 5.6 B/6061 Al, 5.6 B/2024 Al, 5.7 borsic/Ti alloy, and FP/Al composites, where the number in front of B denotes the diameter of boron fiber in mils.

Dropweight testing is another convenient way to give low-velocity impact damage. The impact energy of SiC particulate/6061 Al (SiCp/6061 Al) composite under dropweight was studied by Patterson and Taya [11]. They studied the effect of the volume fraction of SiC particulate (V_f) and impact energy given. Then the damage was assessed in terms of the size of indent on the front surface of the specimen and absorbed energy (residual fracture energy). The results of the residual fracture energy of the impacted composites with three different V_f, $V_f = 0$ (unreinforced matrix), 20% and 30%, are plotted as a function of given impact energy in Fig. 14. It is obvious from Fig. 14 that SiCp/6061 Al composites have modest residual fracture energy compared with the unreinforced matrix, and for the larger impact energy

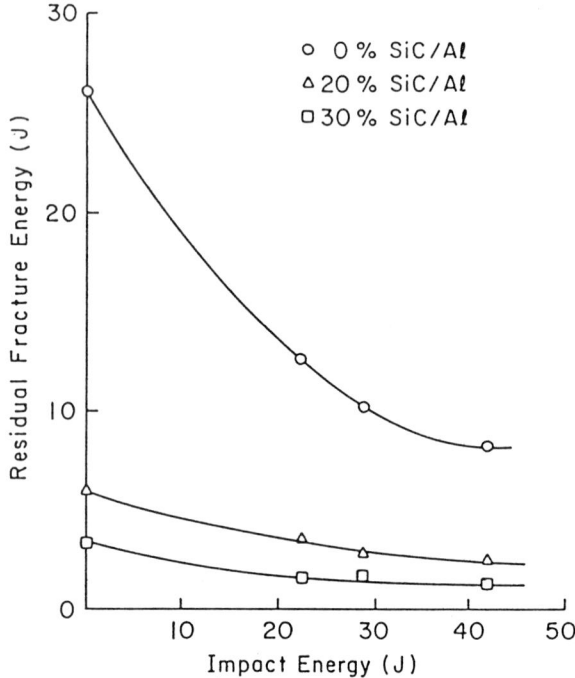

FIG. 14.   Residual fracture energy for SiCp/Al composite with several V_f's as a function of impact energy.

than a certain value (about 20 Joules), the residual fracture energy of the composite did not reduce.

References

1. J. Harding, M. Taya, B. Derby, and S. Pickard, in "ICCM-6/ECCM-2" (F. L. Matthews, N. C. R. Buskell, J. M. Hodgkinson, and J. Morton, eds.), Vol. 2, p. 224. Elsevier Applied Science, London 1987.
2. S. Pickard, B. Derby, J. Harding, and M. Taya, *Scripta Metall.*, **22**, 601 (1988).
3. A. Marchand, J. Duffy, T. A. Christman, and S. Suresh "An Experimental Study of the Dynamic Mechanical Properties of an Al–SiC$_w$ Composite." Brown University Tech. Report ONR-SDO10 N00014-85-K-0687/1, Providence, Rhode Island, December 1986.
4. M. Taya, I. W. Hall, and H. S. Yoon, *Acta Metall.*, **33**, 2143 (1985).
5. D. M. Schuester and R. P. Reed, *J. Comp. Mater.*, **3**, 562 (1969).
6. H. S. Yoon and M. Taya, submitted for publication.
7. T. D. Gray, in "Fatigue of Composite Materials," ASTM STP 569, p. 262. American Society for Testing and Materials, Philadelphia, Pennsylvania, 1975.

8. L. E. Dardi and K. G. Kreider, in "Failure Modes in Composites I" (I. Toth, ed.), p. 231. TMS–AIME, Warrendal, Pennsylvania, 1973.

9. I. Ahmad and J. Barranco, in "Advanced Fibers and Composites for Elevated Temperatures" (I. Ahmad and B. R. Norton, eds.), p. 183. TMS-AIME, Warrendale, Pennsylvania, 1980.

10. K. M. Prewo, in "The Mechanical Behavior of Metal Matrix Composites" (J. E. Hack and M. F. Amateau, eds.), p. 181. TMS-AIME, Warrendale, Pennsylvania, 1983.

11. W. G. Patterson and M. Taya, in "Advances in Aerospace Science and Engineering," (U. Yuceoglu and R. Hesser, eds.), ASME AD-08 Bound Volume, p. 49, The American Society of Mechanical Engineers, New York, 1984.

Creep Behavior of Metal Matrix Composites

M. TAYA

Department of Mechanical Engineering
University of Washington
Seattle, Washington

I. Introduction

Of the high-temperature properties (behavior) of concern of metal matrix composites (MMCs), the creep behavior is considered the most important. This has prompted a number of investigations on this subject. The most fundamental issue of the creep behavior of a MMC is the determination of the mechanism by which the creep rate of the composite is reduced by reinforcing the creeping matrix with less-creeping or noncreeping fibers. Once the mechanism is clarified, one can design a new MMC with higher creep resistance by tailoring constituent parameters of the matrix and fiber phases. Experimentally, the creep data of a MMC are obtained by a "creep test," in which a MMC specimen is loaded in tension with a constant stress at a given temperature, which provides the creep strain (ε_c) as a function of time (t).

II. Creep Behavior Composites

A typical creep strain–time relation is shown in Fig. 1, which is from a 25% V_f SiC whisker/2124 Al composite tested at 573 K and with an applied stress (σ_c) of 48 MPa. Figure 1 illustrates three stages: primary, secondary

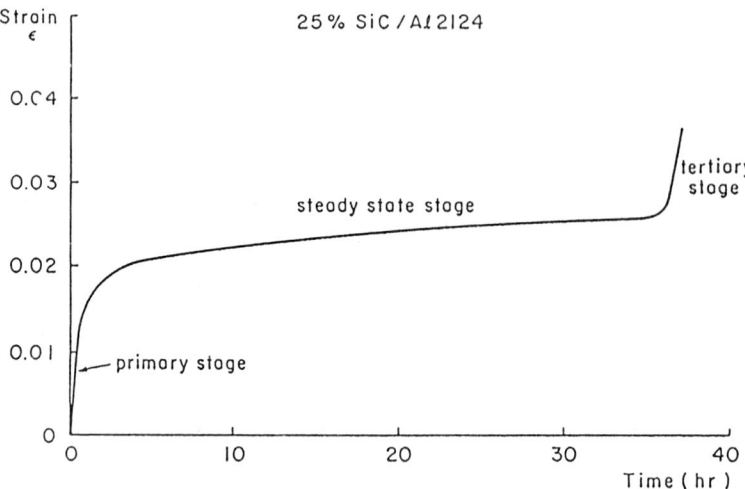

FIG. 1.   A typical creep strain–time curve of 25% V_f SiC$_w$/2124 Al composite at 300°C and applied stress 48 MPa [1].

(or steady state), and tertiary stages. The secondary stage is the longest and provides the minimum creep rate. If no secondary stage, i.e., no steady-state creep rate, is established, then the minimum creep rate is often referred to as the creep rate for a given total creep strain. Thus, the minimum creep rate, $\dot{\varepsilon}_c$, (hereafter simply called "creep rate") has been used extensively as a key parameter in describing the creep behavior of MMCs. The next important parameter is the applied stress, σ_c. This leads to the fact that the relation between $\dot{\varepsilon}_c$ and σ_c for a given temperature (T) has been studied rigorously. In the following sections, the $\dot{\varepsilon}_c$–σ_c relation of selected MMC systems will be reviewed.

The objective of reinforcing a metal with fibers is to reduce the creep rate, thus increasing the creep resistance of the metal matrix composite. This is clearly demonstrated in Fig. 2, where the relation between logarithmic creep rate, log $\dot{\varepsilon}$, and applied creep stress σ of 15% V_f SiC$_w$/6061 Al composite (filled circles connected by a solid line) and the unreinforced 6061 aluminum (open circles connected by a solid line) is shown [2]. Namely, for a fixed applied stress (for example, $\sigma = 55$ MPa), the creep rate of the composite is (ten times) smaller than that of the unreinforced metal. The question arises as to which parameters related to the microstructure of MMCs would have a strong influence on increasing the creep resistance. To answer this question, a number of researchers have proposed analytical models to predict the creep rate of MMCs: continuous-fiber MMCs [3–5] and short-fiber MMCs [1, 6–10].

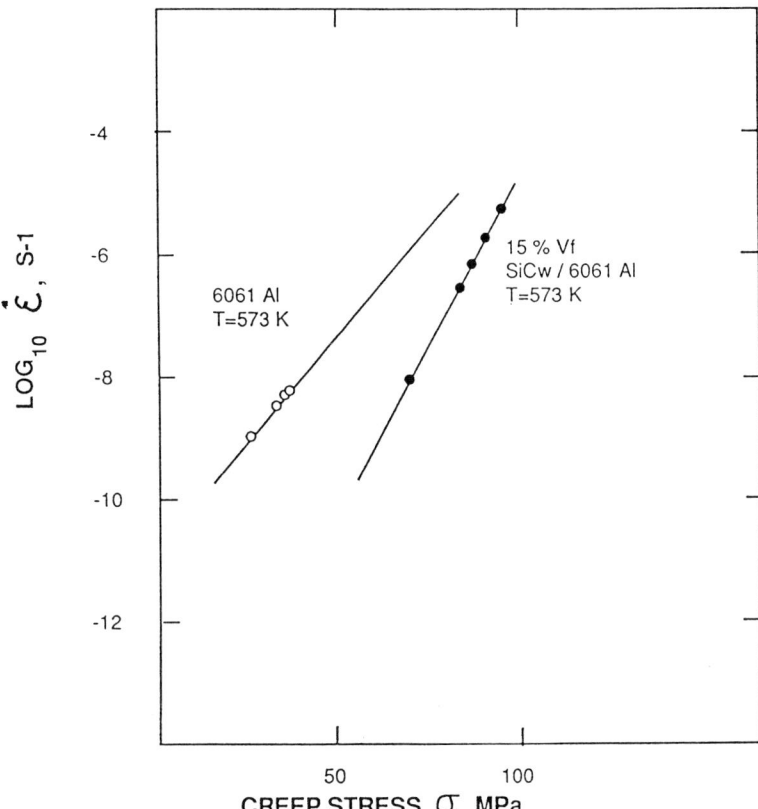

FIG. 2. $\text{Log}_{10}\ \dot{\varepsilon}_c$ (creep strain rate) versus applied stress (σ) for unreinforced 6061 Al (open circles) and 15% V_f SiC$_w$/6061 Al composite (filled circles). Reprinted from [2].

An analytical model for predicting the creep rate of continuous-fiber (unidirectional) MMCs was first constructed by McDanels *et al.* [3], who assumed that both the matrix and fiber creep behaved according to a power law of the type

$$\dot{\varepsilon}_i = A_i \sigma_i^{n_i} \exp\left(-\frac{Q_i}{RT}\right) = B_i \sigma_i^{n_i}, \tag{1}$$

where subscript $i = m$ (matrix) or f (fiber), A_i and n_i, and Q_i are material constants and the activation energy for self diffusion of the ith phase, respectively; R is the gas constant, and T is the absolute temperature. It is also assumed in the McDanels model that the strains in all phases are equal, resulting in the same strain rate for the matrix and the fiber, i.e.,

$$\dot{\varepsilon}_c = \dot{\varepsilon}_m = \dot{\varepsilon}_f. \tag{2}$$

This shows that Eq. (2) is also valid for a short-fiber MMC as far as the steady-state creep behavior is concerned, although it was originally derived by assuming isostrain, which is characteristic of (continuous) unidirectional MMCs. The applied stress σ_c is assumed to be shared by both phases according to the law of mixtures:

$$\sigma_c = \sigma_f V_f + \sigma_m(1 - V_f). \tag{3}$$

From Eqs. (1)–(3), one can obtain the relation between the applied stress (σ_c) and the creep rate ($\dot{\varepsilon}_c$) of the composite as

$$\sigma_c = V_f\left(\frac{\dot{\varepsilon}_c}{B_f}\right)^{1/n_f} + (1 - V_f)\left(\frac{\dot{\varepsilon}_c}{B_m}\right)^{1/n_m}. \tag{4}$$

McLean [5] plotted the results of Eq. (4) in the log $\dot{\varepsilon}_c$–log σ_c graph, shown in Fig. 3, where the volume fraction of the fiber $V_f = 0.1$ was used. Figure

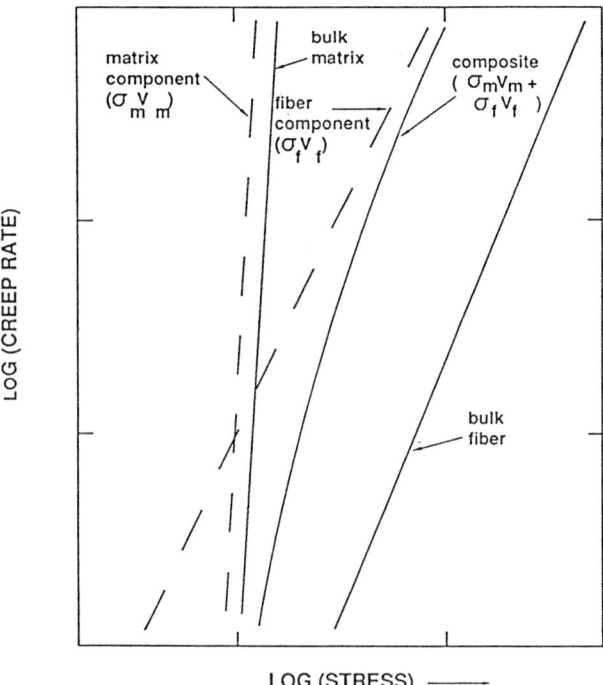

FIG. 3.   Analytical results of the creep rate–applied stress relation where both the matrix metal and fiber obey power-law-type equations. Reprinted from [5].

3 shows that the slope of the composite curve is close to that of the fiber at higher stress level while at lower stress level, it is close to that of fiber.

Although this model of a creeping matrix and a creeping fiber explains the experimental results of tungsten fiber/copper composite [3] reasonably well, it cannot be applied to other MMC systems where fibers are made of ceramics, thus they do not creep at the operating temperatures of the MMCs. The case of a creeping matrix and noncreeping (or elastic) fibers was studied analytically by McLean [5]. In McLean's model, it is assumed that both phases deform at the same rate $\dot{\varepsilon}_c$, although fiber deforms elastically, whereas the matrix creeps according to the power law defined by Eq. (1). Then

$$\text{for the fiber:} \quad \dot{\varepsilon}_c = \frac{1}{E_f} \frac{d\sigma_f}{dt} \tag{5}$$

$$\text{for the matrix:} \quad \dot{\varepsilon}_c = \frac{1}{E_m} \frac{d\sigma_m}{dt} + B_m \sigma_m^{n_m} \tag{6}$$

Equation (6) shows that the first term on the right represents the elastic response of the matrix, whereas the second term is the creep strain rate. Assuming that the composite stress σ_c is obtained by the law of mixtures defined by Eq. (3), one can arrive at the following differential equation,

$$\dot{\varepsilon}_c = \alpha B_m \sigma_c^{n_m} \left(1 - \frac{\dot{\varepsilon}_c}{\varepsilon_\infty} \right)^{n_m}, \tag{7}$$

where

$$\frac{1}{\alpha} = \left[1 + \frac{E_f V_f}{E_m (1 - V_f)} \right] (1 - V_f)^{n_m},$$

$$\varepsilon_\infty = \frac{\sigma_c}{E_f V_f}. \tag{8}$$

In Eq. (8), ε_∞ is the ultimate strain at an equilibrium state where all the applied stress is carried by the fibers. Equation (7) can be solved for $\dot{\varepsilon}_c$ once the initial condition is given, i.e., the strain at $t = 0$, ε_0 is obtained based on elasticity analysis,

$$\varepsilon_0 = \frac{\sigma_c}{E_f V_f + E_m (1 - V_f)}. \tag{9}$$

Then ε_c is solved for two cases:

for $n_m > 1$,

$$\varepsilon_c = \varepsilon_\infty - (\varepsilon_\infty - \varepsilon_0) \left[1 + \left(1 - \frac{\varepsilon_0}{\varepsilon_\infty} \right)^{n_m - 1} \frac{(n_m - 1)\alpha B_m \sigma_c^{n_m} t}{\varepsilon_\infty} \right]^{-1/(n_m - 1)}, \tag{10}$$

for $n_m = 1$,

$$\varepsilon_c = \varepsilon_\infty \left[1 - \left(1 - \frac{\varepsilon_0}{\varepsilon_\infty} \right) \exp\left(- \frac{\alpha B_m \sigma_c t}{\varepsilon_\infty} \right) \right]. \tag{11}$$

Equation (10) represents most of the unidirectional MMC systems including eutectic composites, whereas Eq. (11) may be applicable to the MMCs with linearly viscous metals or metals with fine grain sizes exhibiting diffusional creep. It should be noted in this model that there exists no steady-state creep rate, because ε_c increases monotonically with t; thus, it is convenient to define the creep rates achieved at a given strain, which is needed for construction of the $\dot{\varepsilon}_c$–σ_c relation. McLean applied this model to the case of the eutectic composite γ–γ'–Cr$_3$ C$_2$ at T = 1095 K, and the results of the creep rate at $\varepsilon_c = 0.01$ are plotted as solid curve against the applied stress σ_c in Fig. 4, where the creep rate–stress relation of the matrix only (Eq. (1)) is plotted as a solid line. Also shown in Fig. 4 are the experimental results of γ–γ'–Cr$_3$ C$_2$ composite, which appear to be in good agreement with the prediction.

The case of short-fiber MMCs, the steady-state (or the minimum) creep rate $\dot{\varepsilon}_c$ is usually attained. A number of researchers have proposed analytical models to predict $\dot{\varepsilon}_c$, which can be grouped into so-called shear lag models [1, 2, 7–10] and Eshelby's model [11, 12]. As an example of the shear lag model, the model proposed by Kelly and Street [9] will be reviewed here briefly. Kelly and Street assumed that short fibers of relatively large aspect ratio (l/d) are all aligned along the fiber axis (z-direction), see Fig. 5; and the matrix creep and fiber creep obey power-law-type creep behavior. They considered several cases: creeping matrix/elastic fiber, creeping matrix/creeping fiber, and creeping matrix/elastic fiber with sliding interface. The average shear strain rate $\dot{\gamma}$ in the matrix plays a key role in the Kelly and Street model and can be estimated as

$$\dot{\gamma} = \frac{1}{h} (\dot{u}_m - \dot{u}_i), \tag{12}$$

where the component of shear strain is $r - z$, with r taken perpendicular to the fiber axis (z-axis), $\dot{u}_m$ and $\dot{u}_i$ is the velocity along the z-axis at the mid-point between two short fibers and the point close to the matrix–fiber interface, respectively. In Eq. (12), h is set equal to the average spacing between two fibers, and it is related to the fiber diameter d and the volume fraction of fiber V_f as

$$h = \frac{d}{2} \left[\left\{ \frac{\pi}{2\sqrt{3}V_f} \right\}^{1/2} - 1 \right], \tag{13}$$

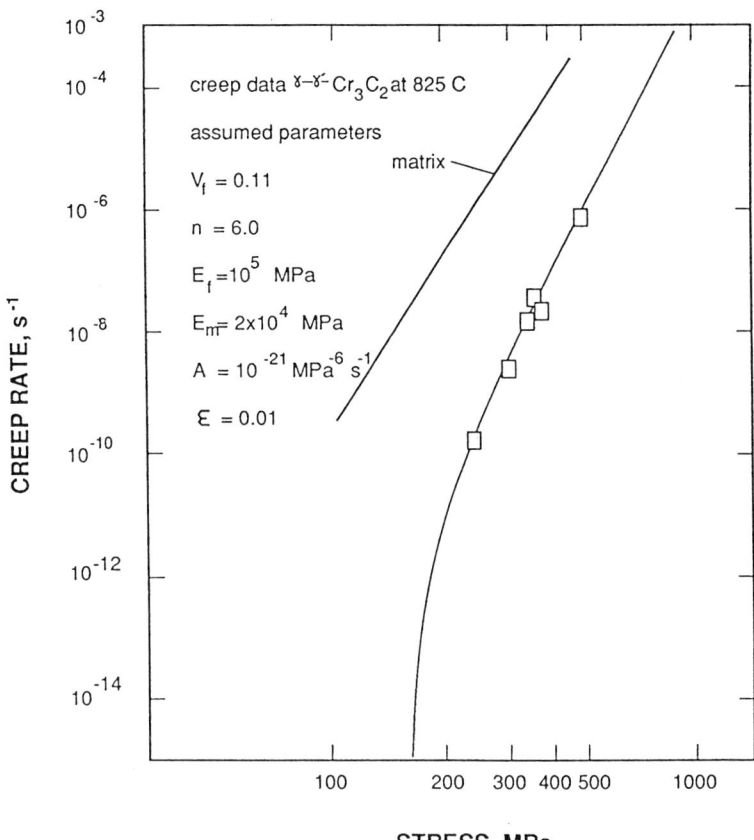

FIG. 4.   Predicted creep rate at 1% strain for an eutectic composite ($\gamma-\gamma'-Cr_3 C_2$) consisting of elastic fibers and creeping matrix. Reprinted from [5].

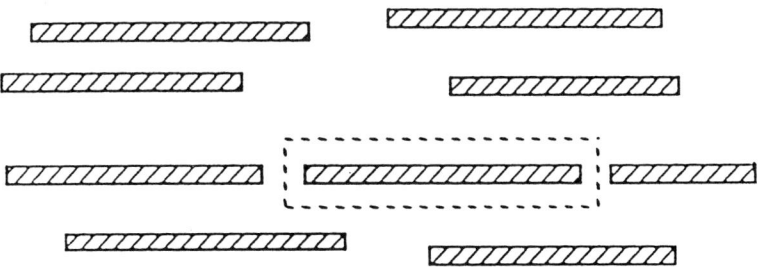

FIG. 5.   A shear lag model for the creep of an aligned short-fiber MMC. Reprinted from [9].

where the packing of short fibers is assumed to be a hexagonal array and the fiber aspect ratio l/d is relatively large. It should be noted here that for the case of short fibers with smaller aspect ratios, Eq. (13) gives rise to errors. The velocity of the matrix near the interface $\dot{u}_i$ is considered to consist of the velocity of the interface sliding $\dot{u}_s$ and that of the fiber $\dot{u}_f$,

$$\dot{u}_i = \dot{u}_f + \dot{u}_s, \tag{14}$$

where $\dot{u}_s$ is assumed to be proportional to the difference $\dot{u}_m - \dot{u}_f$, i.e.,

$$\dot{u}_s = \eta(\dot{u}_m - \dot{u}_f), \tag{15}$$

and where η is a sliding mode parameter ($\eta = 0$ corresponds to a perfectly bonded interface). From Eqs. (12), (14), and (15), the average shear strain rate in the matrix is rewritten as

$$\dot{\gamma} = \frac{1}{h}\{\dot{u}_m - \dot{u}_f - \eta(\dot{u}_m - u_f)\}, \tag{16}$$

where the velocities $\dot{u}_m$ and $\dot{u}_f$ are related to tensile strain rates $\dot{\varepsilon}_m$ and $\dot{\varepsilon}_f$ as

$$\dot{u}_j = \dot{\varepsilon}_j z, \qquad j = m \text{ (matrix) or f (fiber)}, \tag{17}$$

and where $\dot{\varepsilon}_m$ and $\dot{\varepsilon}_f$ are related to the tensile stresses σ_m and σ_f, respectively, by the power-law formulas similar to Eq. (1), i.e.,

$$\dot{\varepsilon}_j = \dot{\varepsilon}_{j0}\left(\frac{\sigma_j}{\sigma_{j0}}\right)^{n_j}, \qquad j = m \text{ (matrix) or f (fiber)}. \tag{18}$$

In Eqs. (17) and (18), the subscript j denotes the phase and should not be summed over, and $\dot{\varepsilon}_{j0}$, σ_{j0}, and n_j are material constants that can be determined by uniaxial tensile creep tests for a fixed temperature. Since the main contribution to the composite creep comes from the matrix shear strain rate $\dot{\gamma}$, one must convert Eq. (18) with $j = m$ (matrix) to the shear deformation type. To this end, Kelly and Street [9] assumed that the Tresca-type yield criterion holds and that the matrix is incompressible in order to arrive at $\sigma_m = 2\tau$ and $\dot{\varepsilon}_m = \frac{2}{3}\dot{\gamma}$. With these, Eq. (18) for the matrix phase is rewritten as

$$\dot{\gamma} = \frac{3}{2}\dot{\varepsilon}_{m0}\left(\frac{2\tau}{\sigma_{m0}}\right)^{n_m}, \tag{19}$$

where τ is the shear stress corresponding to $\dot{\gamma}$ and acting on the matrix–fiber interface. From Eqs. (13), (16), (17), and (18), τ is obtained as

$$\tau = \frac{\beta\sigma_{m0}}{(\dot{\varepsilon}_{m0}d)^{1/n_m}}[\dot{\varepsilon}_m z - \dot{u}_f - \eta(\dot{\varepsilon}_m z - \dot{u}_f)]^{1/n_m}, \tag{20}$$

where

$$\beta = \frac{1}{2}\left(\frac{4}{3}\right)^{1/n_m}\left[\left(\frac{2\sqrt{3}V_f}{\pi}\right)^{-1/2} - 1\right]^{1/n_m}.$$ (21)

Let us examine the case of a creeping matrix/elastic fiber that may represent the case of a ceramic fiber/metal matrix composite. In this case, $\eta = 0$ and $\dot{u}_f = 0$, hence Eq. (20) is reduced to

$$\tau = \beta\sigma_{m0}\left(\frac{\dot{\varepsilon}_m}{\dot{\varepsilon}_{m0}}\right)^{1/n_m}\left(\frac{z}{d}\right)^{1/n_m}.$$ (22)

The shear stress distribution given by Eq. (22) is $\tau = 0$ at $z = 0$ (the center of the fiber) and increases toward the fiber ends. This shear stress at the interface must be in equilibrium along the z-direction with fiber axial stress σ_f, which results in

$$\sigma_f = \int_0^z \frac{4\tau}{d}\,dz.$$ (23)

The average fiber stress $\bar{\sigma}_f$ can be obtained by

$$\bar{\sigma}_f = \frac{2}{l}\int_0^{l/2} \sigma_f\,dz.$$ (24)

The composite stress σ_c is estimated by a law of mixtures (Eq. (3)), where σ_f is replaced by $\bar{\sigma}_f$ and then derived as

$$\sigma_c = \sigma_{m0}\left(\frac{\dot{\varepsilon}_m}{\dot{\varepsilon}_{m0}}\right)^{1/n_m}\left[\phi V_f\left(\frac{l}{d}\right)^{n_m + 1/n_m} + 1 - V_f\right],$$ (25)

where

$$\phi = \left(\frac{2}{3}\right)^{1/n_m}\left(\frac{n_m}{2n_m + 1}\right)\left[\left(\frac{2\sqrt{3}V_f}{\pi}\right)^{-1/2} - 1\right]^{-1/n_m}.$$ (26)

Equation (25) shows that $V_f = 0$ yields the matrix creep law, i.e., Eq. (18) with j = m, and also that the creep rate $\dot{\varepsilon}_m$ decreases with increase in fiber aspect ratio l/d. Kelly and Street [9] also obtained the formula to predict the $\sigma_c - \dot{\varepsilon}_m$ relation for the case of creeping matrix/creeping fibers and compared those predictions with the experiment where phosphor bronze fiber/lead (Pb) composite was used as a model MMC, as shown in Fig. 16. In the case of $l/d = 100$, the model with the creeping matrix/creeping fiber was found to agree better with the experimental data than that with creeping matrix/rigid fiber. The effectiveness of using longer fibers (or fibers with larger fiber aspect ratio) in enhancing the creep resistance is clearly seen in Fig. 6. Shown in the same figure is the prediction based on the Mileiko model [7] (dash–dot line), which seems to agree well with the experiment, although Mileiko used the model with creeping matrix/rigid fiber. It appears from Fig. 6 that the stress exponent n_m is too large ($n_m = 14$) to create an accurate drawing of the line in the log–log plot.

In the case of short-fiber MMCs with a relatively small fiber aspect ratio, the Kelly and Street model [9] tends to give a poor estimate for the creep rate–stress relation, since in this case, the creep behavior in the matrix domain in a MMC consists of two mechanisms, creep by shear stress and creep by normal stress [10], and since the assumption used in deriving Eq. (13) becomes less accurate. Lilholt [10] discussed a general approach to predict the composite creep by using the data of the matrix creep and rigid

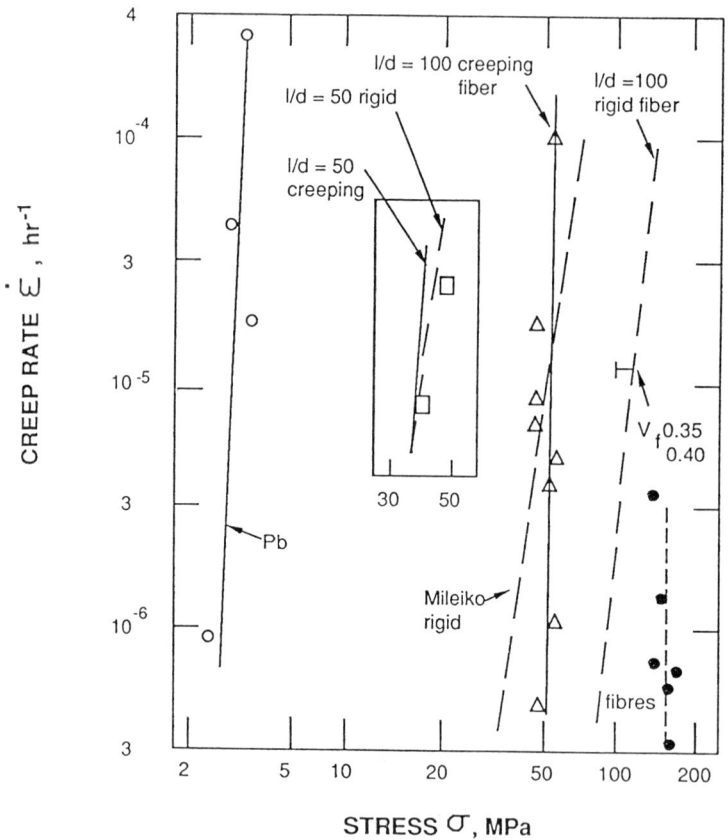

FIG. 6. Minimum creep rates versus applied stress for short phosphor bronze fiber/lead composite for which the analytical results are also shown by lines. Reprinted from [9].

as well as creeping fibers. He emphasized the importance of using the exponential-type creep law in describing the higher stress region in the matrix. The experimental results of a short-fiber MMC with relatively small fiber aspect ratio that have been reported are centered on SiC whisker/ aluminum alloy composites [2, 13–15]. A common observation in these experimental results is that the composite creep line in the log $\dot{\varepsilon}_c$–log σ_c plot or the log $\dot{\varepsilon}_c$–σ_c plot is steeper than the unreinforced matrix, i.e., the composite creep is depending much more strongly on stress than the unreinforced matrix, see Fig. 7 [14].

Motivated by experimental evidence, Taya and Lilholt [1] constructed a model in which fibers do not creep and the matrix creep is assumed to obey the empirical exponential creep law defined by

$$\frac{\dot{\varepsilon}_m}{\dot{\varepsilon}_{m0}} = \exp\left(\frac{\sigma_m}{\sigma_{m0}}\right). \tag{27}$$

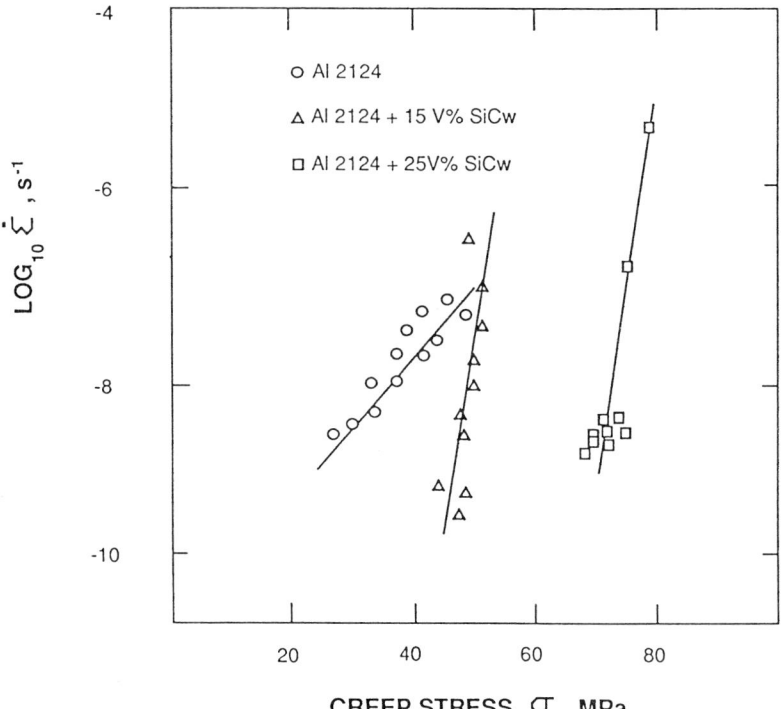

FIG. 7. The creep rate versus applied stress of unreinforced Al 2124 and SiC$_w$/Al 2124 composites, indicating that the exponential-type creep law is valid. Reprinted from [14].

In the Taya–Lilholt model, it is assumed that at low to intermediate stress levels, the matrix–fiber interfaces are perfectly bonded, while at higher stress levels, the interfaces are debonded. The analytical models for the case of perfectly bonded and of debonded interfaces are shown in Fig. 8a and 8b, respectively. These unit cell models are taken from a representative unit dotted in Fig. 5. The case of short fibers with smaller aspect ratio l/d should require a three-dimensional geometry of the unit cell, i.e., use of l_0, d_0, l, and d in Fig. 8. This is in contrast with the Kelly and Street model, where short fibers have large aspect ratio, thus requiring only two-dimensional consideration, leading to Eq. (13). Therefore, in the Taya and Lilholt model, h is given by

$$h = \frac{d}{2}\{\sqrt{(l/l_0)/V_f} - 1\}. \tag{28}$$

Following the Kelly and Street assumption that the matrix obeys the Tresca-type plasticity law and that it is incompressible, one can arrive at the matrix shear stress (τ)–tensile strain rate ($\dot{\varepsilon}_m$) relation,

$$\tau = \tfrac{1}{2}\sigma_{m0} \ln\left\{\frac{\dot{\varepsilon}_m}{\dot{\varepsilon}_{m0}} \eta \frac{z}{d}\right\}, \tag{29}$$

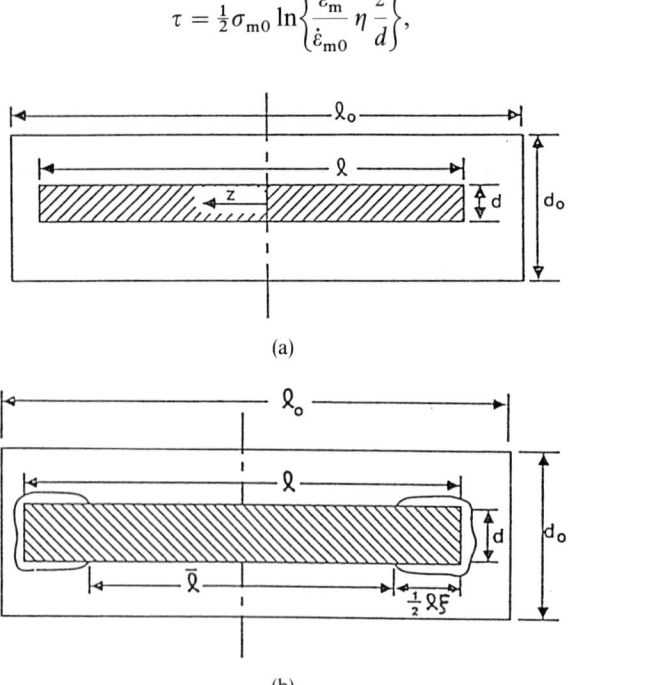

(a)

(b)

FIG. 8. Analytical model of Taya and Lilholt: (a) for perfectly bonded interface, and (b) for partially debonded interface. Adapted from [2, 14].

where z is the axial coordinate with its origin at the fiber center and η is defined by

$$\eta = \frac{4}{3} \frac{1}{[(kV_f)^{-1/3} - 1]}, \tag{30}$$

and k is the ratio of the unit cell aspect ratio to the fiber aspect ratio,

$$k = (l_0/d_0)/(l/d). \tag{31}$$

This k-value can be determined from scanning electron microscope (SEM) photographs. By using Eq. (3) (law of mixtures for stresses), Eq. (23) (load transfer), and Eq. (24) (averaging the fiber stress), one can obtain the σ_c–$\dot{\varepsilon}_m$ relation similar to Eq. (25). Then the composite creep rate $\dot{\varepsilon}_c$ is estimated by use of $\dot{\varepsilon}_m$ as

$$\dot{\varepsilon}_c = (1 - V_f)\dot{\varepsilon}_m. \tag{32}$$

The case of a debonded interface can be formulated in the same manner as the case of a perfectly bonded interface, except for the fiber length l being replaced by the effective fiber length $\bar{l}$, which is related to l by

$$\bar{l} = l(1 - \xi), \tag{33}$$

where ξ is a parameter describing the extent of interface debonding (see Fig. 8b) and its value ranges from 0 (no debonding except for fiber ends) to 1 (complete debonding around the fiber). The sequence of the interface degradation in terms of debonding under increasing applied stress σ_c is schematically shown in Fig. 9 [2]. The composite stress (σ_c) and creep rate ($\dot{\varepsilon}_c$) obtained by Taya and Lilholt for the two cases are

for perfectly bonded interface,

$$\dot{\varepsilon}_c = (1 - V_f)\dot{\varepsilon}_{m0} \exp\left\{\frac{\sigma_c/\sigma_{m0} - C}{1 + V_f(l/d)/2}\right\},$$

$$C = 0.5V_f(l/d)\{\ln(l/d) + \ln\eta - 1.193\}; \tag{34}$$

for debonded interface,

$$\dot{\varepsilon}_c = (1 - V_f)\dot{\varepsilon}_{m0} \exp\left\{\frac{\sigma_0/\sigma_{m0} - C}{1 - V_f + V_f(l/d)(1 - \eta)/2}\right\},$$

$$C = 0.5V_f(l/d)(1 - \xi)\{\ln[(l/d)(1 - \xi)] + \ln\eta - 1.193\}. \tag{35}$$

In Eqs. (34) and (35), η has been defined by Eq. (30).

Based on Eq. (34), Taya and Lilholt [1] examined the effect of the volume

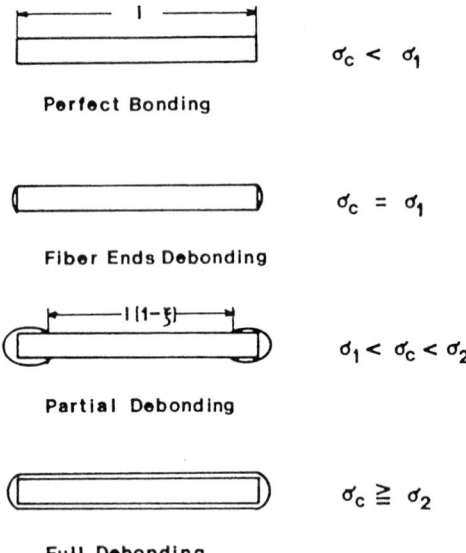

FIG. 9. Sequence of the interface degradation of a short-fiber MMC under creep loading. Reprinted from [2].

fraction of the fiber V_f and the fiber aspect ratio l/d on the overall creep rate, $\dot{\varepsilon}_c$, of a short-fiber MMC. The results of $\dot{\varepsilon}_c$ normalized by the creep rate of the unreinforced matrix $\dot{\varepsilon}_m$ are plotted as a function of V_f as shown in Fig. 10a and of l/d as shown in Fig. 10(b). Figure 10 demonstrates the effectiveness of increasing V_f and l/d when decreasing the composite creep rate, thus increasing the creep resistance. Taya and Lilholt [1] also examined the effect of interfacial debonding by changing parameter ξ, and the results of the composite creep rate normalized by $\dot{\varepsilon}_m$ are shown as a function of ξ in Fig. 11. It is clearly seen from Fig. 11 that the debonding of the interface, even partial debonding, increases the composite creep rate, resulting in the degradation of the creep behavior. Morimoto *et al.* [2] used the above model to explain their experimental results of 15% V_f SiC$_w$/6061 Al composite (Fig. 12). A comparison between the experiment and the prediction based on the Taya–Lilholt model is shown in Fig. 12, where the experimental results are shown by circles (open circles for the unreinforced matrix, filled ones for the composite) connected by solid lines, and the prediction by Eqs. (34) and (35) by dash, dash–dot, dash–double-dot lines. The steeper slope of the $\dot{\varepsilon}_c - \sigma_c$ line of the composite was found to agree with the prediction by the model with debonded interface (Eq. (35)). Thus, they concluded that at higher stress level, the debonding of the interface contributes to the increase in the composite creep rate, thereby resulting in the steeper slope of the $\dot{\varepsilon}_c - \sigma_c$

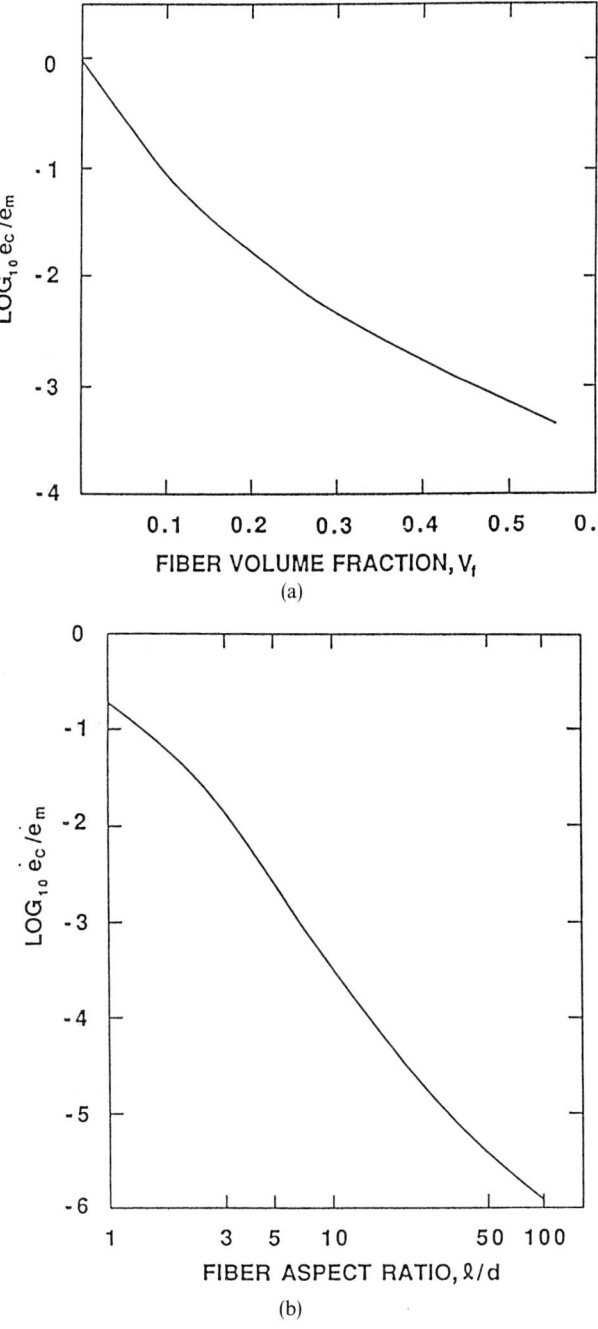

FIG. 10. Dependence of the creep rate of a short-fiber MMC on (a) fiber volume fraction V_f, and (b) fiber aspect ratio l/d [1].

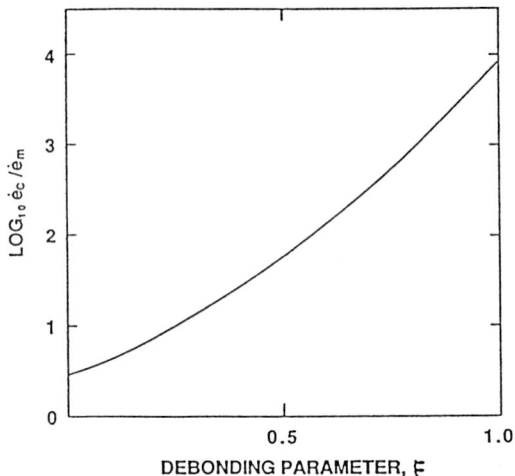

FIG. 11. Effect of interfacial debonding (ξ) on the creep rate of a short-fiber MMC [1].

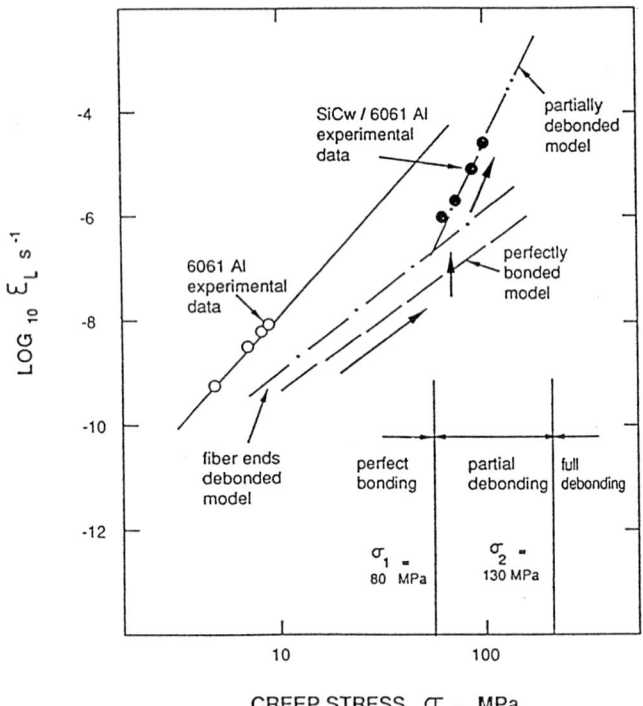

FIG. 12. A comparison between the experimental and analytical results of the creep rate: applied stress relation for a short-fiber MMC. Reprinted from [2].

line. It should be mentioned here that determination of the threshold stresses σ_1 and σ_2 in the Taya–Lilholt model must be carefully assessed in order better to predict the composite creep rate.

The debonded interface that was found to accelerate the composite creep rate is one type of microdamage that would presumably take place during creep loading. Possible other types of microdamage are residual stresses, interfacial sliding, fiber fracturing, and the nucleation and growth of voids and cracks [16, 17]. Among these types of microdamage, McLean [16] claimed that the fiber fracturing, which will reduce the effective fiber aspect ratio, has the strongest influence in enhancing the creep rate, thus weakening the creep resistance of a unidirectional (continuous or discontinuous) fiber MMC. The McLean model for fiber fracturing is similar to the Taya and Lilholt model in that both models are based on the shear-lag-type model, and the fiber used has a smaller aspect ratio due to the fracturing (the McLean model) or partially debonded interface (the Taya and Lilholt model). Goto and McLean [17] have recently analytically examined the effect of weak and strong interfaces on the overall composite creep rate. In the case of long-fiber MMC, they found that the strong (work-hardening) interface has quite an impact on reducing the overall creep rate, whereas the weak interface (slipping at its extreme case) has little effect. In the case of a short-fiber MMC, the weak interface has a very strong effect on the overall creep rate of the composite.

The preceding models are based on several assumptions facilitating the analysis; thus, the problem can be reduced to simple one-dimensional models (shear lag and law of mixtures), yielding a very crude approximation of the composite creep rate. This is particularly true for the case of smaller fiber aspect ratios, where a three-dimensional analysis such as Eshelby's model [18, 19] is needed to predict the creep rate accurately. Eshelby's model, however, is applicable only to linear problems, i.e., the relation between internal stress σ and internal strain ε (or strain rate) $\dot{\varepsilon}$ be linear. Choi et al. [11] proposed an analytical creep model for a specific class of metal matrix composites, i.e., in situ MMCs, where the matrix metal undergoes plastic deformation by dislocation slip with its friction stress k and shear modulus μ, and the fiber undergoes plastic deformation by diffusion. They considered an aligned short-fiber MMC where all the short fibers are aligned along the loading direction (x_3-axis), with the applied stress (σ_A) and the volume fractions of the matrix and fiber being denoted by V_m and V_f, respectively. The plastic deformation of the matrix by dislocation slip was characterized by the plastic strains

$$\varepsilon_{33}^* = \varepsilon_p, \; \varepsilon_{11}^* = \varepsilon_{22}^* = -\varepsilon_p/2. \tag{36}$$

Similarly, the plastic strain of the fiber by diffusion was given by

$$\varepsilon_{33}^* = \varepsilon_d, \; \varepsilon_{11}^* = \varepsilon_{22}^* = -\varepsilon_d/2. \tag{37}$$

These plastic strains (eigenstrains [19]) will induce the internal stresses in the matrix, which along with the applied stress σ_A must satisfy the yield criterion, resulting in

$$\varepsilon_p - \varepsilon_d = \{\sigma_A(1 - C_2 V_f)/V_m - k\}/(C_1 \mu V_f). \tag{38}$$

In the calculation of these internal stresses, the Eshelby's model was used. When ε_d changes by $\delta\varepsilon_d$ due to the fiber diffusion, the change in the total energy F is calculated as

$$\Delta F = -\{C_1 V_m \mu V_f(\varepsilon_p - \varepsilon_d) + C_2 V_f \sigma_A\}\delta\varepsilon_d. \tag{39}$$

In Eqs. (38) and (39), C_1 and C_2 are constants of the constituents, depending on the fiber aspect ratio and the elastic constants of the matrix and fiber. By taking into account the transport of the fiber atoms from the fiber in consideration to itself or the adjacent fibers, Choi and coworkers arrived at the following kinetic equation;

$$\tau \frac{d\varepsilon_d}{dt} = (\varepsilon_p - \varepsilon_d) + \frac{C_2 \sigma_A}{C_1 V_m \mu}. \tag{40}$$

In the case of steady-state creep, the overall creep rate of the composite $\dot{\varepsilon}_c$ is equal to $\dot{\varepsilon}_d$ ($= \dot{\varepsilon}_p$). By using this fact and Eqs. (38) and (39), they obtained the following analytical result;

$$\dot{\varepsilon}_c = \frac{\sigma_A - V_m k}{C_1 V_f V_m \tau \mu}. \tag{41}$$

In Eqs. (40) and (41) τ is the relaxation time and function of several constants related to the fiber atom diffusion. Equation (41) indicates that the steady-state creep of the MMC occurs with the rate linearly depending upon the applied stress σ_A when σ_A is larger than the threshold stress given by $V_m k$. Choi and coworkers then compared the analytical result of Eq. (41) with the experimental results of in situ MMC, Al–Al$_3$Ni, where Al$_3$Ni is considered as an aligned short fiber. Both the analytical and experimental results are shown in Fig. 13, which demonstrates the validity of the Choi et al. model and yields a threshold stress of 4 MPa.

Although the model by Choi et al. [11] can adequately explain the linear dependence of the creep rate on the applied stress, it may not be applicable to the other MMC systems where the creep rate of the composite usually depends nonlinearly upon the applied stress, for example, the power-law type. Sakaki and Taya [12] have recently proposed a model based on Eshelby's model, which is applicable to the case of nonlinear relations of creep rate stress. It is assumed in this model that the plastic (including creep) strain is uniform in each phase, which is equivalent to using the averaged stress and strain in each phase, thus accounting for the interaction between

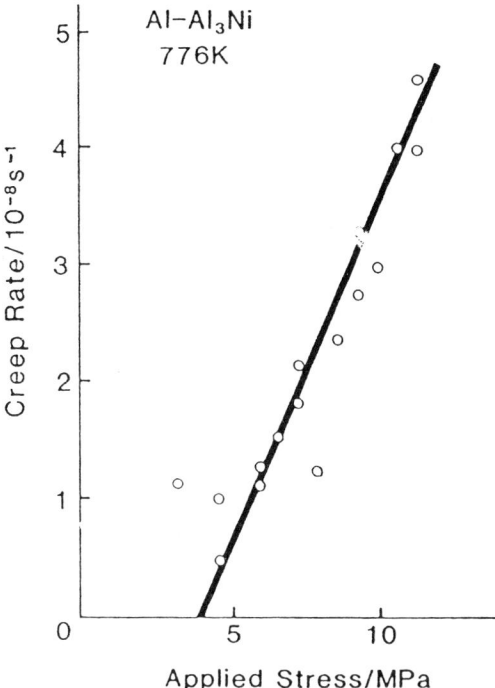

FIG. 13. Stationary creep rate versus applied stress: experimental results (circles) and prediction (solid line). Reprinted from [*11*].

fibers. This model can treat more general cases than the existing models; multiaxial applied stresses, misoriented short fiber MMC, any type of creep law of each phase and transient creep behavior in addition to steady-state creep. The Sakaki and Taya model uses an incremental method of computing the internal stress field caused by inelastic strain increments. In an extreme case, i.e., the steady-state creep of unidirectional MMC where both the matrix and fiber phases undergo the power-law-type creep, the model predicts the same results as given by Eq. (4). In other cases, this model requires step-by-step numerical calculation.

It often becomes important for a MMC component design engineer to evaluate the creep behavior of a candidate MMC subjected to off-axis loading (anisotropic creep behavior). This anisotropic creep behavior is most important in the case of aligned continuous-fiber (unidirectional) MMC, in a marked contrast to isotropic creep behavior of a particulate MMC system. Miles and McLean [*20*] studied the anisotropic creep behavior of a (Co, Cr)$_7$C$_3$ fiber/(Co, Cr) eutectic composite both experimentally and analytically. This eutectic composite is considered a unidirectional MMC, which was creep-loaded at 1098 K. The direction of creep loading (σ_θ) varied from $\theta = 0$ (σ_L) to $\theta = 90°$ (σ_T), shown in Fig. 14. The results of the creep tests are summarized in Fig. 15, where the relation of the minimum creep rate ($\dot{\varepsilon}_c$) versus applied stress (σ_c) is given by the log–log graph with θ as a parameter. Figure 15 illustrates the strong θ-dependence of $\dot{\varepsilon}_c$. To examine the θ-dependence of $\dot{\varepsilon}_c$ more clearly, Miles and McLean plotted the creep data at 1098 K and $\sigma_c = 200$ MPa in terms of the $\dot{\varepsilon}_c$–θ relation in Fig. 16, where open circles denote the experimental results. They then attempted to predict this θ-dependence of the composite creep by a simple mechanical model. In the Miles and McLean model, off-axis creep stress, σ_θ, was resolved into three stresses; normal stress σ_L (applied along the fiber axis, i.e., x_3-axis in Fig. 14), normal stress σ_T (along the x_1-axis) and shear stress τ (acting on the x_1–x_3 plane). These stresses are related to the applied stress σ_θ by

$$\sigma_L = \sigma_\theta \cos^2 \theta,$$
$$\sigma_T = \sigma_\theta \sin^2 \theta, \tag{42}$$
$$\tau = \sigma_\theta \sin \theta \cos \theta.$$

Denoting the strain rates caused by σ_L, σ_T, and τ by $\dot{\varepsilon}_L$, $\dot{\varepsilon}_T$, and $\dot{\gamma}_c$, respectively, one can obtain the creep strain rate $\dot{\varepsilon}_\theta$ of the composite loaded by σ_θ as

$$\dot{\varepsilon}_\theta = \dot{\varepsilon}_L \cos^2 \theta + \dot{\varepsilon}_T \sin^2 \theta + \dot{\gamma} \sin \theta \cos \theta, \tag{43}$$

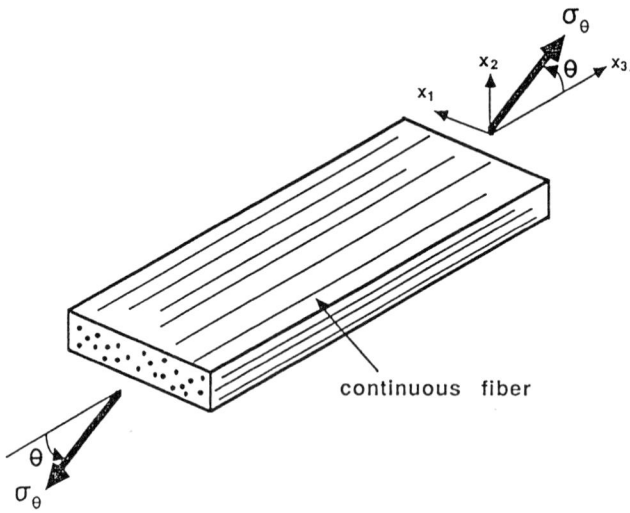

FIG. 14. Off-axis creep loading of a unidirectional MMC.

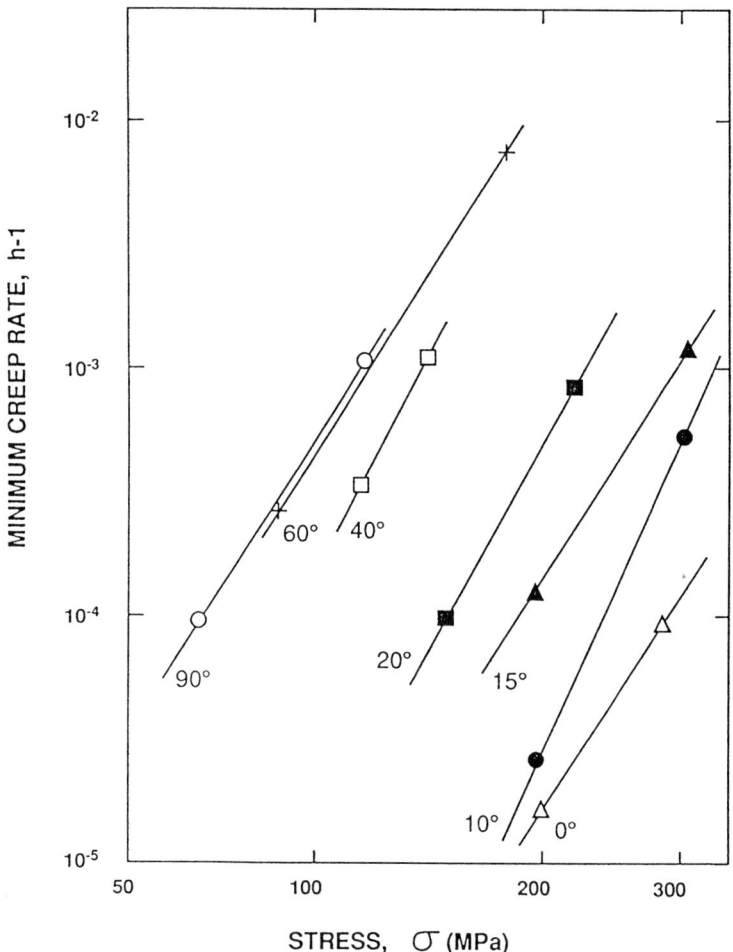

FIG. 15.   Effect of off-axis creep loading with angle θ on the minimum creep rate of eutectic composite. Reprinted from [20].

where $\dot{\varepsilon}_L$, $\dot{\varepsilon}_T$, and $\dot{\gamma}$ are related to σ_L, σ_T, and τ, respectively, by the following power laws,

$$\dot{\varepsilon}_L = A_1 \sigma_L^n,$$
$$\dot{\varepsilon}_T = f_2 A_2 \sigma_T^m, \qquad (44)$$
$$\dot{\gamma} = f_1 A_2 \tau^m,$$

and where f_1 is a parameter for conversion from normal to shear mode.

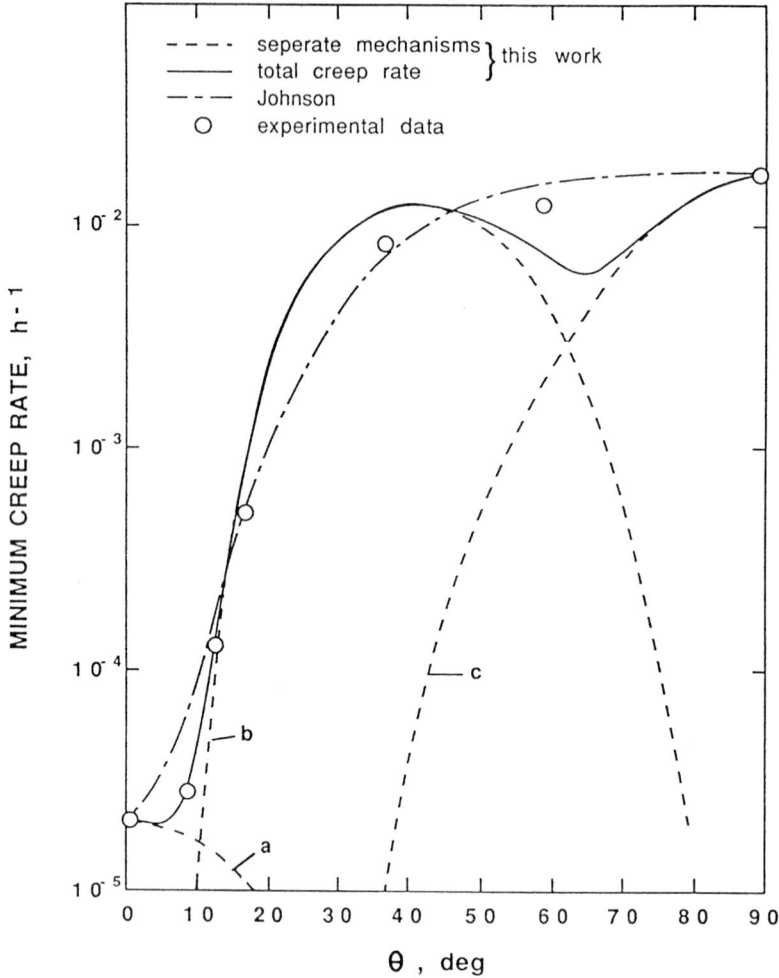

FIG. 16. The minimum creep rate versus the off-axis creep-loading angle θ. Reprinted from [20].

A similar consideration was used to derive Eq. (19) by Kelly and Street [9], and

$$f_1 = \tfrac{1}{2} 3^{(m+1)/2}, \tag{45}$$

whereas f_2 in Eq. (44) is a numerical constant indicating the effectiveness of the conversion from the creep of the unreinforced matrix to that of the composite with fibers located in the direction normal to the loading axis, and $f_2 = 0.25$ was assumed in the Miles and McLean model. From Eqs. (42),

(43), and (44), the total strain rate $\dot{\varepsilon}_\theta$ due to three creep modes is obtained as

$$\dot{\varepsilon}_\theta = \dot{\varepsilon}_L\{\cos^{2(n+1)}\theta - \tfrac{1}{2}\cos^{2n}\theta\sin^2\theta\}$$
$$+ \frac{f_1}{f_2}\dot{\varepsilon}_T(\sin\theta\cos\theta)^{m+1} + \dot{\varepsilon}_T\sin^{2(m+1)}\theta. \tag{46}$$

Johnson [21] proposed a more general approach to predict the anisotropic creep behavior of unidirectional MMC. The Johnson model is based on the concept of a scalar potential, Φ, which takes account of the energy density due to creep. Johnson constructed Φ for anisotropic material, from which three-dimensional creep strain-rate components versus creep stress components were derived. It was found in the Johnson's analysis that the three basic creep modes discussed earlier are good enough for the description of the creep behavior of a transversely isotropic composite such as unidirectional MMC (Fig. 14). In Johnson's formulation, the three basic creep modes are given by

for longitudinal normal loading,

$$\dot{\varepsilon}_L = \lambda^m\sigma_L^{2m-1} \tag{47a}$$

for transverse normal loading,

$$\dot{\varepsilon}_T = \mu^m\sigma_T^{2m-1} \tag{47b}$$

for longitudinal shear loading,

$$\dot{\gamma} = \tfrac{1}{2}v^m\tau^{2m-1} \tag{47c}$$

In these equations, λ, μ, v, and m are material constants that can be determined by three different creep mode tests, for example σ_L, σ_T, and σ_θ loadings. Johnson's model yields the relation between the creep stress (σ_θ) and strain rate ($\dot{\varepsilon}_\theta$) for off-axis loading,

$$\dot{\varepsilon}_\theta = F^{(n+1)/2}\sigma_\theta^n, \tag{48}$$

where n = 2m − 1, and

$$F(\theta) = \lambda\cos^4\theta + (v - \lambda)\sin^2\theta\cos^2\theta + \mu\sin^4\theta. \tag{49}$$

In Fig. 16, the prediction by the Miles–McLean model, Eq. (46) is shown as a solid curve, whereas that by the Johnson model, Eq. (48), is given by the dash–dot curve. Also shown in Fig. 16 are three sets of dash curves, a, b, and c, representing three basic models, i.e., due to σ_L, τ, and σ_T loading, respectively. It follows from Fig. 16 that the composite creep rate can be predicted accurately by both models although the Johnson model appears to agree better with the experiment over the entire range of θ. In fact, Miles and McLean used the values of material constants in the Johnson model that are different from those used by Johnson. If they had used the same values as Johnson, the agreement between the experiment and Johnson's prediction appears to have been even better.

The preceding discussion of the composite creep behavior in terms of creep rate and creep stress may not be useful for a design engineer who wants to assess the durability of a candidate MMC under creep loading. In this case, stress (σ_c) versus rupture time (t) will provide more useful information to the design engineer. The stress versus rupture time data have been reported on selected MMC systems; tungsten/copper (W/Cu) [3], W/superalloy [22–24] and B/Al [25]. Among the MMC systems investigated, the W/superalloy system has been studied extensively by the researchers at NASA Lewis Center [3, 22]. Petrasek and Signorelli reported three superalloy-based MMC

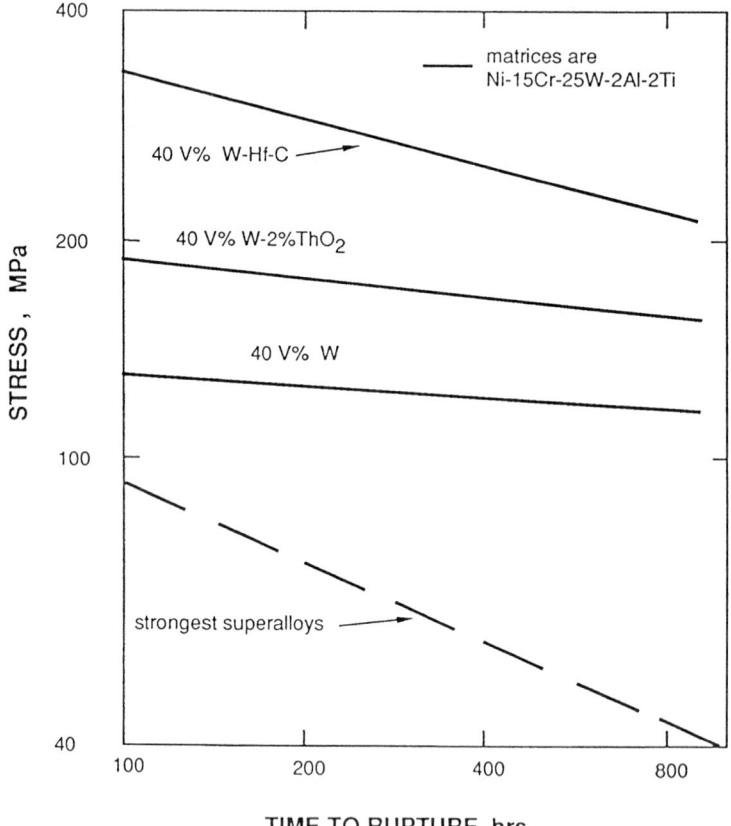

FIG. 17.   Stress versus time-to-rupture comparison for superalloys and their composites. Reprinted from [22].

systems, nickel-, cobalt-, and iron-based MMCs, where a number of W-based fibers were used as a reinforcement. Figure 17 shows some of the results reported by Petrasek and Signorelli, the stress versus time-to-rupture of Ni-based MMC with several types of W-base fibers, tested at a temperature of 1093°C. It appears from this figure that 40% V_f W–Hf–C fiber/Ni–15Cr–25W–2Al–2Ti matrix composite exhibited the best creep resistance. The fiber content (V_f) dependence of rupture strength was studied by Dean [24] for W/Ni–3.7Ti–4.8Al–10Cr–20Co composite, and the results are shown in Fig. 18. Figure 18 shows the linear V_f dependence of the 100-hr rupture strength of this composite tested at 1000°C. The applied stress (σ_c) time-to-rupture (t) relation must be constructed for a fixed temperature (T). To obtain a complete picture of the creep behavior as a function of T and t, one can

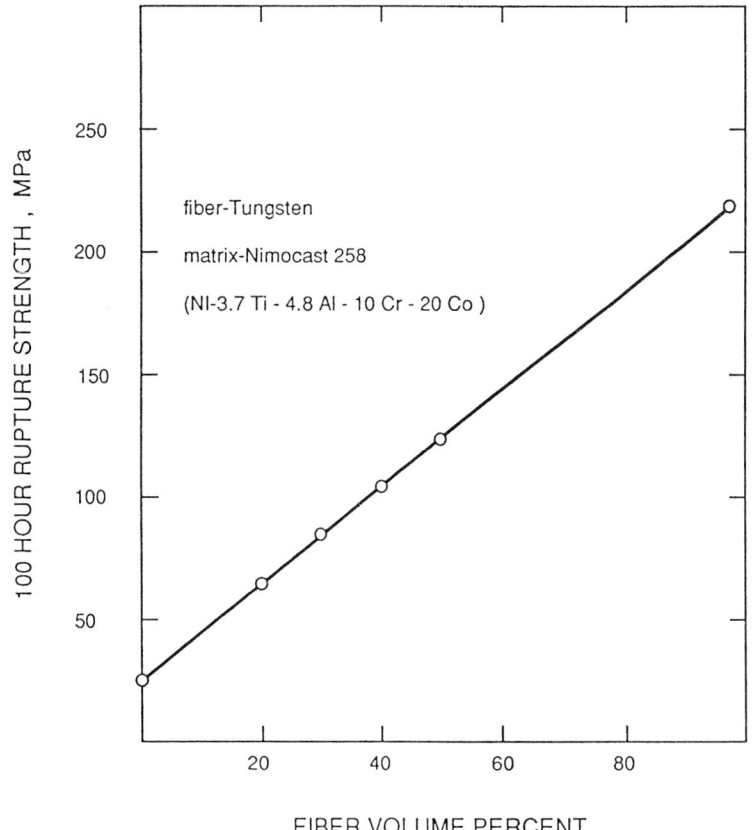

FIBER VOLUME PERCENT

FIG. 18. One-hundred-hour rupture strength versus fiber volume fraction (%) for tungsten/Nimocast 248 composite. Reprinted from [24].

use the applied-stress Larson–Miller parameter relation, where the Larson–Miller parameter (P) is usually defined by

$$P = T(20 + \log t_1),\qquad(50)$$

where T is the temperature in any unit (F, R, C, or K), and t_1 is the time to rupture or to specified elongation, for example, 1% strain and is usually expressed in hours. Then the creep data tested at different temperatures can be converted to a single curve. An example of the σ_c–P relation is shown in Fig. 19 [26], where T is defined by Kelvin, and t (hours) is the time to 3% strain. It should be noted here that the σ_c–P relation of a MMC is valid, provided the microstructure of the fiber and matrix would not change during creep loading.

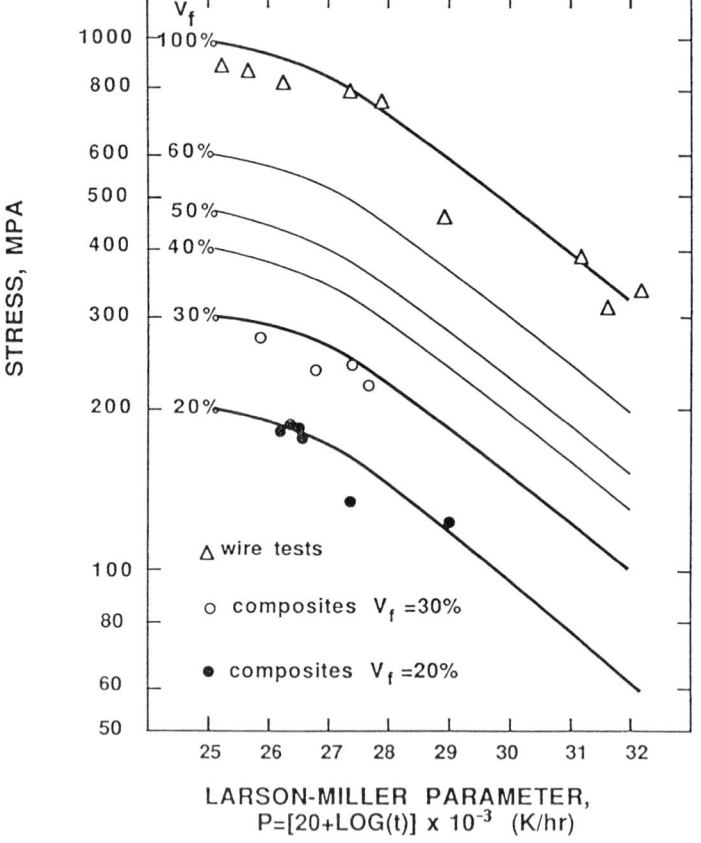

FIG. 19. Stress versus Larson–Miller parameter of W/Cu composite and W wires [26].

References

1. M. Taya and H. Lilholt, in "Advances in Composite Materials and Structures" (S. S. Wang and Y. Rajapakse, eds.), ASME, in press.
2. T. Morimoto, T. Yamaoka, H. Lilholt, and M. Taya, *J. Eng. Mater. Tech.*, **110**, 70 (1988).
3. D. McDanels, R. A. Signorelli, and J. W. Weeton, NASA Rept. No. TN-4173, 1967.
4. E. Bullock, M. McLean and D. E. Miles, *Acta Metall.*, **25**, 333 (1977).
5. M. McLean, "Directionally Solidified Materials for High Temperature Service," Chap. 6. p. 225. The Metal Society, London, 1983.
6. A. R. T. DeSilva, *J. Mech. Phys. Solids*, **16**, 169 (1968).
7. S. T. Mileiko, *J. Mater. Sci.*, **5**, 254 (1970).
8. D. McLean, *J. Mater. Sci.*, **7**, 98 (1972).
9. A. Kelly and K. N. Street, *Proc. Roy. Soc. Lond.*, **A328**, 283 (1972).
10. H. Lilholt, "Fatigue and Creep of Composite Materials" (H. Lilholt and R. Talreja, eds.), pp. 63. Risø National Laboratory, Roskilde, Denmark.
11. B. H. Choi, K. Wakashima, and T. Mori, in "Proc. 8th Risø Intl. Symp. on Constitutive Relations and Their Physical Basis (S. I. Anderson, J. B. Blide Sørensen, N. Hansen, T. Leffers, H. Lilholt, O. B. Pedersen, and B. Ralph, eds.), p. 259. Risø National Laboratory, Denmark, 1987.
12. T. Sakaki and M. Taya, unpublished work.
13. T. G. Nieh, *Metall. Trans.*, **15A**, 139 (1984).
14. H. Lilholt and M. Taya, in "Proc. ICCM6/ECCM-2" (F. L. Matthews, N. C. R. Buskell, J. M. Hodgkinson, and J. Morton, eds.), Vol. 2, p. 234. Elsevier Sci. Pub., 1987.
15. T. G. Nieh, K. Xia, and T. G. Landon, *J. Eng. Mater. Tech.*, **110**, 77 (1988).
16. M. McLean, in "Proc. Materials and Engineering Design—The Next Decade" (B. F. Dyson and D. R. Hayhurst, eds.). The Inst. of Metals, London, 1989.
17. G. Goto and M. McLean, *Scripta Metall.*, in press.
18. J. D. Eshelby, *Proc. Roy. Soc. London*, **A241**, 376 (1957).
19. T. Mura, "Micromechanics of Defects in Solids," 2d ed. Martinus–Nijhoff Publishers, The Hague (1987).
20. D. E. Miles and M. McLean, *Metall. Sci.*, **11**, 563 (1977).
21. A. F. Johnson, *J. Mech. Phys. Solids*, **25**, 117 (1977).
22. D. W. Petrasek and R. A. Signorelli, "Ceramic Eng. Sci. Proc." p. 739. American Ceramics Society, July–Aug., 1981.
23. I. Ahmad, D. N. Hill, J. Barranco, R. Warenchak, and W. Herffernan, "Advanced Fibers and Composites for High Temperatures" (I. Ahmad and B. R. Norton, eds.), p. 156–174. The Metallurgical Society of AIME, Walendale, Pennsylvania, 1980.
24. A. V. Dean, *J. Inst. Metall.*, **95**, 79 (1967).
25. E. M. Breinan and K. G. Kreider, *Metall. Trans.*, **1** (1970).
26. L. O. Larsson and R. Warren, "Advanced Fibers and Composites for High Temperatures" (I. Ahmad and B. R. Norton, eds.), pp. 108–125. The Metallurgical Society for AIME, Walendale, Pennsylvania, 1980.

7

Fracture Toughness of Particulate
Metal Matrix Composites

D. L. DAVIDSON

Southwest Research Institute
San Antonio, Texas

I. Introduction

Although there are benefits to be derived from particulate-reinforced composites, there are also some drawbacks in many of the materials that have been produced to date. One of the properties that has been of the most concern is fracture toughness. This property, one of the principal factors used in damage-tolerant design of structures, is, in general, not as large as for unreinforced matrix materials. Materials engineers have, therefore, attempted to increase the fracture toughness of these composites to satisfy design criteria, which has led to investigations of the origins of fracture toughness. The goal of this research has been to understand fracture sufficiently well to predict fracture toughness from the fundamental physical and mechanical properties of the composite constituents. The purpose of this section is to examine the factors that contribute to the fracture toughness of materials in general, and specifically to particulate composites, and to assess our level of understanding of these factors.

217

Fast fracture of a flawed body occurs when the driving force imposed on the crack exceeds the ability of the material to resist rapid crack extension. Fracture toughness is thought to be a material property, and the American Society for Testing and Materials has expended considerable effort standardizing tests for this property [1, 2]. A proper fracture toughness test measures the load and crack length at the point of unstable crack growth for a specimen with a manufactured flaw. The fracture toughness of a material is often given the symbol K_C or J_C, although other terms are also used to specify fracture toughness determined under special conditions, such as the fracture toughness of thin sheets, K_Q, or that for thick sections, K_{IC} and J_{IC}. The loading rates used for these tests are quasistatic; if rapid loading is used, the dynamic fracture toughness is determined. Fracture toughness can be defined as the condition required for initiation of material tearing (stable crack extension), J_C, if the material exhibits that behavior, but it is defined as the stress intensity factor required for rapid, sustained (uncontrolled or unstable) crack growth, K_{IC}, if the material does not tear stably. The materials considered in this section are relatively brittle and are best described by the latter behavior; thus, fracture toughness will be considered to be described adequately by K_C. Whether or not the values of fracture toughness measured by various investigators can be considered to be the plane strain value of fracture toughness K_{IC} is not known. Often, toughness measurements were made on specimens of varying size and shape, and this has been particularly true of these composites because these materials are in their developmental stage. A review of reported values of fracture toughness for these materials, as of mid-1988, may be found in the work of Goolsby and Austin [3]. Also, an interesting discussion of fracture toughness in composites in general, including continuous-fiber and whisker reinforcement, as well as particulate reinforcements, has been given by Friend [4]. The main point made by Friend is that no single parameter seems to be capable of describing the toughness response of metal matrix composites, and that both initiation and propagation resistance must be considered in a proper assessment of fracture toughness of composites.

II. Description of Materials

The composites of interest in this section have matrices of ductile alloys (elongations to fracture of 5 to 10%) and particles not small enough to be considered as dispersoids (larger than 1 μm), yet not exceeding about 100 μm in diameter. Chemical composition of the reinforcements are well defined and controlled, unlike the constituent particles found in some commercial materials, e.g., the aluminum alloys. The moduli of reinforcements thought to be useful in composites are higher than those of the matrix, which, to date, has meant that the particulate has been ceramic, or ceramiclike. Therefore, the composite is composed of particles that deform elastically in a matrix that is capable of plastic deformation. Typical microstructures of these materials are shown in Figs. 1 and 2.

Characterization of these composites microstructurally is difficult ‹and requires a significant effort. Important particulate characteristics include composition, size distribution, volume fraction and dispersion. Matrix characteristics of interest are typical of the usual metallurgical description of microstructure, and may include such factors as composition, grain size,

FIG. 1.   Metallography of the mechanically alloyed composite IN-9052 + 15 vol-% silicon carbide at progressively higher resolution. (a) and (b) secondary electron images in the scanning electron microscope, and (c) transmission electron microscope image. Dark objects in (c) are silicon carbide particles.

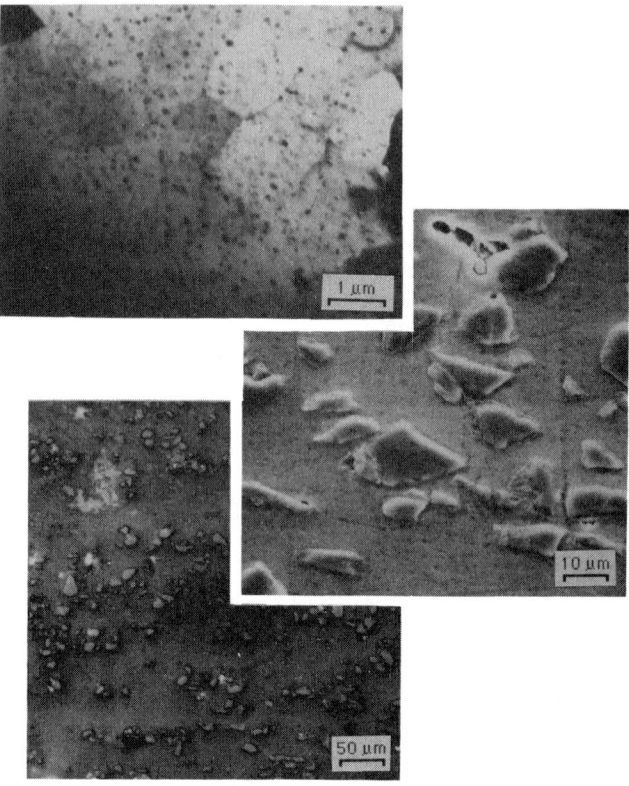

FIG. 2. Metallography of the cast and extruded composite 2014 + 15 v/o silicon carbide at progressively higher resolution: (a) and (b) secondary electron images in the scanning electron microscope, with (a) having reverse contrast to (a), and (c) transmission electron microscope image. Dark objects on the left and right borders of (c) are silicon carbide particles.

precipitate and dispersoid content and volume fraction, dislocation density, and subgrain size. Optical microscopy, and scanning and transmission electron microscopy are all required for a proper characterization of these composites. Machine processing of images made over a wide range in magnifications is usually necessary to characterize the particulate, whereas transmission electron microscopy is almost always required for matrix characterization.

In addition to the microstructural description, these materials also have interfaces that require characterization, both chemically and mechanically. However, the techniques for interfacial characterization are not nearly as advanced as are those for microstructure. Auger electron spectroscopy may be used for elemental analysis, which is only a partial description of interfacial chemistry. Analytical transmission and scanning transmission electron microscopy may also be used for structural and elemental analysis, but these techniques are difficult and sample only minute amounts of available material [5]. Determinations of interfacial bonding strength are difficult and techniques for measurement of this property are still being developed.

These composites have residual stresses of considerable magnitude attributed to the difference in coefficient of thermal expansion between the matrix and reinforcement. These residual stresses are three-dimensional in nature, are difficult to calculate and measure, and methods are still being evolved to assess their importance. However, to a first approximation, residual stresses are of less importance to fracture toughness than to other composite properties (i.e., yield stress), although they may cause secondary effects such as increasing the dislocation density in undeformed material.

III. Mechanisms of Fast Fracture

Fracture toughness specimens are usually precracked by initiating a fatigue crack from a sharp notch. The conditions that initiate crack tearing and rapid growth are dependent on a number of factors. The load level at which initiation occurs depends on fatigue-crack growth parameters, and the thickness of the specimen is also an important factor. These important details of testing are related to the magnitude of the fracture toughness values determined, and they must be considered carefully in interpreting experimental data. However, they will not be considered further here. Rather, toughness of the composite in the context of this section is considered as the work of fracture—the dissipation of the elastic energy imposed in the specimen by the externally applied load per unit of new crack surface formed.

As a crack grows rapidly through a particulate-reinforced composite, many events occur that are potentially related to the fracture toughness, and each of these either make it easier or harder for the crack to extend. The events that have been identified as being related to resisting crack growth are shown in Fig. 3. Each of these is a mechanism for absorbing energy, which, in turn, requires more work to be done by the external load.

The energy absorbed by each mechanism is not the same, but to quantify each requires a detailed knowledge of the mechanism, and that does not exist currently. Some estimates of the energy absorbed by several of the mechanisms will be given in later sections.

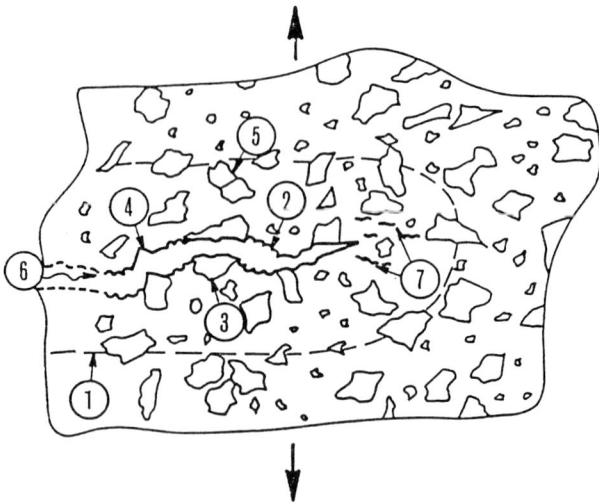

FIG. 3. Energy absorption mechanisms for metal matrix composites with discontinuous reinforcement: (1) deformation within plastic zone; (2) formation of voids along fracture surface. Voids have not been found to form within plastic zone away from fracture path; (3) fracture of SiC particles along crack path; (4) interfacial separation between matrix and SiC; (5) fracture of SiC within plastic zone; (6) tortuous fracture path increases fracture surface area; (7) matrix cracks near, but not continuous with, the main crack.

IV. Origins of Fracture Toughness

Mechanistic descriptions of fracture toughness go back to Griffith and his research on the fracture of glass. He was able to relate the stress, σ_{L}, required to break glass to the size of the flaw, a, the energy of new surface created, γ, and the modulus E, given by the relation

$$\sigma_{\mathrm{L}} = (2E\gamma/\pi a)^{1/2}. \tag{1}$$

Since glass has no measurable plasticity at ambient temperature, fracture is related solely to the breaking of atomic bonds, so if this concept is to be used in conjunction with a deformable material, it must be modified. This was done by Irwin and Orowan by extending the definition of surface energy to include the energy absorbed by plastic processes in the creation of new fracture surface—an "effective surface energy." A well-written review of the equivalence between mechanistic theories of fracture and those based on thermodynamics for purely elastic and elastic–plastic solids may be found in the text of Lawn and Wilshaw [6].

Fracture toughness, K_C, is related to the work done within the material per unit area of new surface created, G_e, by

$$G_e E = K_c^2, \qquad (2)$$

where the magnitude of G_e is determined by all of the energy-dissipating mechanisms that are operating as the crack grows, as listed in the previous section. Another interpretation of G_e is that it is the same as J_c at the initiation of stable tearing.

The energy absorptive mechanisms of Fig. 3 are listed approximately in the order of their magnitude for relatively brittle materials, to the extent that they can be estimated at this time. These mechanisms will now be taken in turn, and their energy absorptive capacity estimated.

A. Deformation within the Plastic Zone

Most, but not all, of the energy absorbed within the plastic zone of the growing crack occurs within the matrix. A small amount of energy is absorbed within the reinforcement. However, this energy is elastic and will be recovered as the crack passes and the stress concentration associated with the crack tip is diminished. Therefore, the energy *dissipated* as the crack tip passes will occur entirely within the matrix.

To estimate the level of plastic-zone energy dissipation, it is necessary to follow the load–unload sequence of an element of material within the plastic zone of the passing crack. A model for this [7] is depicted schematically in Fig. 4, where a crack of length c is growing under the application of a remotely applied stress field. Figures 4a and b show the motion of the plastic zone as the crack tip extends from c_1 to c_3. By considering the steady-state condition of crack growth, some of the complexity of the rise and fall in strain can be eliminated. If the strain in an element of material just ahead of c_2 is followed, it goes through the entire stress–strain history shown in Fig. 4c, dissipating an energy W, which is the area beneath the curve. Material elements between c_1 and c_2 will be unloaded along the path 3 to 4 as the crack tip moves from c_1 to c_3, and material elements between c_3 and c_4 will be loaded along the path 1 to 3. By considering only the steady state, it is not necessary to consider any of these elements, because loading of the elements between c_3 and c_4 just offset those that occurred between c_1 and c_2. Stated in another way, energy dissipation within the plastic zone c_1 to c_2 just offsets that in c_3 to c_4 when steady-state crack growth is considered.

To sum up the contributions of all elements within the plastic zone, it is necessary to know the strain distribution *parallel to the loading axis*, in the

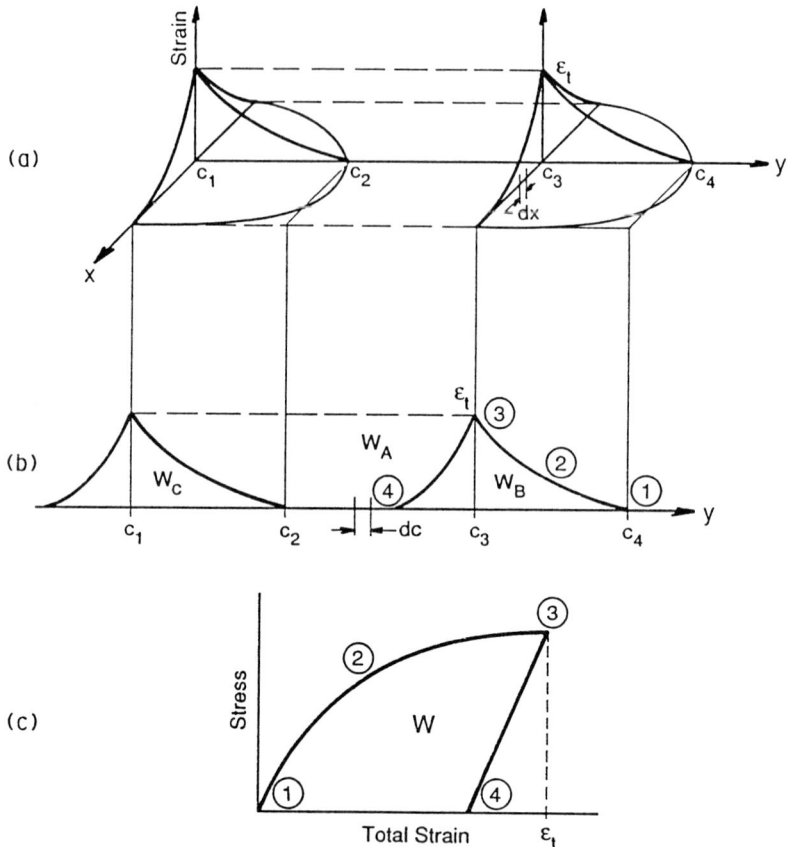

FIG. 4. Method for calculating the work adsorbed in the plastic zone during extension of the crack from c_1 to c_3: (a) is in strain-coordinate space, and (b) is a similar view showing the location of the points in the strain history of an element of material along the crack path; (c) shows those same points on a tensile stress–strain curve.

x-direction in Fig. 4, which is approximately perpendicular to the direction of crack growth. This strain distribution might be a power function as given theoretically by the Hutchinson–Rice–Rosengren analysis [8] for a stationary crack, or a logarithmic distribution, which is given by the theoretical analysis for a growing crack [9]. Examination of these predicted distributions by experiment cannot be found in the literature, and even if such work had been done for unreinforced materials, there might be an effect of reinforcement that is unaccounted for in the theories. Therefore, measurements were made on composites to determine this strain distribution.

   Several typical measured strain distributions are shown in Fig. 5 for fatigue cracks that were being cycled at a ΔK level just below K_c. Strain distributions having the form of both logarithmic and power function have been found experimentally, but a power function having a slope of approximately -1 has been found the most often. Use of either of these strain distributions in an estimate of the work done in the plastic zone gives about the same value. The important factor in this calculation is the magnitude of the strain at the

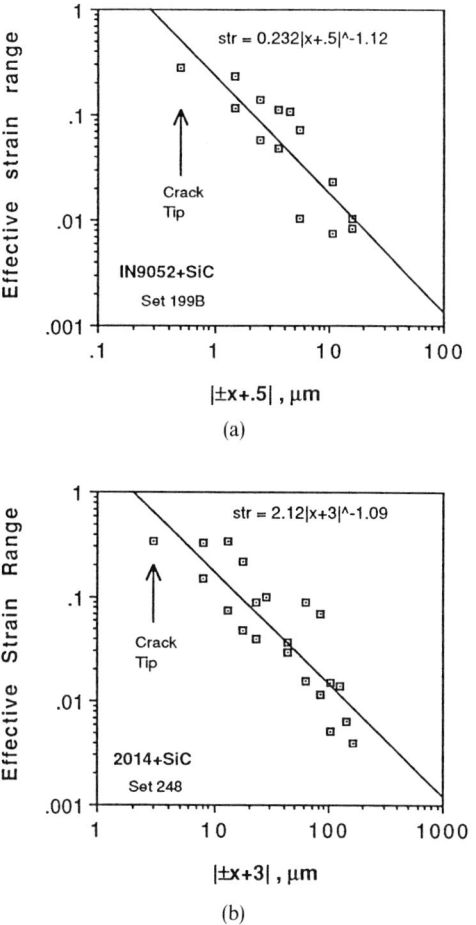

FIG. 5.   Strain distributions parallel to the loading axis for cracks in two metal matrix composites: a crack in (a) mechanically alloyed IN–9052 + 15 v/o SiC, and (b) peak aged 2014 + 15 v/o SiC. By addition of a small increment to the axis, the crack-tip strain value may be shown on this distribution. The function that describes the least-squares line fit to this data is shown on the figure.

crack tip because this specifies the limit of integration on the stress–strain curve as well as the size of the plastic zone. For a given level of crack-tip strain, the strain distribution function alters only the plastic-zone size, but that has much less influence on the work done during fracture than the crack-tip strain. The work done in the outer part of the plastic zone is much less than that done near the crack tip even though the volume of material is larger.

The hypothesis that fracture toughness can be estimated from the energy dissipated in the plastic zone during fast fracture has not been tested enough experimentally at this time to determine its validity for relatively-low-toughness materials. For the aluminum alloy composite IN–9052 + 15 v/o (vol-%) SiC, as made by powder metallury techniques, there was agreement between hypothesis and experiment [6]. However, for a cast-extruded composite of 2014 + 15 v/o SiC, the estimate of plastic-zone energy dissipation significantly underestimated that measured [10], but this material indicated a considerable increase in K_C as compared with other, similar materials, that has not been explained. This hypothesis has been applied also to fracture of an unreinforced titanium aluminide alloy and found to be compatible with measured toughness; however, the strain at the point of fracture was not measured for this case. In fact, experiments to determine the strain distribution at the onset of final fracture have proven difficult to perform; thus, experimental examination of the hypothesis is incomplete.

B. Formation and Growth Microvoids

This is a very important energy absorption mechanism. For ductile metals, such as annealed copper and austenitic stainless steels, the growth and coalescence of microvoids during fracture is by far the dominant energy-absorbing mechanism. Voids that form during fracture are found as dimples on the fracture surface, where their diameters may be directly measured. The models that exist for computing energy absorption by this mechanism are elementary and full of assumptions that have not been substantiated experimentally. The energy absorbed in void coalescence and growth is directly dependent on the size of the voids and their aspect ratio, or depth. Measurements of void size and shape are difficult because they are made usually from fracture surfaces, which are rough, thereby requiring extensive analysis, such as that done by photogrammetry of stereopairs.

Evaluation of the importance of the void mechanism to fracture toughness is begun by determining the coverage and size of dimples from the fracture surfaces generated by rapid crack growth. For particulate-reinforced composites, this factor varies widely with material, with the coverage being nearly zero for some materials [*11*], but a large fraction of the surface for other materials [*12*]. The causes for this level of variation in these composite materials are not known, but they are probably due to matrix composition and processing; no established correlation between these factors and microvoid formation characteristics currently exists.

All of the composite materials that this investigator has had the opportunity to examine have evidenced only a small fractional coverage by microvoids, and the size of the microvoids found have been very small [*7, 10*]. Examples of these microvoids and their coverage of the fracture surface are shown in Fig. 6. Typical of the microvoids found in the materials that I

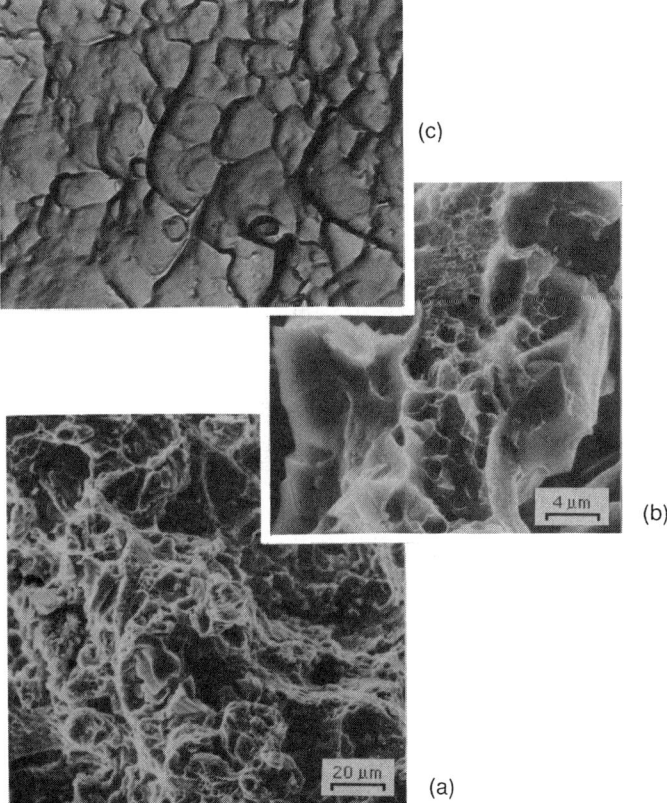

FIG. 6.   Fractography of fast fracture in metal matrix composites: (a) shows patches of very small dimples surrounded by more brittle-looking regions; (b) shows a similar surface at higher magnification; no arrows on photographs, therefore leave out phrase; (c) is a two-stage plastic-carbon replica of a dimpled region at much higher resolution.

have examined are those shown in Fig. 6b. The energy absorbed during the formation of these microvoids was estimated [7] to be $6\text{--}13 \times 10^{-5}$ MJ/m^2, which is a small fraction of the energy dissipated within the plastic zone.

Microvoids are often considered to be initiated by particles in the material, such as intermetallics and dispersoids. The interface between these particles and the matrix fails at some point in the stress history of the material in their vicinity; whether this interfacial failure is stress- or strain-controlled is not well known. The result is that the particle that initiated the microvoid is found at the bottom of a dimple on one of the fracture surfaces; thus, about half the dimples in microvoids so initiated should contain particles. The microvoids shown in Figs. 6b and 6c have no such particles and were thought to have been initiated at subgrain boundaries. Subgrains of about the same size as dimples were found in the material of Fig. 6c by transmission electron microscopy [7]. Likewise, other particulate-reinforced composites showing dimples on the fracture surface have little evidence of particle-initiated microvoids. Tear ridges also are found on fracture surfaces of these composites; a tear ridge is a large, curvilinear region that may or may not close on itself, and which has deformed down to approximately a sharp edge in the final fracture process. Arrows indicate tear ridges in Fig. 6. If a tear ridge described a small circle, it would be said to be the edge of a dimple. The energetics of tear-ridge formation is not as well assessed as for dimples. Energy is expended in the formation of tear ridges, but it is presently neglected in the analysis of fracture toughness because of an inability to describe the parameters associated with this mechanism.

Clearly, one of the best methods of ensuring an acceptable level of fracture toughness for particulate-reinforced composites is to manufacture material that creates relatively large microvoids over much of the surface during fracture. The methods by which this is to be accomplished are still to be determined.

As an example of the importance of the void-forming mechanism to fracture, the energy absorption estimates discussed earlier for void formation and plastic-zone formation were estimated for IN–9052 + 15 v/o SiC and are shown in Fig. 7. This analysis emphasizes the importance of fractional coverage and void size in determining the dominant mechanism for fracture toughness.

C. Debonding of Interfaces to Cause Microcracks near the Main Crack

Several theoretical continuum analyses [13, 14] have hypothesized that the presence of a large number (a "mist") of microcracks near the main crack would effectively toughen the material because these microcracks have the

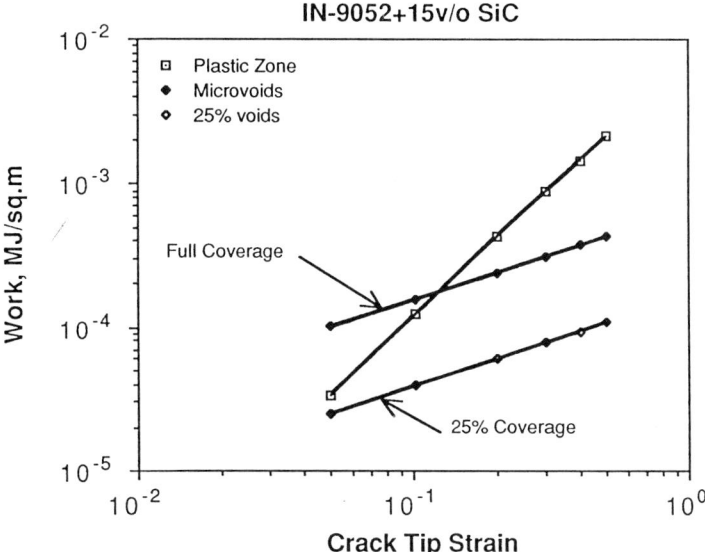

FIG. 7. The work dissipated during fracture, computed by using a model [7], as a function of crack tip strain. Trends are shown for work done in the plastic zone and in formation of a void sheet (at two levels of coverage) for small voids typical of those illustrated in Fig. 6. If voids fully cover the surface, then the dominant mechanism depends upon how much strain can be sustained at the crack tip.

effect of blunting the main crack, in the sense that the stress field experienced at the main crack tip is lowered. Thus, the stress intensity at the main crack tip caused by an applied external stress would be lowered by the formation of these microcracks. In some particulate composites, microcracks have been found to form near the main crack tip, at least at some levels of stress intensity factor, and at least some of these cracks are in the interface between particle and matrix. Are these microcracks beneficial, as they might be thought to be from continuum theory?

As has been shown by Rose [15] and Kachanov [16], the location of specific microcracks relative to the main crack tip is very important to the influence of that particular microcrack on the stress intensity factor of the main crack tip. Experimental examination of this effect [10] has shown that the microcracks that actually form do, in fact, have a negligible effect on crack-tip stress intensity factors because they formed too far ahead of the crack tip. If anything, these cracks were found to be antishielding because they formed in the zone ahead of the crack tip. This result agrees with the analysis of Shang and Ritchie [17], on what they termed "crack bridging." Furthermore, microcracks were found to be just as likely to form in the matrix, apparently unassociated with particles, as they were to form at particle–matrix interfaces. The concept of weakening the interface in these particulate composites to promote microcracking is, therefore, questionable; this effect might be used to promote microvoid intitiation and growth rather than microcracking if the particles are small enough and are well dispersed.

D. Formation of New Surface—The Effect of Fracture Surface Roughness

Fracture surface energy is the dominant contribution to fracture toughness for materials that do not plastically deform. For most materials, the magnitude of surface energy is ≈ 2 J/m^2. This level of energy absorption for deformable materials, such as the metal matrix composites, is several orders of magnitude less than the other fracture mechanisms. But the hypothesis has been forwarded [18] that an increase in the roughness of the fracture surface might have a large effect on fracture toughness.

For particulate metal matrix composites, this concept has been experimentally examined by measuring the fracture surface roughness and comparing it to the fracture toughness [11]. Roughness was quantified by using the surface roughness factor defined by Underwood, and the fractal characteristics of these fracture surfaces were also measured. Neither the surface roughness factor nor the fractal dimension could be correlated with the fracture toughness for the particulate composites studied. This result is not surprising considering the many mechanisms of fracture contributing to fracture toughness and the relatively small magnitude of the surface energy.

V. Effects of Particle Reinforcement on Fracture

As discussed in the foregoing sections, there are a number of factors contributing to the fracture characteristics of particulate-reinforced composites that are not well understood. The addition of particles to ductile matrices can result in at least the following changes to the material behavior.

(A) Slip characteristics of the matrix alloy are altered. Particles will have the primary effect of blocking slip lines, but there may also be secondary effects, such as limiting the operation of secondary dislocation sources and limiting the extent of cross slip. These effects will have such consequences as limiting the size of the plastic zone surrounding a crack tip, thereby decreasing the volume of deforming material within it. Since the volume of deforming material is at least partly responsible for the magnitude of the fracture toughness, this would be a major effect. However, a detailed assessment has shown that plastic-zone size is not changed very much for fatigue cracks in composites at relatively large ΔK (approaching K_c) when compared with unreinforced aluminum alloys.

(B) The presence of particles near the crack tip could also limit the strain to fracture at the crack tip. This, combined with limitations on plastic-zone size, might alter the distribution of strain within the plastic zone; however, this has not been found in the limited work done to date, as evidenced by strain distributions such as those shown in Fig. 5.

Strain at the crack tip does appear to be limited, in comparison with unreinforced material [7]. The observed brittle nature of the matrix when reinforced with particles, as compared with unreinforced material, may be evidence of the slip-limiting characteristics of the particles. It has been found that the fatigue-crack threshold stress intensity factor, ΔK_{th}, may be computed from the mean free slip of dislocations through the matrix and the yield stress [19]. This is considered to be further evidence of the slip-limiting characteristics of adding particles to the matrix, but it is not completely clear that this limitation continues with increasing K level up to K_C.

(C) The anticipated effect of adding stiff particles to a matrix is an increase in yield stress for the composite. For some particle-reinforced aluminum composites, this effect has been found, whereas for others it has not. These disparate results have not been satisfactorily explained, but there are a number of factors that could influence a change in yield stress besides just the presence of particles. These include matrix residual stresses and an increase in dislocation density in the matrix due to thermal expansion mismatch between particulate and matrix, alteration of the recrystallization characteristics of the matrix during thermal treatment, which would control the size of grains and formation of subgrains, and the alteration of precipitation kinetics and distribution of precipitate particles. Conversely, several of these factors may have little influence on the deformation of the matrix during final fracture, e.g., residual stress and precipitation kinetics. Yield stress will have some influence on fracture toughness through alteration of plastic-zone size, but a relatively large change in yield stress would be required to cause much effect.

(D) The debonding of particles near the crack tip to form microcracks not directly connected to the main crack is another mechanism that has the potential to influence the fracture toughness. The magnitude of this effect may be largely controlled by the size of the particles and their bonding with the matrix. Location of the microcracks is also important. If microcracking is confined to the particle–matrix interface, and if the particles are small, the effect may be promotion of microvoid initiation, which has a potent influence on fracture toughness.

The increase in fracture toughness of high-impact polystyrene (HIPS) occurs because of the microcracking mechanism. Debonding between matrix and the dispersed rubber particles causes numerous cracks to occur that lower the stress intensity factor at the crack tip. This effect is illustrated in Fig. 8. Whether or not the conditions necessary for this type of fracture-toughening mechanism can be created at a fast moving crack tip in a metal matrix composite has yet to be answered.

(E) The influence of particulate fracture on composite fracture toughness has been the subject of considerable debate. Hard, brittle particles within the plastic zone of a growing crack deny the plastic zone of material that could plastically deform. The work-to-fracture ratio should be decreased in proportion to the volume fraction of the particulate. The elastic distortional energy of these particles is released once the crack tip has passed. However, if the particle should break, then, at least, the energy of new surface formation is absorbed. But, because this energy is so low, the effect on fracture toughness is negligible. If the particle is small and contains a large enough flaw to break far enough ahead of the crack tip to form a microvoid, then the energy effect would be greater. For most of the particulate systems thus far examined, particle size has been too large to enhance void formation, and the number of fractured particles has been relatively small. Therefore, the overall effect of particle fracture on these composites has been negligible.

(F) It might be argued that the effect of particle fracture would be to increase surface roughness, and that would, in turn, enhance fracture tough-

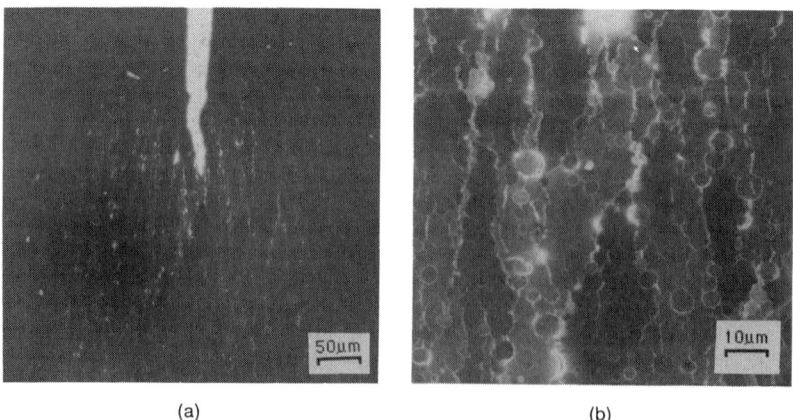

(a)   (b)

FIG. 8.   Multiple cracking induced at the tip of a fatigue crack in high-impact polystyrene, showing the energy-absorbing characteristics of a "mist" of microcracks, (a) showing the extent of the cracking, and (b) showing the sizes of the cracks immediately at the crack tip (at top center of photograph). This mechanism, although a possible method for increasing fracture toughness in particulate-reinforced composites, has not yet been employed.

ness because of the increase in fracture surface area created. Evidence to date indicates that the surface roughness does not have much influence on fracture toughness [11].

In summary, it appears as though the major influence of particles on fracture toughness is related to the effects that these particles have on the slip characteristics of the matrix under large-strain conditions. This issue has not been explored sufficiently either theoretically or experimentally for an understanding to emerge.

VI. Predicting Fracture Toughness

One of the objects of materials science is to understand and predict material behavior from fundamental considerations. This goal has not been achieved yet for predicting fracture toughness of particulate composites, specifically the ceramic-reinforced aluminum alloy materials, nor has it been achieved for any composites [4]. This section has attempted to identify the components that may contribute to fracture toughness and estimate their effects. For most of these mechanisms, neither adequate experimental nor theoretical assessment is available. It should be kept in mind that there is not even good agreement amongst investigators about the best test method to use for determining a material parameter termed "fracture toughness"; in fact, this "property" is not adequately defined by anyone, not even by the standards organizations.

References

1. ASTM Standard Test Method E399-83 "Plane-strain fracture toughness of metallic materials," Annual Book of Standards, Vol. 02.02. Am. Soc. for Testing and Mater., Philadelphia, Pennsylvania.
2. ASTM Standard Test Method E813-87 "J_{IC}, A measure of fracture toughness," Annual Book of Standards, Vol. 03.01. Am. Soc. for Testing and Mater., Philadelphia, Pennsylvania.
3. R. D. Goolsby and L. K. Austin. "Fracture Toughness of Discontinuous SiC Reinforced Aluminum Alloys," Proc. ICF-7, Vol. 4, Pergamon Press, pp. 2423–2435, 1989.
4. C. M. Friend, "Toughness in metal matrix composites," *Mater. Sci. Tech.*, **5**, 1–7.
5. R. W. Carpenter, "High Resolution Studies of Interfaces in Materials Science," pp. 400–403. Proc. Electron Micro. Soc. Am., San Francisco Press, 1986.
6. B. R. Lawn and T. R. Wilshaw, "Fracture of Brittle Solids," pp. 46–90. Cambridge University Press, Cambridge, 1975.

7. D. L. Davidson, "Fracture characteristics of Al-4 Pct Mg mechanically alloyed with SiC," *Metall. Trans. A,* **18A**, 2115–2128 (1987).

8. J. W. Hutchinson, J. R. Rice, and G. F. Rosengren, *J. Mech. Phys. Solids,* **16**, 1–31 (1968).

9. J. R. Rice, "Mathematical analysis of mechanics of fracture" in *Fracture,* Vol. II, eq. 287, p. 287. Academic Press, 1975.

10. D. L. Davidson, "Micromechanisms of fatigue crack growth and fracture toughness in metal matrix composites." Technical Report to the Office of Naval Research, Arlington, Virginia, 1989.

11. D. L. Davidson, Fracture surface roughness as a gauge of fracture toughness: aluminum-particulate SiC composites, *J. Mater. Sci.,* Vol. 24, pp. 681–687, 1989.

12. C. R. Crowe, R. A. Gray, and D. F. Hasson, "Proc. Fifth Int. Conf. on Composite Materials," ICCM-V, p. 843. TMS–AIME, Warrendale, Pennsylvania, 1985.

13. R. G. Hoagland and J. D. Embry, *J. Amer. Ceram. Soc.,* **63**, 404 (1980).

14. J. W. Hutchinson, *Acta Metall.,* **35**, 1605 (1987).

15. L. R. F. Rose, *Int. J. Fracture,* **31**, 233 (1986).

16. M. Kachanov, *Int. J. Fracture,* **30**, R65 (1986).

17. J. Shang and R. O. Ritchie, *Mater. Sci. Eng.,* (in press) (1989).

18. B. B. Mandlebrot, D. E. Passoja, and A. J. Paullay, *Nature,* **308**, 721 (1984).

19. D. L. Davidson, "The growth of fatigue cracks through particulate SiC reinforced aluminum alloys," *Eng. Fracture Mech.,* Vol. 33, pp. 965–977, 1989.

8

Fatigue of Continuously Reinforced Metal Matrix Composites

K. K. CHAWLA

Department of Materials and Metallurgical Engineering
New Mexico Institute of Mining and Technology
Socorro, New Mexico

I. Introduction

Fatigue is the phenomenon of mechanical property degradation leading to failure of a material or a component under cyclic loading. Understanding the fatigue behavior of composites of all kinds is of vital importance, because without such an understanding, it would be virtually impossible to gain acceptance of the design engineers. Many high-volume applications of composite materials involve cyclic-loading situations, e.g., automobile components. It would be a fair admission that this understanding of the fatigue behavior of composites has lagged behind that of other aspects such as the elastic stiffness or strength. The major difficulty in this regard is that the application of conventional approaches to fatigue of composites, for example, the stress versus cycles ($S–N$) curves or the application of linear elastic fracture mechanics (LEFM), is not straightforward. The main reason for this is the inherent heterogeneity and anisotropic nature of the composites. This

235

results in damage mechanisms in composites that are very different from those encountered in conventional, homogeneous, or monolithic materials. Despite these limitations, conventional approaches have been used and are therefore described briefly before describing some new approaches to the problem of fatigue in composites. Two types of continuously reinforced metal matrix composites (MMCs) are described: unidirectional, fiber-reinforced MMCs and laminated MMCs. This is followed by a description of hybrid composites. Finally, a brief description is made of the important problem of thermal fatigue in MMCs.

II. Types of Continuous Reinforcement in Metal Matrix Composites

There are two types of continuously reinforced MMCs:

(i) unidirectional, fiber-reinforced composites, and
(ii) laminated metal matrix composites.

We describe briefly these two types of MMCs.

A. *Unidirectional, Fiber-Reinforced Metal Matrix Composites*

This type of composite has fibers aligned in one direction. Figure 1 shows a longitudinal section of an alumina (FP) fiber-reinforced aluminum–2% lithium alloy matrix composite. They are frequently made by casting techni-

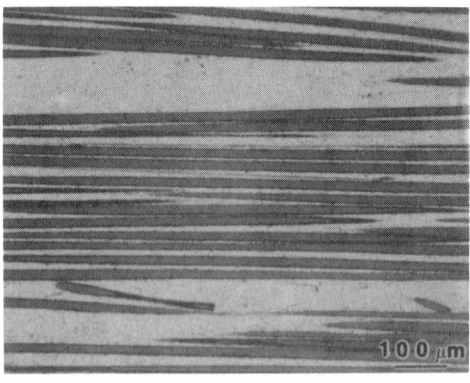

FIG. 1.   A longitudinal section of a unidirectionally aligned alumina (FP) fiber reinforced Al–2% Li matrix composite.

ques, although there are other techniques as well [*1*]. Not unexpectedly, this kind of reinforcement results in maximum strength and modulus improvement in the direction of reinforcement. This has its consequences in the fatigue behavior as well, as we shall describe in the following.

B. Laminated Metal Matrix Composites

There are two types of laminated MMCs:

(i) Metallic matrix containing fibers oriented at different angles in different layers. This type of MMC is similar to the polymeric laminates. One can vary the fiber orientation from ply to ply and stack the plies in any desired sequence.

(ii) Two or more different metallic sheets bonded to each other. In this type of MMC, one gets a bi-dimensional reinforcement. There are two types of arrangements of sheets, viz., crack divider and crack arrest, Fig. 2. In the crack-divider arrangement, at any given instant the crack front is shared by the two components, say Al 1100 and Al 2024, as shown in Fig. 2a. In the crack arrest configuration, the crack front resides in any one component at any given instant, Fig. 2b. The microstructure of an aluminum 1100/stainless-steel-type 304 laminated MMC is shown in the scanning electron micrograph in Fig. 3a. SS and Al indicate the stainless steel and aluminum, respectively. Figure 3b is a bright field transmission electron micrograph of an aluminum

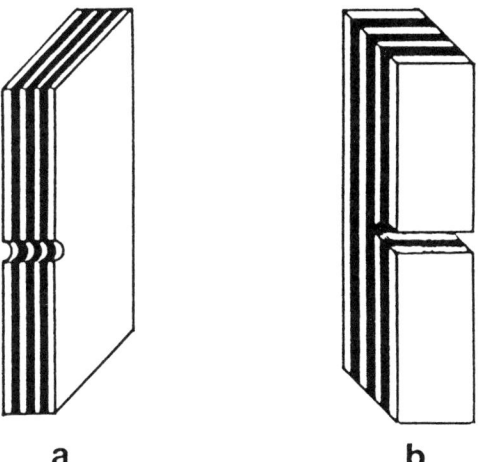

a   **b**

FIG. 2.   Crack-divider and crack-arrest geometries in sheet laminates.

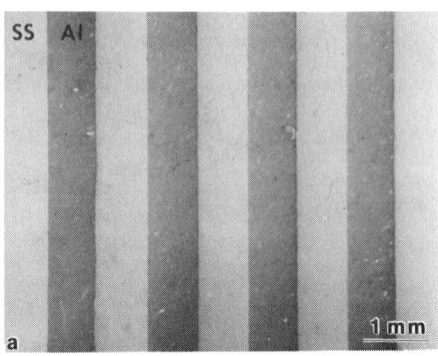

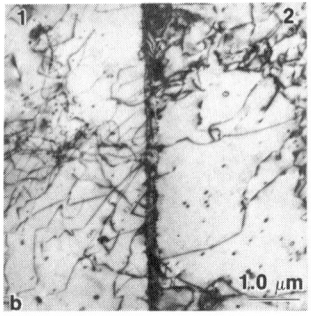

FIG. 3. (a) An aluminum 1100/stainless steel-type 304 laminated composite made by roll bonding (SEM). SS and Al indicate the stainless steel and aluminum, respectively. (b) Bright-field TEM micrograph of an aluminum 1100/aluminum 2024 laminate composite made by roll bonding. 1 and 2 designate Al 1100 and Al 2024, respectively. Note the integrity of the interface and a larger dislocation density in 1 than in 2.

1100/aluminum 2024 composite, indicated by 1 and 2, respectively. Both composites were made by roll bonding. Note the integrity of the interface between the two components and a larger density of dislocations in Al 1100 than in Al 2024.

III. Fatigue Behavior of Continuously Reinforced Metal Matrix Composites

There are many material variables that play an important role in the fatigue of MMCs; for example, the modulus, strength, ductility, and work-hardening characteristics of the constituents (fiber and matrix) as well as the characteristics of the interface.

The conventional approaches to fatigue involve $S-N$ curves and fatigue crack propagation studies. A novel approach involves the monitoring of damage accumulation. Specifically, measurement of modulus loss as a function of cycles has shown good results in polymer matrix composites. In the following, we describe some results of the conventional as well as not-so-conventional approaches to the problem of fatigue of MMCs.

A. S-N Curves

A very popular conventional fatigue testing technique used with metals involves determination of the so-called $S-N$ curves, where S is the stress amplitude and N is the number of cycles to failure. In general, for ferrous metals, one obtains a fatigue limit or endurance limit. For stress levels below this endurance limit, theoretically, the material can be cycled infinitely. If, however, such an endurance limit does not exist, for example for nonferrous metals, then one can arbitrarily define a certain number of cycles, say 10^6, as the cutoff value. The fatigue behavior of unreinforced metals can be conveniently divided into two stages: crack initiation and propagation. In high cycle–low stress fatigue of metals, most of the property changes occur in what has been called the small process zone at the crack tip. The rest of the material away from the crack does not seem to be affected seriously by the phenomenon of fatigue. In the low cycle–high stress fatigue of metals, the phenomena of work hardening and work softening are observed. Quite frequently, however, a rule-of-thumb approach in metal fatigue is to increase its monotonic strength, which concomitantly results in an increase in its cyclic strength. This rule-of-thumb assumes that the ratio of fatigue strength to tensile strength is about constant. It should also be noted that the maximum efficiency in terms of stiffness and strength gains in fiber-reinforced composites occurs when the fibers are continuous, uniaxially aligned and the properties are measured parallel to the fiber direction. As we go off-angle, the strength and stiffness drop sharply. Also, at off-angles, the role of the matrix becomes more important in the deformation and failure processes. One major drawback of studies of fatigue behavior of a material using this $S-N$ approach is that no distinction can be made between the crack initiation phase and the crack propagation phase.

Incorporation of fibers generally improves the fatigue resistance of any fiber-reinforced composite in the fiber direction. One would not expect high strength, brittle fibers such as carbon or boron to show fatigue behavior similar to that of metals. Not surprisingly, therefore, composites containing these fibers aligned along the stress axis, and in large volume fractions, do show high monotonic strength values that are translated into high fatigue strength values as well. Typically, the S–N curves of such composites are very flat. Figure 4 shows the S–N curves in tension–tension for unidirectionally reinforced boron (40 v/o)/Al6061, alumina–FP (50 v/o)/Al, and alumina–FP (50 v/o)/Mg composites [2]. The cyclic stress is normalized with respect to the monotonic ultimate tensile stress. Note the rather flat S–N curves in all the cases and the fact that the unidirectional MMCs show better fatigue properties than the matrix when loaded parallel to the fibers. For example, at 10^7 cycles, the fatigue-to-tensile-strength ratio of the composite is about 0.77, almost double that of the matrix.

Gouda *et al.* [3] observed crack initiation early in the fatigue life at defects in boron fibers in unidirectionally reinforced B/Al composites. These cracks then grew along the fiber–matrix interface and accounted for a major portion of the fatigue life, as would be the case in a composite with high fiber-to-matrix-strength ratio. In composites with low fiber-to-matrix-strength ratio, crack propagation may be a major portion of fatigue life, but the crack would be expected to grow across the fibers, and a poor fatigue resistance will result. This simply confirms the observation that in unidirectional composites, the fatigue resistance will be maximum along the fiber direction, and the greatest efficiency will be achieved if the fibers have uniform properties, are as much as possible defect free, and are much stronger than the matrix. Similar results have been obtained by other researchers. For example, McGuire and Harris [4] studied the fatigue behavior of tungsten-fiber-reinforced aluminum–4% copper alloy under tension–compression cycling ($R = \sigma_{min}/\sigma_{max} = -1$). They found that increasing the fiber volume fraction from 0 to 24% resulted in increased fatigue resistance. This

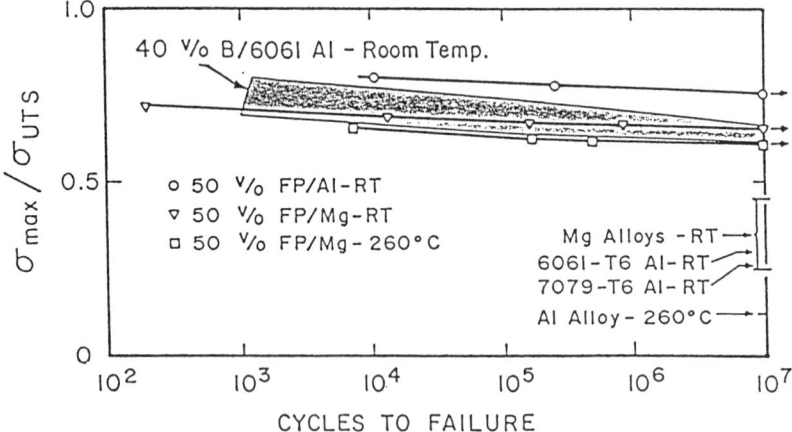

Fig. 4. S–N curves for some continuous fiber MMCs (after [2]).

was a direct result of increased monotonic strength of the composite as a function of the fiber volume fraction. The reader should note that due to the highly anisotropic nature of the fiber-reinforced composites in general, the fatigue strength of off-axis MMCs, just like that of any fibrous composite, will be expected to decrease with increasing angle between the fiber axis and the stress axis. This has been confirmed by studies involving S–N behavior of alumina (FP) fiber-reinforced magnesium composites [5, 6]. It was found that the S–N behavior followed the tensile behavior. Increased fiber volume fractions resulted in enhanced fatigue lifetimes in the axial direction, but little or no improvement was observed in the off-axis directions. Fatigue-crack initiation and propagation occurred primarily through the magnesium matrix. Thus, alloy additions to increase the strength of the matrix and fiber–matrix interface were tried. The alloy additions did improve the off-axis properties but decreased the axial properties. The reason for this was that while the alloy additions resulted in matrix and interface strengthening, they decreased the fiber strength.

There have been few fatigue-related direct structural observations on continuously reinforced MMCs. One such study has involved a single-crystal copper matrix containing tungsten fibers [7]. Tungsten/copper is quite an unusual metal matrix composite system. The two metals are mutually insoluble, yet molten copper wets tungsten. This allows for a strong mechanical bond between the tungsten fiber and the copper matrix, without any attendant chemical complications at the interface. Dislocation etch-pitting technique was used to make structural observations on the single-crystal copper matrix in these W/Cu composites. The composites were made by liquid metal infiltration of fibers in vacuum. It was observed that the process of fabrication involving cooling from a temperature $> 1080°C$ (melting point of copper) to room temperature resulted in thermal stresses large enough to deform the copper matrix plastically [8]. The dislocation density in the matrix was higher $(> 10^8 \text{ cm}^{-2})$ near the fiber–matrix interface than away from the interface. The dislocation distribution had a cellular structure, with the cell structure being better defined and smaller near the fiber than away from the fiber. This very heterogeneous distribution became a homogeneous one on stress cycling. The copper matrix structure after cycling had a high dislocation density distributed in a more or less uniform cellular structure.

B. Fatigue-Crack Propagation Tests

Fatigue-crack propagation tests are generally conducted in an electrohydraulic closed-loop testing machine on notched samples. The results are presented as log da/dN (crack growth per cycle) versus log ΔK (the alternating stress intensity factor). Crack growth rate, da/dN, is related to the cyclic stress intensity factor range, ΔK, according to the power-law relationship first formulated by Paris and Erdogan [9],

$$\frac{da}{dN} = A(\Delta K)^m \tag{1}$$

where A and m depend on the material and test conditions. The applied cyclic stress intensity range is given by

$$\Delta K = Y \Delta \sigma \sqrt{a},$$

where Y is a geometric factor, $\Delta \sigma$ is the cyclic stress range, and a is the crack length.

The major problem in this kind of test is to make sure that there is *one and only* one dominant crack that is propagating. This is called the *self-similar* crack growth, i.e., the crack propagates in the same plane and direction as the initial crack. Fatigue-crack propagation studies, under conditions of *self-similar* crack propagation, have been made on metallic sheet laminates [10–13] and unidirectionally aligned fiber-reinforced MMCs [14, 15]. For the crack arrest geometry, with the crack growing perpendicular to the thickness of the composite, the mechanism proposed by Cook and Gordon [16] seems to be valid. According to this model, if the interface is weak, then the crack bifurcates and changes its direction when reaching the interface, and thus the failure of the composite is delayed. The improved fatigue crack propagation resistance in crack divider geometry has been verified by a number of researchers [10–12]. This improved performance has been attributed either to interfacial separation, which removes the triaxial state of stress, or to an interfacial *holding back of crack* in the faster crack-propagating component by the slower crack-propagating component. Generally, a relationship of the form of Eq. (1) describes the fatigue crack propagation behavior. Typical *self-similar* fatigue-crack propagation curves for metallic sheet laminates such as Al 1100/Al 2024 as well as for those of the individual components are shown in Fig. 5. [17]. An example of similar results obtained in the case of boron fiber (coated with B_4C)/Ti–6Al–4V matrix composite and the parent matrix alloy is shown in Fig. 6 [18]. Note that the MMC has a higher threshold value and a much lower rate of crack growth through most of the fatigue life than the unreinforced matrix alloy.

In general, the fibers provide a crack-impeding effect, but the nature (morphology, rigidity, and fracture strain) of the fiber surface, the fiber interface, and/or any reaction-zone phases that might form at the interface can have great influence. Soumeldis *et al.* [19], for example, observed lower fatigue-crack growth rates in fiber-reinforced Ti–6Al–4V compared with that in the unreinforced alloy. Long isothermal exposures at 850°C, however,

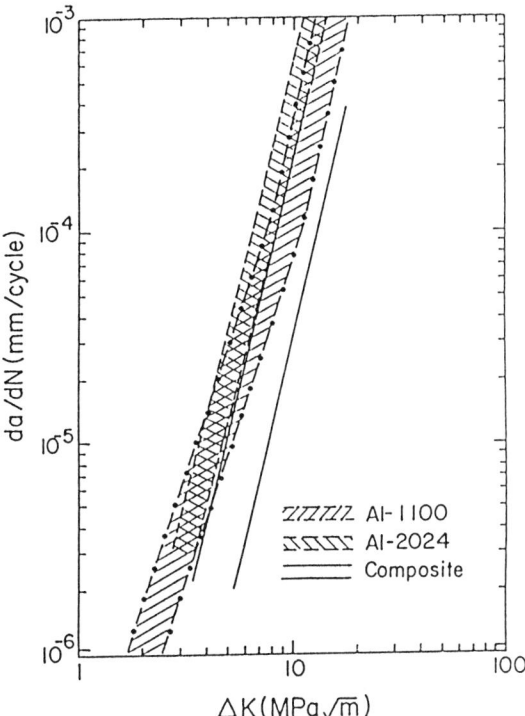

FIG. 5. Fatigue-crack propagation rate as a function of alternating stress intensity factor for aluminum 1100/aluminum 2024 laminate as well as monolithic Al 1100 and Al 2024. Note the lower crack growth rate in the composite than in either one of the components for most of the fatigue life [17].

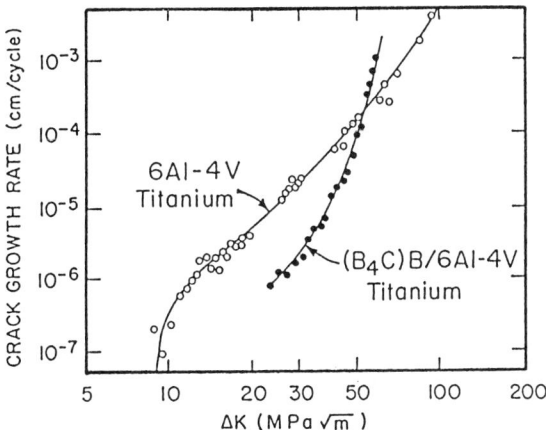

FIG. 6. Fatigue-crack propagation in boron fiber (coated with B_4C)/Ti–6Al–4V matrix composite. For most of the fatigue life, the crack growth rate is lower in the composite than that in the matrix alone [18].

resulted in a reduced crack growth resistance of the MMC. This was due to fiber degradation, fiber–matrix debonding, and an increase in matrix brittleness. Short-time isothermal exposures (up to about 10 h for B/Ti–6Al–4V, 30 h for B(B$_4$C)/Ti–6Al–4V, and 60 h for SiC/Ti–6Al–4V) improved the fatigue-crack resistance. This was attributed to energy-dissipating mechanism of fiber microcracking in the vicinity of the crack tip.

Fatigue-crack propagation studies have been done on aligned eutectic or in situ composites as well. Since many of these in situ composites are meant for high-temperature applications in turbines, their fatigue behavior has been studied at temperatures ranging from room temperature to 1100°C. The general consensus is that the mechanical behavior of in situ composites, static, and cyclic strengths, is superior to that of the conventional cast superalloys [20].

It should be emphasized that only fatigue-crack propagation rate data obtained under conditions of self-similar propagation can be used for comparative purposes. In a laminated MMC consisting of plies having different fiber orientation, in general, the self-similar mode of crack propagation will not be obtained. A possible approach under such circumstances is the one using the strain energy density formulation due to Sih [21]. He defines a strain energy density function S as follows,

$$S = a_{11}k_1^2 + 2a_{12}k_1k_2 + a_{22}k_2^2 + a_{33}k_3^2,$$

where a_{11}, etc., are functions of the angle θ made by the crack growth direction and the stress direction.

Sih's strain energy density formulation says that under mixed-mode crack extension conditions, which are commonly encountered in fiber-reinforced composites, the crack will propagate at a certain angle θ determined by the strain energy density field around the crack. From this, one can define an effective stress intensity factor, K_{ef} for cracks subjected to simultaneous shear and tensile loading (i.e., mixed-mode cracking) as

$$K_{ef} = S^{1/2}.$$

Preliminary verification of the applicability to metal matrix composites has been done [14, 15]. Clearly, more work needs to be done on mixed-mode crack propagation in MMCs.

C. Stiffness Loss

It was mentioned in Section I that the complexities in composites lead to the presence of many modes of damage, such as matrix cracking, fiber fracture, delamination, debonding, void growth, multidirectional cracking,

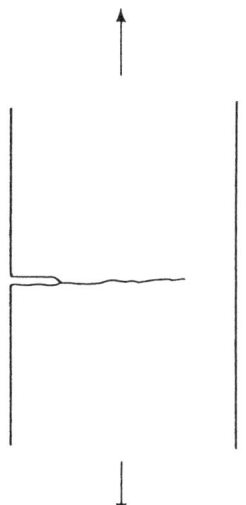

FIG. 7. Damage zone in a conventional, homogeneous, monolithic material (isotropic).

and so on. These modes appear rather early in the fatigue life of composites. Figure 7 shows schematically the type of damage zone formed in an isotropic material (e.g., a metal, ceramic, or polymer) whereas Fig. 8 shows the damage zone in a fiber-reinforced composite, which is an anisotropic material. In the case of the isotropic material, a single crack propagates in a direction perpendicular to the cyclic loading axis (mode-I loading), i.e., *self-similar* mode of crack propagation, Fig. 7. In contradistinction to this, in a fiber-reinforced composite, a variety of subcritical damage mechanisms generally come into play (Fig. 8a, b, c). And these subcritical damage-accumulation mechanisms come into play rather early in the fatigue life, well before the fatigue limit as determined in an S–N test, and a highly diffuse damage zone

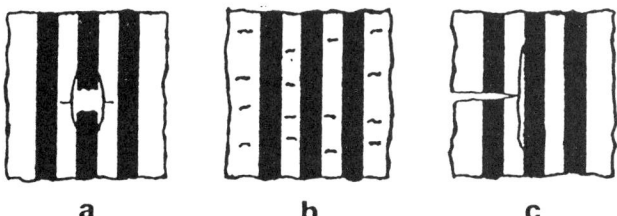

a   **b**   **c**

FIG. 8. Diffuse damage zone in a fiber-reinforced composite (anisotropic): (a) fiber break and local debonding, (b) matrix cracking, deflection of the principal crack along a weak fiber–matrix interface.

is formed. One manifestation of such damage is the stiffness loss as a function of cycling. In general, one would expect the scatter in fatigue data of composites to be much greater than that in fatigue of monolithic, homogeneous materials. This is because of the existence of a variety of damage mechanisms in composites, to wit, random distribution of matrix microcracks, fiber–matrix interface debonding, fiber breaks, and so on. Thus, with continued cycling, there occurs an accumulation of damage. This accumulated damage results in a reduction of the overall stiffness of the composite laminate. Measurement of stiffness loss as a function of cycling has been shown to be quite a useful technique of assessing the fatigue damage in polymer matrix composites. Information useful to designers can be obtained from such curves [1]. In MMCs, the fatigue behavior of boron fiber and silicon carbide fiber-reinforced aluminum and titanium alloy matrix composite laminates having different stacking sequences has been examined using the stiffness-loss measurement technique [22–24]. It was observed that on cycling below the fatigue limit but above a distinct stress range, ΔS_{SD}, the plastic deformation and cracking (internal damage) in the matrix led to a reduced modulus. Figure 9 shows the response of a boron/aluminum composite subjected to a constant cyclic stress range (225 MPa) but to varying values of S_{max}, the maximum stress. The modulus drop occurred only when S_{max} was shifted upward. Johnson [23] has proposed a model that envisions

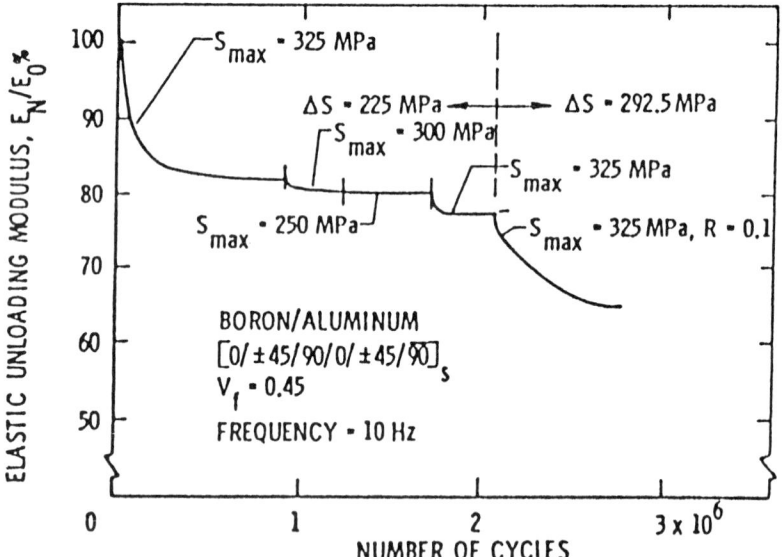

FIG. 9. Change in elastic unloading modulus under a variable loading program. Reprinted from [23].

that the specimen reaches a "saturation damage state" (SDS) during constant amplitude fatigue testing.

Gomez and Wawner [25] also observed stiffness loss when subjecting silicon carbide/aluminum composites to tension–tension fatigue ($R = 0.1$) at 10 Hz. Periodically, the cycling was stopped and the elastic modulus was measured. Figure 10 shows a typical modulus loss curve for unidirectional silicon carbide/aluminum composites. Modulus at N cycles, E_N, normalized with respect to the original modulus, E_0 is plotted against the log (number of cycles, N). These authors used a special type of silicon carbide fiber, called the SCS-8 silicon carbide fiber, which is a silicon carbide fiber with a modified surface to give a strong bond between the fiber and the aluminum matrix. The SCS coating broke off at high cycles, and the fracture surface showed the coating clinging to the matrix.

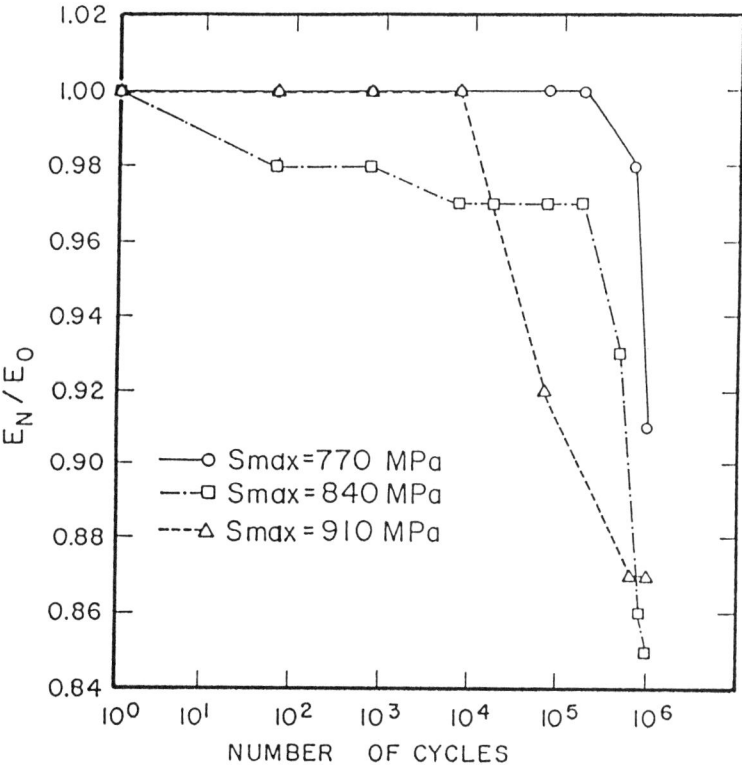

FIG. 10. Modulus loss in axially aligned SiC/Al composites [25].

IV. Hybrid Composites

Composites containing more than one type of fibers are called hybrid composites. Such composites, by using two or more type of fibers, extend the idea of tailor-making a composite material to meet specific property requirements. Partial replacement of expensive fibers by cheaper but adequate fiber types is another attractive feature of hybrid composites. Additionally, there is the possibility of obtaining a synergistic effect in the fatigue behavior of hybrid composites.

An interesting type of hybrid composite is made of alternating layers of high-strength aluminum alloy sheets and layers of unidirectional aramid fibers in an epoxy matrix. It is called the aramid aluminum laminate (ARALL). Improved fatigue resistance of ARALLs over that of monolithic aluminum structures is the main attractive feature. Cracks can grow only a short distance before being blocked by the aramid fibers spanning the crack tip. Figure 11 shows the slow fatigue-crack growth characteristics of

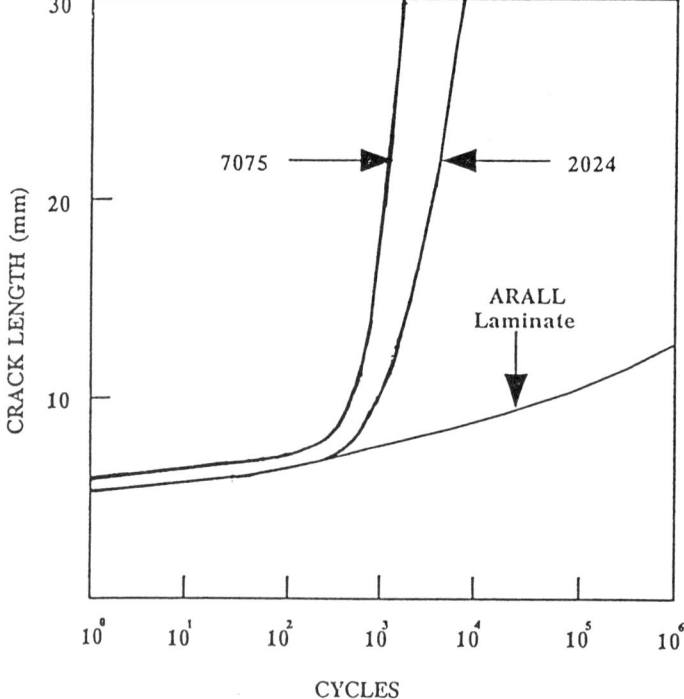

FIG. 11. ARALL composite shows slow fatigue-crack growth characteristics compared with the monolithic aluminum alloys. Reprinted by permission of the Society for the Advancement of Materials and Process Engineering, L. B. Mueller and M. Gregory, paper presented at First Annual Metals and Metals Processing Conf. of SAMPE. Cherry Hill, New Jersey, 1988.

ARALLs compared with two monolithic aluminum alloys [26]. Applications of ARALLs are envisaged in tension-dominated fatigue structures such as aircraft fuselage, lower-wing, and tail skins. It should be mentioned that the use of ARALLs will result in 15–30% weight savings over conventional construction.

Another version of hybrid laminated MMCs is a family of layered composites consisting of metallic outer skins with a viscoelastic core material (for example, polyethylene, nylon, polypropylene, paper, or cork). Such composites will be useful where sound and vibration dampening are required. The viscoelastic layer provides a high loss factor, i.e., a high capacity to convert vibrational energy to heat.

V.   Thermal Fatigue

There exists a very fundamental physical incompatibility between the fiber and the matrix, to wit, the difference in their thermal expansion (or contraction) coefficients. This problem of thermal expansion mismatch between the components of a composite is a very general and serious one [1]. Thermal stresses arise in composite materials because of the large differences in the thermal expansion coefficients (α) of the fiber and the matrix. It should be emphasized that thermal stresses in composites will arise even if the temperature change is uniform throughout the volume of the composite. Such thermal stresses can be introduced in composites during cooling from high fabrication, annealing, or curing temperatures or during any temperature excursions (inadvertent or by design) during service. Turbine blades, for example, are very much susceptible to thermal fatigue. The U.S. National Aeronautics and Space Administration has a program involving the use of tungsten fiber-reinforced superalloy (TFRS) composite as a turbine blade material.

The magnitude of thermal stresses in composites is proportional to $\Delta\alpha\Delta T$, where $\Delta\alpha$ is the difference in the expansion coefficients of the two components, and ΔT is the amplitude of the thermal cycle. In MMCs, the matrix generally has a much higher coefficient of thermal expansion than the fiber. Rather large internal stresses can result when fiber-reinforced composites are heated or cooled through a temperature range. When this happens in a repeated manner, we have what is called the phenomenon of *thermal fatigue*, because the cyclic stress is thermal in origin. Thermal fatigue can cause cracking in the brittle polymeric matrix or plastic deformation in a ductile metallic matrix [27–29]. Cavitation in the matrix and fiber–matrix debonding are the other forms of damage observed due to thermal fatigue in composites

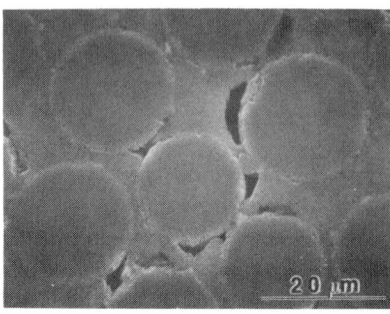

FIG. 12.   Cavitation damage at the fiber–matrix interface due to thermal fatigue (2000 cycles between room temperature and 300°C) in alumina (FP) fiber/magnesium alloy composite.

[28–31]. Figure 12 shows a scanning electron micrograph of an alumina-fiber-reinforced magnesium alloy matrix that was subjected to 2000 thermal cycles between room temperature and 300°C. Note the cavitation at the fiber–matrix interface.

It is possible to obtain a measure of the internal stressses generated when subjecting a composite to thermal cycling. A computer-controlled, servo-hydraulic thermal fatigue system was used to perform tests on alumina (FP)/Al–Li alloy composite. Thermal fatigue testing in this case involved cycling the temperature of the sample while its gage length was kept constant. This constraint resulted in a stress on the sample, which was measured. Such a test thus provides the stress required to keep the specimen gage length constant as a function of thermal cycles. Figure 13 shows the results of cycling between 300°C and 500°C for two samples of alumina (FP)/Al–Li [32]. The alumina fibers were unidirectionally aligned parallel to the stress axis, and the fiber volume fraction was 35%. The graph shows the variation of the maximum stress in tension and compression as a function of number of cycles. The initial rise in the maximum tensile stress curve is due to the work hardening of the aluminum alloy matrix caused by the thermal stresses arising from the thermal mismatch between the fiber and the matrix. With continued cycling, the microstructural damage sets in, and a plateau in the maximum tensile stress-versus-cycles curve is obtained as long as the strength increase due to work hardening of the matrix is balanced by the strength decrease due to microstructural damage, e.g., voiding at the interface. Figure 14 shows the interfacial damage in this system after 490 cycles. The maximum compressive stress, however, decreases with cycling to a plateau value. This may be due to the Bauschinger effect, i.e., a higher strength in tension results in a concomitant lowering of the strength in compression. Eventually, the alumina fibers suffer fracture, which causes the tensile stress curve in Fig. 13 to fall.

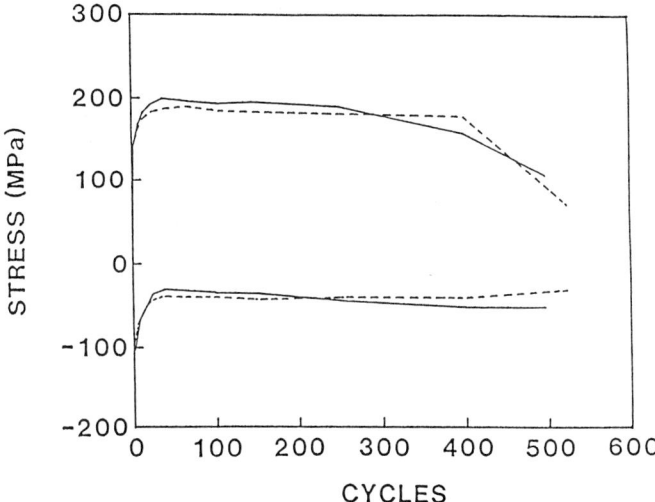

FIG. 13. Maximum stress in tension and in compression as a function of the number of thermal cycles for two samples of alumina (FP)/Al–Li. The alumina fibers (35% volume fraction) were aligned parallel to the stress direction [*32*].

In general, one can reduce the damage in the matrix by choosing a matrix material that has a high yield strength and a large strain to failure (i.e., ductility). The eventual fiber–matrix debonding can only be avoided by choosing the components such that the difference in the thermal expansion characteristics of the fiber and the matrix is low.

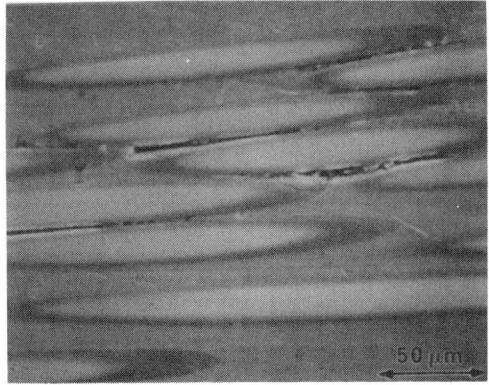

FIG. 14. Cavitation damage along the interface in an alumina (FP)/Al–Li sample after 490 cycles [*32*].

VI. Summary

In general, the fatigue resistance of a given metal can be enhanced by reinforcing it with continuous fibers or by bonding two different metals judiciously selected to give the desired characteristics. Not unexpectedly, the improvement is greatest when the fibers are aligned parallel to the stress direction.

While conventional approaches such as stress-versus-cycle (S–N) curves or fatigue-crack propagation tests under conditions of self-similar crack propagation can be useful for comparative purposes and for obtaining information on the operative failure mechanisms, they do not provide information useful to the designers. Fatigue-crack propagation under mixed-mode cracking conditions should be analyzed analytically and experimentally. Novel approaches such as those epitomized by the measurement of stiffness reduction of the composite as a function of cycles seem to be quite promising.

Since many applications of metal matrix composites do involve temperature changes, it is important that thermal fatigue characteristics of these composites be evaluated in addition to their mechanical fatigue characteristics.

References

1. K. K. Chawla, "Composite Materials: Science & Engineering", Springer-Verlag, New York, 1987.
2. A. R. Champion, W. H. Krueger, H. S. Hartman, and A. K. Dhingra, in "Proc. 1978 Intl. Conf. Composite Materials (ICCM/2)," p. 883, TMS-AIME, New York, 1978.
3. M. Gouda, K. M. Prewo, and A. J. McEvily, in "Fatigue of Fibrous Composite Materials," p. 101, ASTM STP, **723**, Amer. Soc. Testing and Materials, Philadelphia, 1981.
4. M. A. McGuire and B. Harris, J. Phys., Appl. Phys., **7**, 1788 (1974).
5. J. E. Hack, R. A. Page, and G. R. Leverant, Metall. Trans. A, **15A**, 1389 (1987).
6. R. A. Page, J. E. Hack, R. Sherman, and G. R. Leverant, Metall. Trans. A, **15A**, 1397 (1987).
7. K. K. Chawla, Fiber Sci. Tech., **8**, 49 (1975).
8. K. K. Chawla and M. Metzger, J. Mater. Sci., **7**, 34 (1972).
9. P. C. Paris and F. Erdogan, J. Basic. Eng. Trans. ASME, **85**, 528 (1963).
10. R. F. McCartney, R. C. Richard, and P. S. Trozzo, Trans. ASM, **60**, 384 (1967).
11. L. G. Taylor and D. A. Ryder, Composites, **1**, 27 (1976).
12. N. J. Pfeiffer and J. A. Alic, J. Eng. Mater. Tech., **100**, 32 (1978).
13. K. K. Chawla and P. K. Liaw, J. Mater. Sci., **14**, 2143 (1979).
14. C. R. Saff, D. M. Harmon, and W. S. Johnson, J. of Metals, **40**, 58 (1988).
15. D. S. Mahulikar, Y. H. Park, and H. L. Marcus, in "Fracture Mechanics: Fourteenth Symposium, Vol. II, Testing and Applications," ASTM STP 791, p. 579. American Society for Testing and Materials, Philadelphia, 1983.

16. J. Cook and J. E. Gordon, *Proc. Roy. Soc. Lond.*, **A282**, 508 (1964).
17. L. B. Godefroid and K. K. Chawla, in "3rd Latin American Colloquium on Fatigue and Fracture of Materials," Rio de Janeiro, Brazil, 1988.
18. D. M. Harmon, C. R. Saff, and C. T. Sun, "Durability of Continuous Fiber Reinforced Metal Matrix Composites," AFWAL-TR-87-3060. Air Force Wright Aeronautical Labs., Dayton, Ohio, 1987.
19. P. Soumelidis, J. M. Quenisset, R. Naslain, and N. S. Stoloff, *J. Materials Sci.*, **21**, 895 (1986).
20. N. S. Stoloff, in "Advances in Composite Materials," p. 247. Applied Sci. Pub., London, 1978.
21. G. C. Sih, "Handbook of Stress Intensity Factors for Researchers and Engineers." Inst. of Fracture and Solid Mech., Lehigh University, Bethlehem, Pennsylvania, 1973.
22. W. S. Johnson, in "Damage in Composite Materials," ASTM STP 775, p. 83, American Society for Testing and Materials, Philadelphia, 1982.
23. W. S. Johnson in "Mechanical and Physical Behavior of Metallic and Ceramic Composites, 9th Riso Intl. Symp. on Metallurgy and Materials Science," Risø Nat. Lab., Roskilde, Denmark, 1988.
24. W. S. Johnson and R. R. Wallis, in "Composite Materials: Fatigue and Fracture," ASTM STP 907, p. 161. American Society for Testing and Materials, Philadelphia, 1986.
25. J. P. Gomez and F. W. Wawner, personal communication, September, 1988.
26. L. R. Mueller and M. Gregory, paper presented at First Annual Metals and Metals Processing Conf. of SAMPE. Cherry Hill, New Jersey, August 1988.
27. K. K. Chawla, *Metallography*, **6**, 155 (1973).
28. K. K. Chawla, *Philos. Mag.*, **28**, 401 (1973).
29. K. K. Chawla, in "Grain Boundaries in Eng. Materials, Proc. 4th Bolton Landing Conf.," p. 435. Claitor's Pub. Div., Baton Rouge, Louisiana, 1975.
30. C. S. Lee and K. K. Chawla, in "Proc. Industry-University Adv. Mater. Conf.," p. 289. TMS-AIME, Warrendale, Pennsylvania, 1987.
31. C. S. Lee, K. K. Chawla, J. M. Rigsbee, and M. Pfeifer, in "Cast Reinforced Metal Composites," p. 301. ASM Intl., Metals Park, Ohio, 1988.
32. L. K. Kwei and K. K. Chawla, paper presented at the TMS Spring meeting, Las Vegas, Nevada, Feb. 1989.

Suggested Readings

1. L. J. Broutman (ed.), "Composite Materials," Vol. 5, Fracture and Fatigue, Academic Press, New York, 1974.
2. K. K. Chawla, "Composite Materials: Science & Engineering", Springer-Verlag, New York, 1987.
3. V. Gerold, in "Mechanical and Physical Behavior of Metallic and Ceramic Composites, Proc. 9th Riso Intl. Symp. on Metallurgy and Materials Sci., p. 35. Risø Nat. Lab., Roskilde, Denmark, 1988.
4. J. R. Hancock, in "Composite Materials: Fracture and Fatigue," Vol. 5, p. 371. Academic Press, New York, 1974.

Fatigue of Discontinuously Reinforced Metal Matrix Composites*

JIAN KU SHANG† AND R. O. RITCHIE

Department of Materials Science and Mineral Engineering
University of California
Berkeley, California

I. Introduction

In recent years, the development of metallic alloys reinforced with high-strength ceramic phases, such as graphite, silicon carbide, and alumina, has led to a series of metal matrix composites that have provided important material alternatives to traditional engineering alloys, such as precipitation-hardened aluminum alloys. The inclusion of such phases, in general, confers improved stiffness, higher strength and superior wear and elevated-tempera-

* Work supported by the Air Force Office of Scientific Research under the University Research Initiative Contract No. F49620-87-C-0017 to Carnegie Mellon University.
† Presently at The University of Illinois at Urbana-Champaign, Urbana, IL 61801.

ture properties compared with the constituent matrix material, although the use of aligned continuous fibers or laminated sheets as the reinforcement results in highly directional properties [1–4]. Discontinuous reinforcements, in the form of chopped fibers, whiskers, platelets or particles, conversely, show essentially isotropic properties and have the added advantages that they are more economical and can be fabricated by using conventional metallurgical techniques, such as casting, powder metallurgy, extrusion, and hot-forming [3].

Although improved processing techniques are still required to overcome the poor ductility and fracture-toughness properties displayed by many of these composites [5–9], studies on the initiation and *subcritical* growth of incipient cracks in particulate– and whisker-reinforced aluminum alloys have shown improved endurance strengths and high-cycle fatigue resistance, and a potential for superior fatigue-crack growth properties, compared with the unreinforced alloys [10–23].

It is the objective of this chapter to review what is currently known about the cyclic fatigue properties of such discontinuously reinforced metal matrix composites, with particular emphasis on aluminum composites. It must be remembered, however, that a mechanistic understanding of fatigue in these materials is far from complete at this time; the intent here is simply to summarize current knowledge so that directions for future research and material development can be identified to best serve the ultimate applications of these materials.

II. Cyclic Deformation

Limited data on the cyclic deformation properties of metal matrix composites are available in the literature, although cyclic stress–strain relations have been reported for powder-metallurgy (P/M) and cast SiC-whisker-reinforced aluminum alloys (SiC$_w$/Al) [10, 11]. Williams and Fine's [10] data on SiC$_w$/2124 are typical; results indicate that the cyclic yield strength (σ_{yc}) of the composite is significantly higher than that of the unreinforced alloy at the same aging condition (Fig. 1). However, compared with properties measured under monotonic loading, the composite behaves similarly to unreinforced 2124; both materials show comparable degrees of cyclic hardening (~16% increase in yield strength under cyclic loading), little change in elastic modulus (E), and 6 to 7% higher cyclic yield strengths in compression compared with tension. Typical data, taken from [10] for 2124 with 20% SiC whiskers, are shown in Table I.

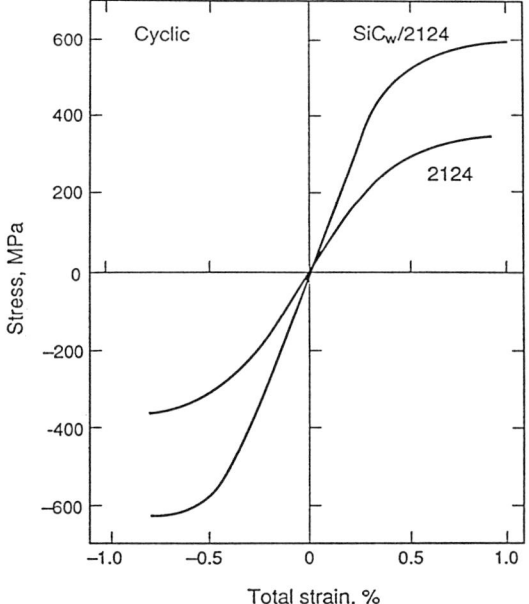

FIG. 1. Cyclic stress/strain behavior for unreinforced and SiC-whisker-reinforced Al-Cu 2124 alloys. Reprinted with permission from the Proceedings of the *5th International Conference on Composite Materials* (ICCM-V), edited by W. C. Harrigan, James Strife, and A. K. Dhingra, 1985. The Metallurgical Society, 420 Commonwealth Drive, Warrendale, Pennsylvania 15086.

TABLE I
CYCLIC PROPERTIES OF 20% SiC$_w$/2124 COMPOSITE

Alloy	Monotonic properties		Cyclic properties		
	E (GPa)	σ_y (MPa)	E$_c$ (GPa)	σ_{yc}[a] (MPa)	σ_{yc}[b] (MPa)
20% SiC$_w$/2124	123	484	128	560	600
2124-T6	74	268	76	310	330

[a] Tension
[b] Compression

Reprinted with permission, one table from the article "Quantitative Determination of Fatigue Microcrack Growth in SiC Whisker Reinforced 2124 Al Alloy Composites," by D. R. Williams and M. E. Fine, a part of *5th International Conference on Composite Materials (ICCM-V)*, edited by W. C. Harrigan, Jr., James Strife, and A. K. Dhingra, 1985, The Metallurgical Society, 420 Commonwealth Drive, Warrendale, PA 15086.

III. Low-Cycle and High-Cycle Fatigue

Hassen *et al.* [*12*] studied stress-life (*S–N*) fatigue behavior in several P/M particulate- and whisker-reinforced 6061 aluminum alloys in various environments. Their results, plotted in Fig. 2a as stress amplitude σ_a as a function of number of cycles to failure *N* in room-temperature air, show that the two forms of reinforcement have a similar effect, the composites are superior in high-cycle fatigue (HCF) but offer no particular advantage over the unreinforced alloy in low-cycle fatigue (LCF). The superior HCF properties appear to result from the higher stiffness of the composites, as differences between the composite and unreinforced alloys are diminished by replotting the data in Fig. 2a in terms of strain amplitude (σ_a/E).

Not all reinforced alloys, however, show superior fatigue properties to their constituent matrices. For example, in cast Al–Cu–Mg composites reinforced with "saffil" (short alumina fibers), Harris [*13*] showed that where the yield strength of the composites exceeded that of the unreinforced alloy, their fatigue strength for both HCF and LCF was also superior. However, where the saffil fibers were bonded to the matrix too weakly to provide strengthening under monotonic loading, the weakened interfaces provided preferential sites for crack initiation; this results in the unreinforced alloy having the superior low-cycle fatigue strength, although high-cycle fatigue strength was unaffected.

Such dependence of the fatigue strength on monotonic behavior is not uncommon. Hurd [*11*], in his study of cast saffil-reinforced Al–Mg–Si and Al–Si composites under cyclic strain control, found that in the Al–Si alloys, which are weakened by the addition of the saffil fibers, the LCF properties of the composite were inferior to the unreinforced alloy; conversely, in the Al–Mg–Si alloys, which are strengthened by fiber additions, the LCF properties were comparable (Fig. 3).

Microscopically, the initiation of cracks during low-cycle fatigue has been found to occur much later in the fatigue life of many composites compared with their constituent matrices. Williams [*14*], for example, found that visible (> 2.5 μm) fatigue cracks appeared in $SiC_w/2124$ composites only after 70 to 80% of the life (at $\sigma_a = 300$ MPa), whereas in the unreinforced materials, larger cracks appeared after only 5% of the life at the same stress amplitude. He noted that although the SiC_w–matrix interface was a preferential site for crack initiation in the composites, subsequent small-crack growth was generally inhibited. Such beneficial fatigue properties are apparently not shared by cast saffil-reinforced Al–Si alloys, where the very weakly bonded saffil fibers readily promote interfacial crack initiation, such that a large fraction of cracks is invariably present throughout the life [*11*].

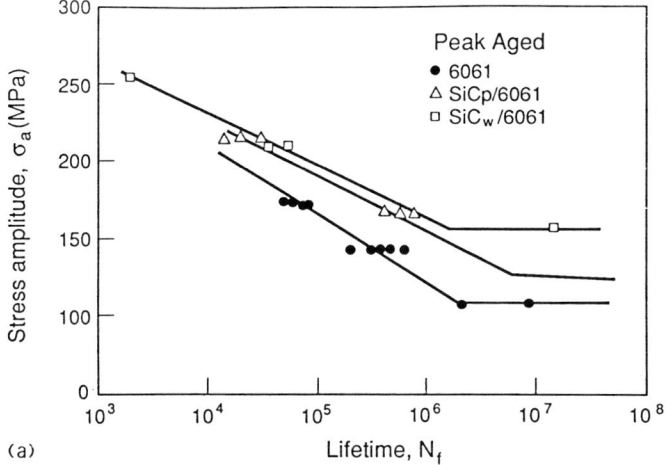

(a)

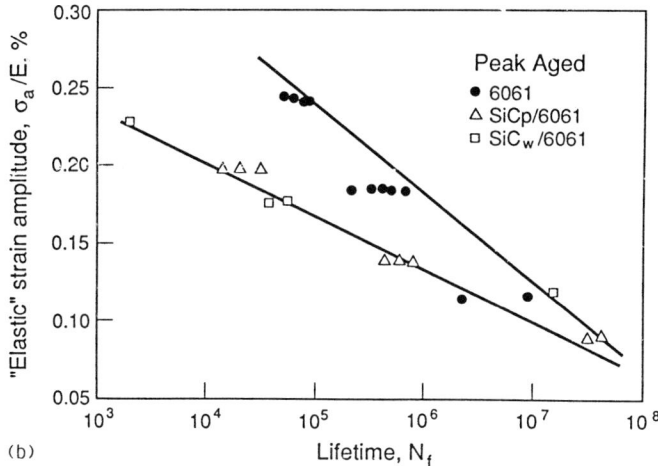

(b)

FIG. 2. Stress amplitude versus life (*S–N*) curves for unreinforced, SiC-particulate-reinforced and SiC-whisker-reinforced Al–Mg–Si 6061 alloy in the peak-aged condition. Fatigue lifetime data are presented as a function of both (a) stress amplitude σ_a and (b) elastic strain amplitude σ_a/E. Adapted and reprinted with permission from Cambridge University Press, D. F. Hassen, D. R. Crowe, J. S. Ahearn, and D. C. Cook, in "Failure Mechanisms in High Performance Materials," *Proc. of 29th Mechanical Failures Prevention Group Meeting*, 1984, NBS (J. G. Early, T. Robert Shives, and J. H. Smith, eds.), 1985, p. 147.

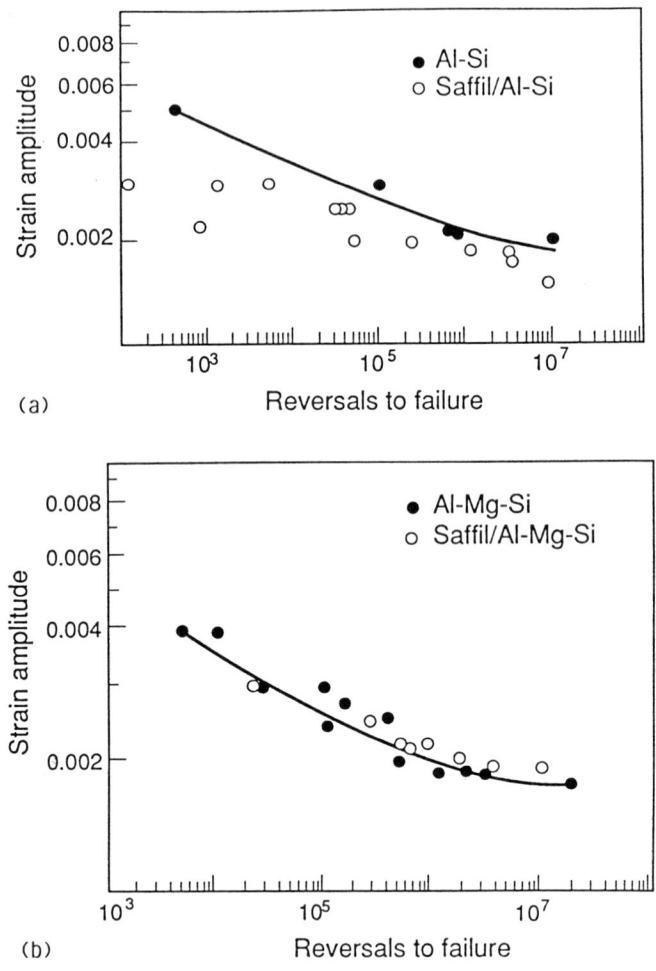

FIG. 3.   Strain amplitude versus life curves for cast unreinforced and saffil-fiber reinforced-aluminum alloy composites. Plotted are results for (a) Al–Mg–Si alloys, where the composite has the lower tensile strength, and (b) Al–Si–alloys, where the matrix material has the lower strength. Reprinted with permission from the Institute of Metals, N. J. Hurd, *Mater. Sci. Tech.* **4** (1988).

IV.   Fatigue-Crack Propagation

Fatigue-crack propagation behavior is generally characterized in terms of fracture mechanics; crack-growth rate (da/dN) data are displayed as a function of the (applied) stress-intensity range ($\Delta K = K_{max} - K_{min}$, where K_{max} and K_{min} are the extremes of stress intensity in the fatigue cycle). Such results are often described by equations of the form [25]

$$da/dN = C\Delta K^m,$$ (1)

where C and m are experimentally determined scaling constants. The growth-rate curve, however, is generally sigmoidal in shape because growth rates are accelerated at high ΔK levels as K_{max} approaches instabililty or the material's fracture toughness, K_{IC}, and tend toward a threshold value, ΔK_{TH}, at low ΔK levels, below which long-crack growth is presumed dormant [26].

Since the K_{IC} values for many metal matrix composites are of the order of 15 MPa$\sqrt{m}$ and ΔK_{TH} values of the order of 3–5 MPa$\sqrt{m}$, unlike most monolithic metallic alloys, the crack-growth rate curves in composites generally exist over a narrow range of ΔK. An example of such data is shown for overaged SiC-particulate-reinforced Al–Zn–Mg–Cu composites in Fig. 4; results are plotted at a load ratio (R = K_{min}/K_{max}) of 0.1 and are

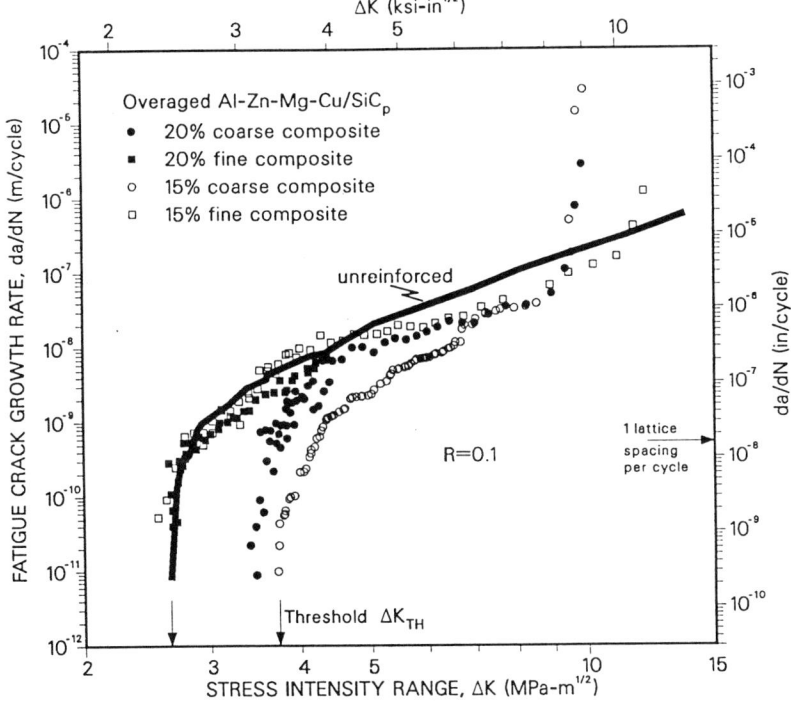

FIG. 4. Variation in fatigue-crack propagation rates, da/dN, with the stress-intensity range, ΔK, for overaged (15 hr at 171°C) SiC-reinforced Al–Zn–Mg–Cu alloys at room temperature, showing behavior at a load ratio of 0.1. Plotted are data for 15 and 20% reinforcement with either fine or coarse SiC particulate (nominal size 5 and 16 μm, respectively). Results for the unreinforced alloy (solid line) are shown for comparison [21].

compared with behavior in the unreinforced matrix alloy after identical aging treatments [21].

Crack growth rate data in particulate-reinforced composites, as shown in Fig. 4, often display several distinct regimes of behavior, each characterized by a dominant crack-extension mechanism, as illustrated schematically in Fig. 5. At near-threshold levels, below typically 10^{-9} m/cycle, the crack often follows a path that tends in general to avoid the reinforcement particles; this results in superior crack-growth resistance in many composites, compared with their constituent matrices, due primarily to crack-closure [18] and crack-trapping [21] mechanisms. In fact, the magnitude of the fatigue threshold is often found to be dependent upon mean particle size, an observation which has been shown to be consistent with a crack-trapping mechanism [21]. At higher growth rates, generally between 10^{-9} and 10^{-6} m/cycle, crack-growth resistance can again be somewhat improved in the composite alloys; in this regime, limited fracture of reinforcement particles *ahead* of the main crack tip can lead to the development of uncracked ligaments which act as crack bridges [20]. Finally, at very high crack-growth rates above $\sim 10^{-6}$ m/cycle, where K_{max} approaches K_{IC}, fatigue-crack propagation

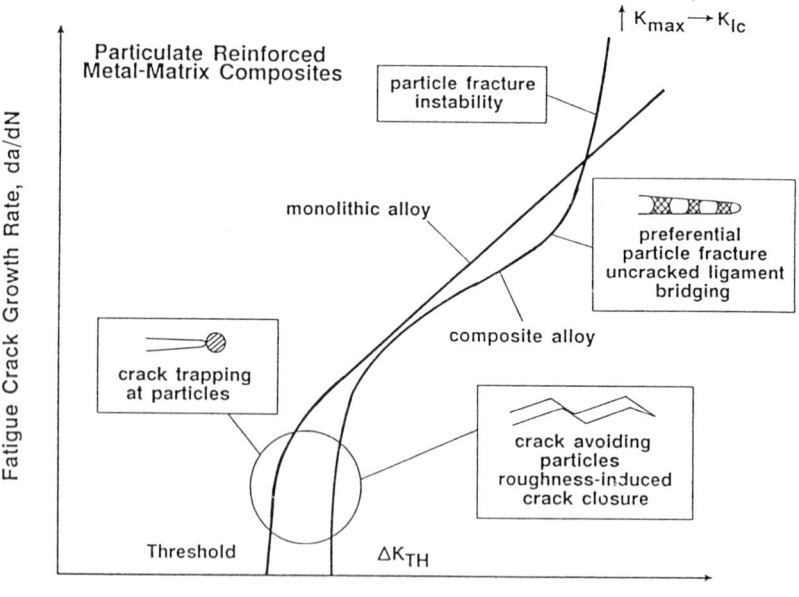

FIG. 5. Schematic illustration showing a comparison between the variation in fatigue-crack propagation rates with ΔK for unreinforced and particulate-reinforced metal matrix composites; shown are the salient micromechanisms associated with crack growth in each regime.

properties are invariably superior in the unreinforced alloys, because the much lower toughness of the composites leads to characteristically accelerated growth rates [*18*].

Such conclusions are largely reflected in reported results for whisker-reinforced composites, specifically SiC_w-reinforced 6061 and 2124 [*15, 16*] and cast saffil-reinforced Al–Mg alloys [*23*]. For example, data [*15*] for the $SiC_w/6061$ composite, shown in Fig. 6a, indicate slower fatigue-crack growth

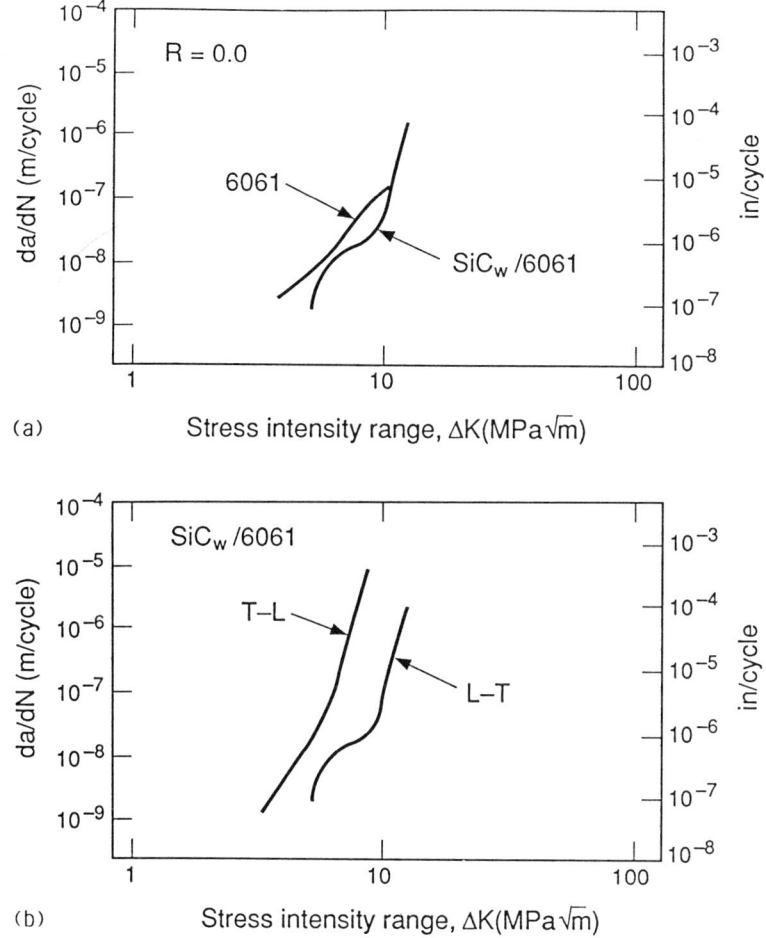

(a)

(b)

FIG. 6.   Fatigue-crack growth behavior, as a function of ΔK, for unreinforced and SiC-whisker-reinforced 6061 aluminum alloys at a load ratio of 0. Plotted are a comparison (a) between the composite and constituent matrix alloys, and (b) between the T–L and L–T orientations in the composite material. Reprinted with permission from Elsevier, Sequoia, S. A., S. S. Yau, and G. Mayer, *Mater. Sci. Eng.* **82** (1986).

rates in the composite at near-threshold and intermediate growth rates; however, the advantage offered by reinforcement is lost as ΔK levels approach instability. Similarly, fatigue thresholds for many whisker-reinforced composites exceed those for the coresponding unreinforced alloys [15, 19], although in other cases, ΔK_{TH} values are similar [16, 18]. One major difference in the properties of the different reinforcements is that, unlike particulate additions, the majority of the whiskers in whisker-reinforced alloys are oriented close to the extrusion direction; these composites thus generally display some orientation dependence [15, 24], as shown by the superior crack-growth properties of the $SiC_w/6061$ alloy in the L–T, rather than T–L, orientation (Fig. 6b [15]).

V.   Near-Threshold Fatigue Behavior

A.   Crack-Closure Mechanisms

Similar to many monolithic alloys, near-threshold fatigue-crack growth in metal matrix composites shows a strong dependence on the load ratio, principally due to the development of significant levels of crack closure, i.e., premature contact of the crack surfaces during the loading cycle [e.g., 27, 28]. Examples of this load-ratio effect are presented in Fig. 7 for several aluminum-alloy composites, reinforced with SiC or saffil whiskers or SiC particulate [7, 14, 20]. The role of crack closure can be appreciated from compliance measurements during the fatigue cycle on cracked specimens at near-threshold ΔK levels (Fig. 8). At high load ratios (R = 0.75), a linear relationship exists between applied load, P, and (back–face) displacement, δ, whereas at R = 0.1, a marked change in compliance can be detected on unloading once the crack surfaces first come into contact.

The degree of crack closure is generally quantified by measuring the closure stress intensity, K_{cl}, which is determined at first contact of the fracture surfaces during unloading [29]. Typical data [18] for a P/M Al–Zn–Mg–Cu composite (similar to 7091) with either fine (~ 5 μm) or coarse (~ 16 μm) SiC-particle sizes are shown in Fig. 9. Plotted are measured values of K_{cl}, normalized with respect to K_{max}, as a function of the applied ΔK. Increasing values of K_{cl} tend to inhibit crack advance by reducing the *local*, or effective, stress-intensity range ($\Delta K_{eff} = K_{max} - K_{cl}$) actually experienced at the crack tip (crack-tip shielding). Closure K_{cl}/K_{max} levels are typically increased with decreasing ΔK due to the smaller crack-tip opening displacements (CTODs). Moreover, by subtracting out the influence of closure

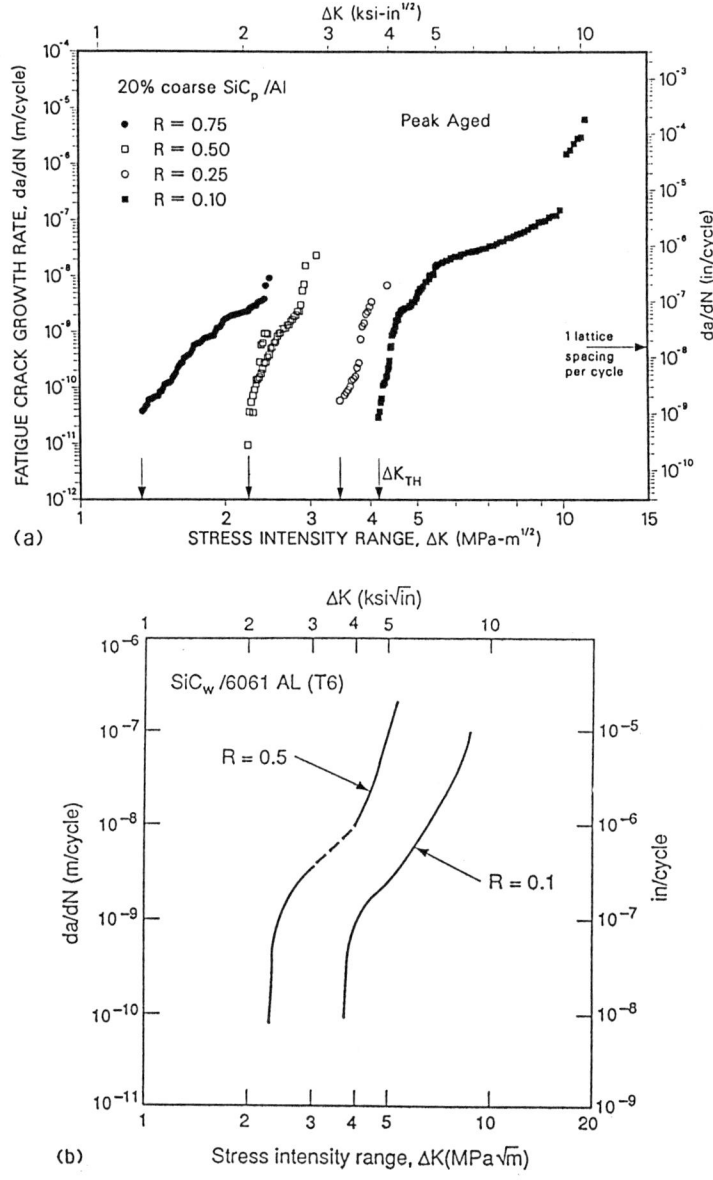

FIG. 7. Variation in fatigue-crack propagation rates in metal matrix composites with load ratio ($R = K_{min}/K_{max}$). Data are shown for (a) peak-aged (24 hr at 121°C) Al–Zn–Mg–Cu alloy reinforced with 20 vol-% SiC particulate, (b) peak-aged 6061 aluminum alloy reinforced with SiC whiskers. Reprinted with permission from *Eng. Fract. Mech.* **24** (5), W. A. Logsdon and P. K. Liaw, © 1986, Pergamon Press, plc. and (c) 10 vol-% saffil-fiber-reinforced Al-2Mg alloy. Copyright ASTM. Reprinted with permission. S. Preston, A. Melander, H. L. Groth, and A. F. Blom in "Thermal and Mechanical Behavior of Ceramic and Metal Matrix Composites," ASTM STP, 1990.

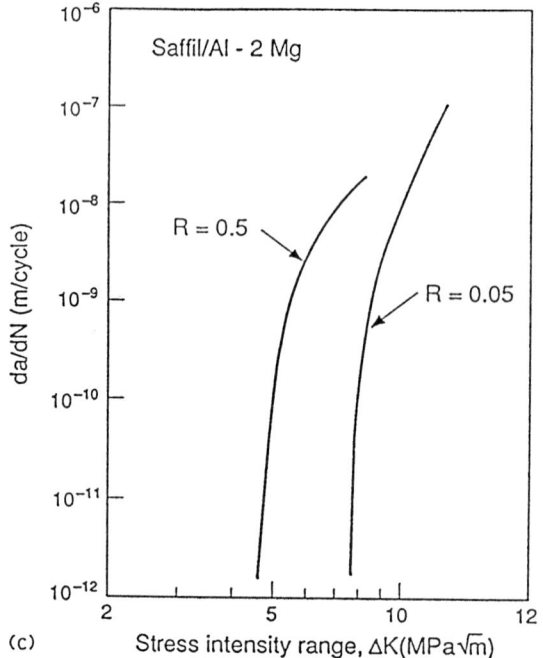

(c)

FIG. 7. (*Concluded*)

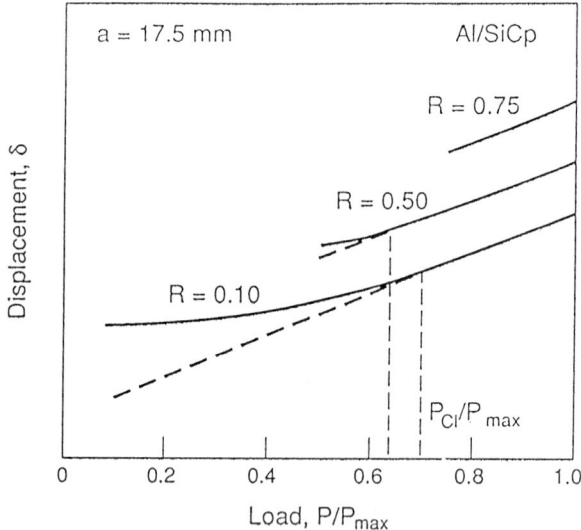

FIG. 8. Compliance curves of crack-mouth opening displacement, δ, as a function of applied load, P, for double-cantilever bend specimens, of SiC_p-reinforced Al–Zn–Mg–Cu alloy, containing a fatigue crack. Note how the compliance for crack growth at low load ratio ($R = 0.1$) shows an inflection at P_{cl}, indicative of crack-surface contact, or crack closure; no closure in the fatigue cycle is apparent at $R = 0.75$.

through a characterization of growth rates in terms of ΔK_{eff}, differences in near-threshold behavior at low and high load ratios become minimal (Fig. 10); differences between in the behavior of coarse and fine particle composites (e.g., Fig. 4), however, remain, but are reduced [20].

In the SiC-particulate-reinforced Al–Zn–Mg–Cu alloy, it is clear that the magnitude of the closure is enhanced with coarser particle sizes (Fig. 9), and only marginally affected by aging condition (Table II) [21]. Such observations are consistent with the primary source of crack closure in SiC_p/Al composites originating from the wedging action of fracture-surface asperities (roughness-induced closure [30–32]), as is common in many monolithic alloys [28], although studies by Davidson [22] on $SiC_p/2014$ purport to dispute this. On the one hand, metallographic studies [18] in SiC_p/Al–Zn–Mg–Cu, however, show that the crack tends to avoid the SiC particles at near-threshold levels (Fig. 11), and since the computed CTOD levels at K_{cl}, δ_{cl}, are several orders of magnitude larger than the SiC-particle size, $\bar{D}_p$, the presence of

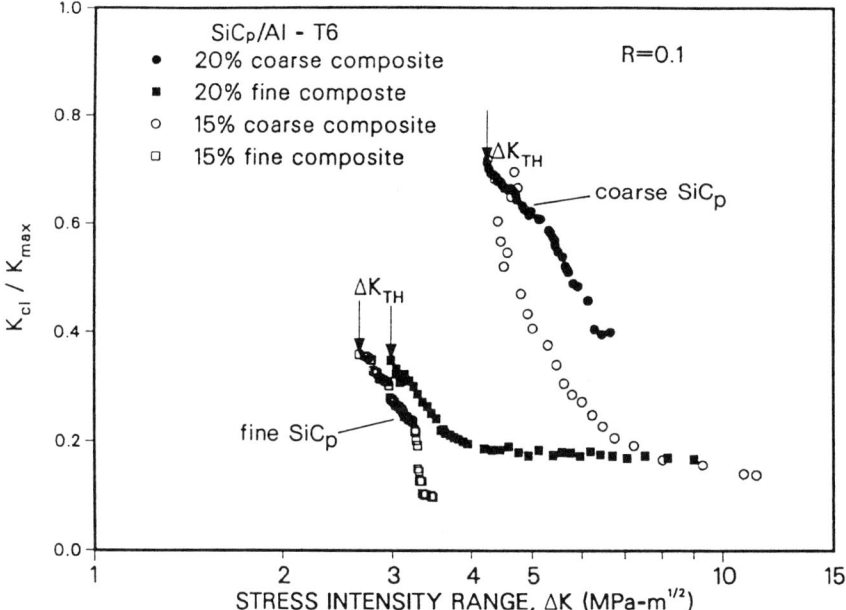

FIG. 9. Variation in crack closure, plotted as the ratio of closure stress intensity to maximum stress intensity, K_{cl}/K_{max}, as a function of ΔK for peak-aged Al–Zn–Mg–Cu alloy reinforced with fine (5 μm) or coarse (16 μm) SiC particulate. Note how the coarse particle distributions promote crack closure, particularly at the smaller crack-tip opening displacements associated with crack advance at low ΔK levels. Reprinted with permission from *Acta Metall.* **37** (8), Jian Ku Shang and R. O. Ritchie, © 1989, Pergamon Press, plc.

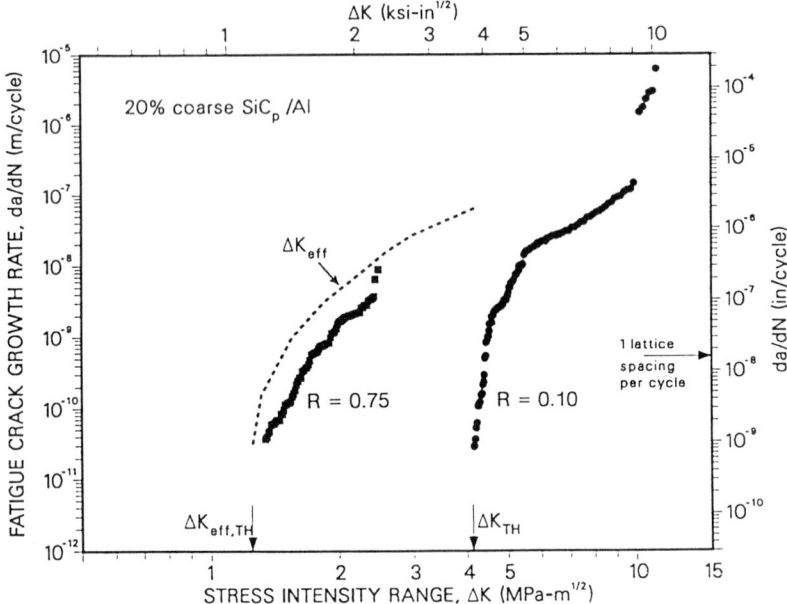

FIG. 10. Influence of load ratio R on fatigue-crack propagation rates in Al–Zn–Mg–Cu alloy reinforced with 20 vol-% SiC particulate, showing increased growth rates and lower threshold ΔK_{TH} values at high R. Note, however, that the effect of load ratio can be normalized by accounting for the role of crack closure, i.e., by characterizing crack-growth rates in terms of the effective stress-intensity range ($\Delta K_{eff} = K_{max} - K_{cl}$), rather than ($\Delta K = K_{max} - K_{min}$).

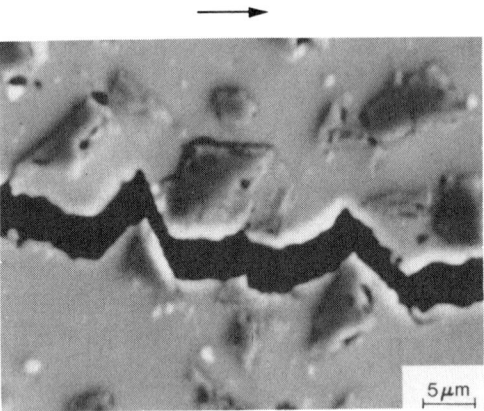

FIG. 11. Mechanisms associated with the interaction of the fatigue crack with reinforcement particles at near-threshold levels in SiC-reinforced Al–Zn–Mg–Cu alloy, showing crack deflection and consequently rougher fracture surfaces as the crack in general seeks to avoid the SiC particles. Arrow indicates direction of crack growth [*21*].

such reinforcement particles on the fracture surface must provide a potent source of wedging [21]. With increasing particle sizes, the larger particles would be expected to induce more crack meandering and hence are more effective in generating high closure levels at low load ratios. On the other hand, the lineal roughness of the fracture surfaces of the composites with fine and coarse grain particle distributions are often similar (~ 1.07 to 1.10), suggesting that our understanding of this mechanism is far from complete.

The extent of such roughness-induced closure can be assessed theoretically from simple geometric modelling [32] of the wedging of asperities (height h, width w) inside a crack deflecting through angle θ (where $\theta = \tan^{-1}\{2h/w\}$); this gives the closure stress intensity K_{cl} in terms of θ, K_{max}, and mismatch X, as

$$K_{cl} = K_{max}[X \tan \theta/(1 + X \tan \theta)]^{1/2}, \qquad (2)$$

where X is the ratio of mode-II to mode-I unloading displacements (u_{II}/u_{I}). Using this relationship for $R = 0.1$, predicted values of K_{cl}/K_{max} for angles of θ between 20 and 75° range from 0.29 to 0.70 for a mismatch of 0.25, which is consistent with the measured maximum closure ratios of 0.28 to 0.73 listed in Table II.

TABLE II
FATIGUE-THRESHOLD DATA FOR SiC_p/Al–Zn–Mg–Cu[1] COMPOSITE ALLOYS

Alloy	Aging[2] condition	$\bar{D}_p$ (μm)	ΔK_{TH} (MPa $\sqrt{m}$)	$\Delta K_{eff,TH}$ (MPa $\sqrt{m}$)	K_{cl}/K_{max}	K_{cl} (MPa $\sqrt{m}$)	δ_{cl}[3] (nm)
20% coarse	PA	10.5	4.2	1.3	0.73	3.4	112
SiC_p/Al	OA		3.5	1.4	0.65	2.5	68
	VO		3.7	1.3	0.69	2.8	85
20% fine	PA	6.1	3.0	1.9	0.42	1.4	26
SiC_p/Al	OA		2.7	1.9	0.35	1.0	15
	VO		3.0	1.1	0.67	2.2	68
15% coarse	PA	11.4	4.3	1.3	0.73	3.5	127
SiC_p/Al	OA		3.9	1.5	0.65	2.8	88
	VO		3.5	1.2	0.69	2.7	98
15% fine	PA	4.5	2.6	1.9	0.34	1.0	11
SiC_p/Al	OA		2.3	1.9	0.28	0.7	6
	VO		2.4	1.3	0.53	1.5	26

[1] The Al–Zn–Mg–Cu matrix of this composite is similar to 7091.
[2] PA $\equiv$ aged at 121°C (24 h), OA $\equiv$ 171°C (15 h), VO $\equiv$ 171°C (50 h).
[3] computed from $\delta_{cl} = \frac{1}{2}K_{cl}^2/E\sigma_y$ [33].
Reprinted with permission from Jian Ku Shang and R. O. Ritchie, *Acta Metall.*, **37**(8), 1989, 2267–2278.

B. Crack-Trapping Mechanism

Measured fatigue-crack growth threshold stress-intensity ΔK_{TH} values below which (long) cracks are presumed dormant are generally found to be relatively insensitive to the matrix-aging condition in both SiC whisker- and particulate-reinforced aluminum alloys; moreover, they are similar for several different matrices (2014, 2024, 7475) and SiC volume fractions (15–25%) [22]. In particulate alloys, however, the size of the reinforcement particles appears to be important; in general, threshold values are increased, and near-threshold growth rates decreased with increasing particle size (Fig. 12), where the average size, $\bar{D}_p'$, of particles "sampled" by the crack front is the appropriate measure [21]. At such near-threshold levels, crack-path morphology studies show a strong tendency for the crack to avoid the particles; in coarse SiC_p/Al–Zn–Mg–Cu composites, a fraction of only $\sim 13\%$ SiC particles are intersected by the crack path ($\sim 9\%$ are cracked, $\sim 4\%$ decohere), far less than the ~ 22 vol-% of SiC in the microstructure [18].

As the crack tends to avoid SiC particles at low ΔK levels, the particles can be considered to impede crack advance; their presence at the crack tip acts to lower *local* stress intensities, and hence their function can be

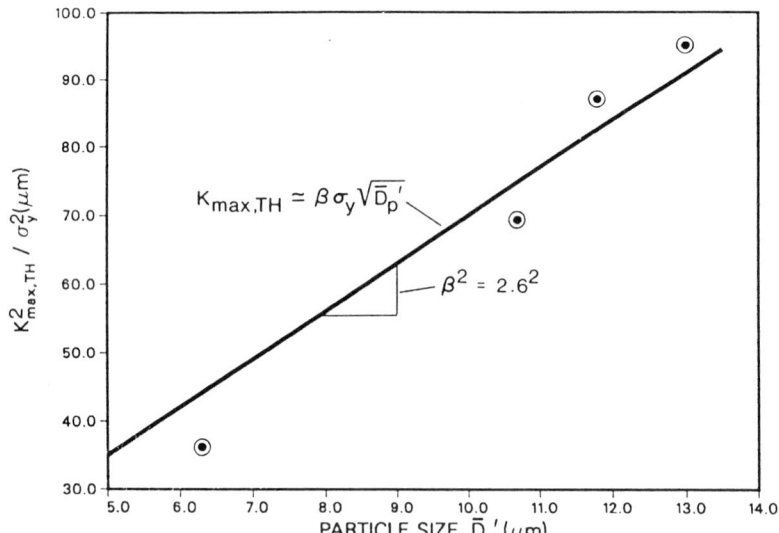

FIG. 12.   Experimental results (data points) and predicted variation (Eq. (5)) for the relationship between the fatigue threshold maximum stress intensity, $K_{max, TH}$, plotted as $K_{max, TH}^2/\sigma_y^2$, and the effective mean particle size, $\bar{D}_p'$, for Al–Zn–Mg–Cu alloys reinforced with SiC particulate, tested at $R = 0.1$. Reprinted with permission from *Acta Metall.* **37**(8), Jian Ku Shang and R. O. Ritchie, © 1989, Pergamon Press, plc.

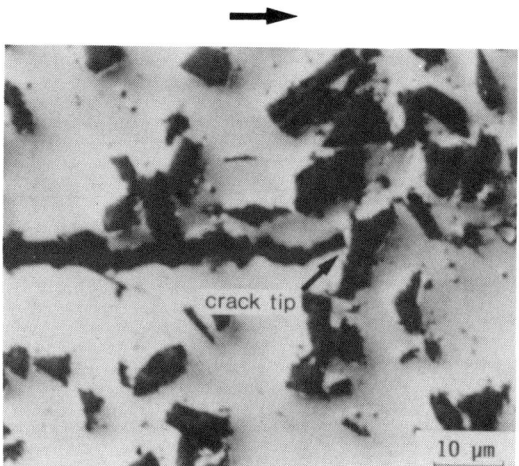

FIG. 13. Mechanisms associated with the interaction of the fatigue crack with reinforcement particles at the threshold stress intensity in SiC-reinforced Al–Zn–Mg–Cu alloy, showing crack trapping at SiC particles. Arrow indicates direction of crack growth. Reprinted with permission from *Acta Metall.* **37**(8), Jian Ku Shang and R. O. Ritchie, © 1989, Pergamon Press, plc.

envisioned as crack traps. Based on this notion, the fatigue threshold in particulate-reinforced composites has been modelled in terms of the role of a nondeforming particle at the crack tip (Fig. 13) in affecting the *local* "driving force" required for crack extension [21]. It is significant in these alloy systems that *at* ΔK_{TH}, maximum plastic zone sizes ($r_{y,\max}$) are comparable with the average particle size $\bar{D}'_p$ (Fig. 14), implying that crack hindrance at particles occurs because of limited local plasticity, which impedes fracture or bypassing of the particle. Accordingly, a limiting condition for fatigue-crack advance in particulate-reinforced composites can be defined in terms of the matrix stress beyond the particle at the crack tip exceeding the yield strength σ_y of the material, i.e., $r_{y,\max}$ must exceed $\bar{D}'_p$ [21].

The presence of a nondeforming inclusion at the crack tip significantly reduces the matrix stresses. Analyses by Atkinson [34, 35] for a finite crack and by Rubinstein [36] for a semi-infinite crack, show that the crack-tip stress intensity decreases (i) with decreasing distance between the inclusion and the tip, and (ii) with an increasing ratio of the modulus of the inclusion and the matrix; the stress intensity approaches zero as the inclusion approaches the tip. Recent studies [21] for a SiC carbide particle in an aluminum alloy (with a ratio of crack size to particle size of ∼ 1250) gives the matrix tensile stress distribution just beyond the inclusion as (Fig. 15),

$$\sigma_{yy} = CK/\sqrt{\bar{D}'_p}, \tag{3}$$

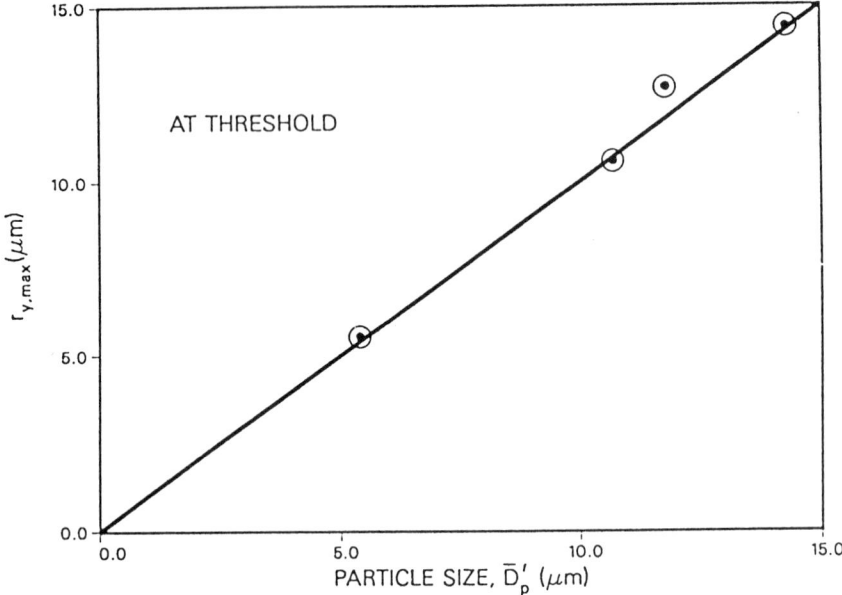

FIG. 14. Plot showing the equivalence in SiC-particulate-reinforced aluminum-alloy composites of the maximum plastic-zone size, $r_{y,\max}$, and the effective particle size, $\bar{D}'_p$, at the fatigue threshold stress intensity.

where the constant C equals 2.4^{-1} and only depends on material constants for a crack that is large compared with particle size [21].

For a fatigue crack trapped at a SiC particle to propagate, this matrix stress must exceed the matrix yield strength to induce plastic flow; this sets a limiting condition for the threshold as [21],

$$\sigma_{yy} = \sigma_y, \quad \text{over} \quad r = r_{y,\max} \approx \bar{D}'_p, \tag{4}$$

such that

$$K_{\max,\text{TH}} \approx 2.4\sigma_y\sqrt{\bar{D}'_p}. \tag{5}$$

Equation (5) implies that for particulate-reinforced composites where near-threshold crack advance is influenced primarily by crack trapping, the fatigue threshold is proportional to the yield strength and the square root of the particle size, which is consistent with experimental observations (Fig. 12).

An alternative model for the threshold in particulate-reinforced composites has been proposed by Davidson [37], based on observations that near-threshold crack growth involves significant Mode-II displacements. Here the threshold condition is associated with the generation of a single slip band ahead of the crack tip; crack growth is stopped once the stress at the tip of

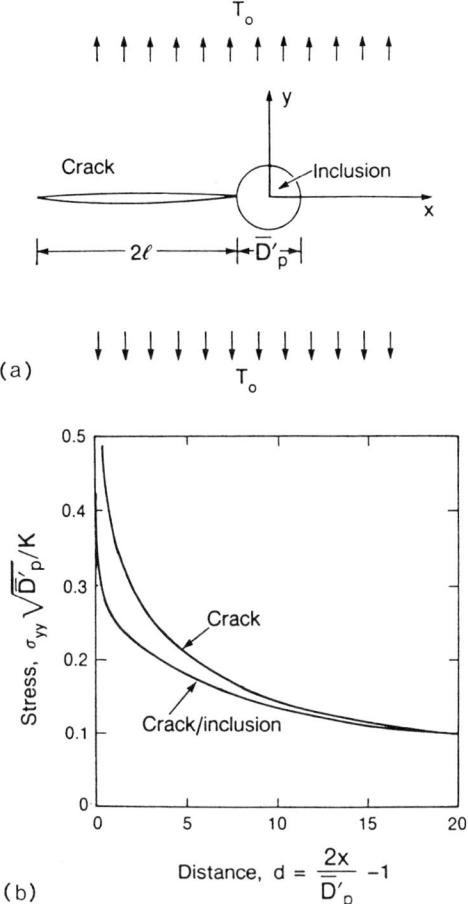

FIG. 15.   (a) Schematic illustration of the crack/inclusion interaction, showing (b) predicted distribution of normalized tensile stress in the matrix, σ_{yy}, as a function of normalized distance ahead of the crack/inclusion, compared to the crack-tip stress distribution in a homogeneous solid. Calculations are for a ratio of crack size to inclusion size of over 10^3, and a ratio of shear modulus of inclusion and matrix of 3. Reprinted with permission from *Acta Metall.* **37** (8), Jian Ku Shang and R. O. Ritchie, © 1989, Pergamon Press, plc.

the band is reduced to σ_y. For SiC-particulate composites, the limiting slip distance is assumed to be equal to the dislocation mean free path; this gives an expression for ΔK_{TH} in terms of the particle size, $\bar{D}_p$, and volume fraction, f_p, as [37]

$$\Delta K_{TH} \approx \sigma_f \left[\frac{4\pi}{3} \bar{D}'_p \left(\frac{1 - f_p}{f_p} \right) \right]^{1/2}, \tag{6}$$

where σ_f is the flow stress at the end of the slip band. Although the dependency on volume fraction has not been reported, the model does predict threshold values in a limited number of particulate-reinforced composites to within $\sim 18\%$.

VI. Higher Growth-Rate Behavior

A. General Considerations

At very high stress intensities where K_{max} approaches the fracture toughness, K_{IC}, fatigue-crack growth rates in the composites are invariably faster than in their unreinforced constituent matrix alloys (e.g., Fig. 4). This is primarily of consequence for the much lower toughness of the composites, which leads to accelerated growth rates, typically above $\sim 10^{-6}$ m/cycle, as conditions approach instability [18].

Conversely, at intermediate growth rates, typically between $\sim 10^{-9}$ and 10^{-6} m/cycle, the fatigue-crack growth resistance of particulate-reinforced composites is generally somewhat superior to the unreinforced matrix alloys (Fig. 4), although the degree of superiority appears to be dependent on interparticle spacing, d_p. In alloys with coarse reinforcement particle distributions, e.g., the 15% coarse SiC_p/Al–Zn–Mg–Cu alloy where $d_p \approx 56$ μm, growth rates are up to an order of magnitude slower in the composite; in the 20% fine SiC_p/Al–Zn–Mg–Cu alloy, conversely, where $d_p \approx 23$ μm, the effect is much smaller (Fig. 16) [18, 20].

It appears that in this intermediate regime, compared with near-threshold behavior, there is an increasing tendency in many particulate-reinforced composites for the crack path to interact with the particles. For example, in the coarse SiC_p/Al–Zn–Mg–Cu alloy at a ΔK of 7 to 10 $MPa\sqrt{m}$, a fraction of $\sim 21\%$ SiC particles is intersected by the crack path ($\sim 18\%$ are cracked, $\sim 3\%$ decohere), a figure which is comparable to the ~ 22 volume % of SiC in the microstructure, but large compared to the $\sim 13\%$ "sampled" at near-threshold levels [18]. Moreover, rather than trapping the crack or promoting crack deflection as at near-threshold levels, the role of the reinforcement phase is now principally to cause *limited* fracture *ahead* of the main crack tip. In fact, there is increasing evidence of cracked particles on the fatigue fracture surfaces with increasing ΔK [18]. The process occurs by the cracking of specific particles (Fig. 17a), or by the formation of voids in the matrix at the sharp corners (poles) of elongated particles (or whiskers) or in constrained material between closely spaced particles (Fig. 17b). Extension of these microcracks and their coalescence with the main crack

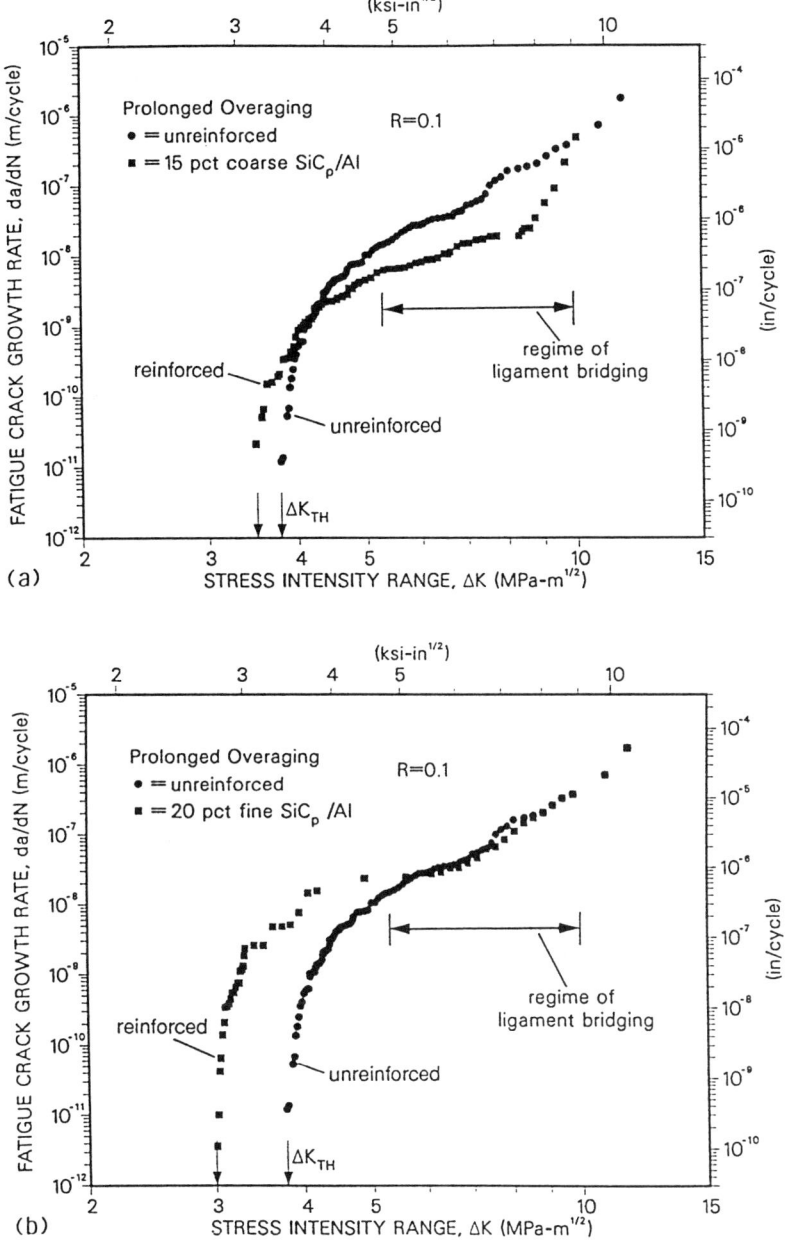

FIG. 16. Fatigue-crack growth rates at $R = 0.1$ in overaged SiC-particulate-reinforced Al–Zn–Mg–Cu alloys, compared to behavior in the unreinforced matrix alloy. Crack bridging via uncracked ligaments predominates over the range $\sim 10^{-9}$ to 10^{-6} m/cycle. Such bridging is effective in (a) the alloy containing coarse (16 μm) SiC particles, where bridges are overlapping; it is less effective in (b) the alloy with fine (5 μm) particles, where bridges are coplanar. Reprinted with permission from TMS-AIME, Jian Ku Shang and R. O. Ritchie, *Mettall. Trans. A* **20A** (1989).

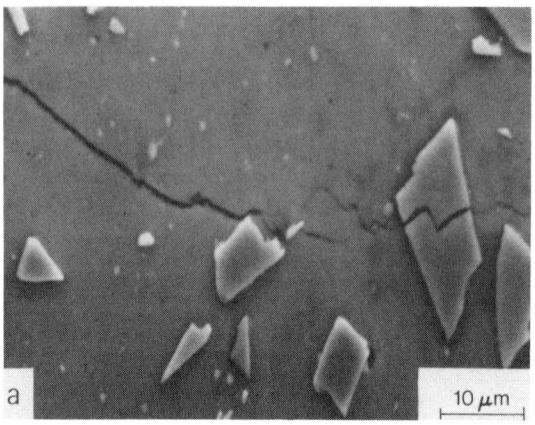

FIG. 17. Micrographs illustrating the generation of microcracks/voids at, or near, SiC-reinforcement particles *ahead* of the main crack tip at intermediate stress-intensity ranges in SiC$_p$/Al composites, showing (a) limited cracking of specific SiC particles, and (b) formation of microvoids in constrained matrix material between closely spaced particles.

results in the nonuniform advance of the main crack front and the consequent formation of small uncracked regions (in any two-dimensional section) along the crack length (Fig. 18). Such uncracked ligaments act to restrain crack opening, and thus shield the crack tip by a mechanism of crack bridging [20].

Evidence for such uncracked ligaments can be seen from metallographic sections perpendicular to the crack plane; in composites with smaller interparticle spacing, the ligaments are essentially coplanar (Fig. 18b), whereas with coarser particle distributions they have an overlapping morphology

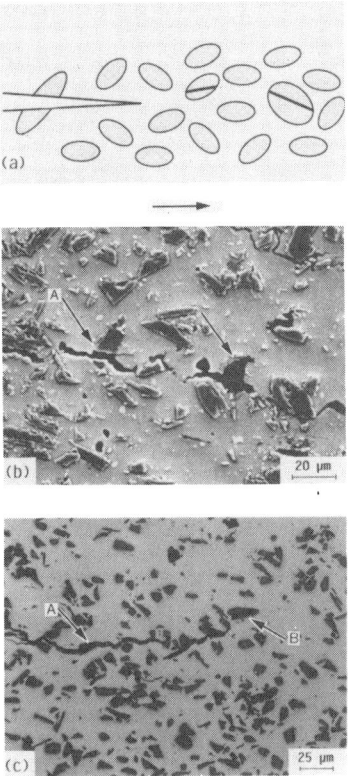

FIG. 18. Evidence for the creation of uncracked ligaments in particulate-reinforced metal-matrix composites, resulting from limited particle fracture, both ahead (location B) and in the wake (location A) of the crack tip (indicated by inclined arrows). Shown are (a) schematic illustration of the cracking process, and bridging from (b) coplanar uncracked ligaments in a fine SiC_p/Al-alloy composite, and (c) "overlapping" ligaments in a coarse SiC_p/Al-alloy composite. Horizontal arrow indicates general direction of crack growth. Reprinted with permission from TMS-AIME, Jian Ku Shang and R. O. Ritchie, *Metall. Trans. A* **20A** (1989).

(Fig. 18c), as illustrated schematically in Fig. 19. The latter type of uncracked ligaments, which exist typically over a bridging zone of some 500 μm behind the crack tip, represents the more potent source of crack-tip shielding, as shown by the experimental data in Fig. 16a (*c.f.,* Fig. 16b) and confirmed by the models described below.

It should be noted that the cracking of specific SiC particles ahead of the crack tip is both mechanically and statistically analogous to the cracking of carbides during the cleavage fracture of low-carbon steels [*38, 39*]. Since the process is stress-controlled, it is triggered at sites of weak ("eligible") particles

UNCRACKED LIGAMENT BRIDGING

coplanar ligaments   overlapping ligaments

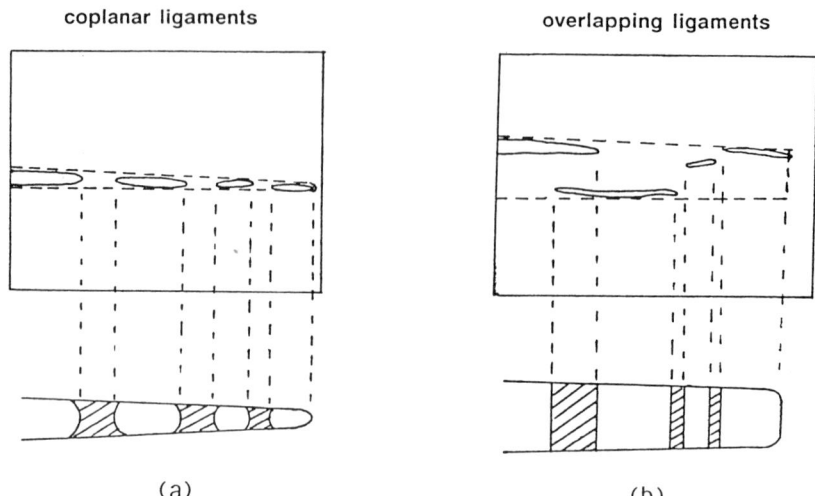

(a)   (b)

FIG. 19.   Schematic illustration of uncracked ligament bridging from (a) coplanar and (b) overlapping ligaments.

(generally the largest ones) *ahead* of the crack tip where the stresses are highest (due to crack-tip blunting). The location of these eligible particles is a function of the crack-tip stress distribution and the statistical distribution of particle strengths (inversely related to particle size) within the plastic zone ("sampling" volume), as described in [38, 39]. At low ΔK levels where the plastic zone, and hence the region of high stresses ahead of the tip, are small, the probability of finding a crackable particle is very low (it is essentially zero close to ΔK_{TH}, where the plastic-zone size is no larger than the particle size); with increasing ΔK levels, conversely, the plastic-zone volume and hence sampling region increases rapidly, such that the probability of finding an eligible particle is substantially enhanced [18].

B. Models for Crack-Bridging Mechanisms

In general terms, the dominant contribution from bridging reinforcements can be expressed solely as the product of the volume fraction f_b of the bridging phase with the area under the stress/strain curve, i.e., in terms of the yield strength σ_y, radius r_b, and f_b [40],

$$G_c \approx C\sigma_y r_b f_b, \tag{7}$$

where C depends on the ductility of the reinforcement phase and the extent of interface debonding. To quantify G_c, however, it is necessary to derive the appropriate constitutive relationship for the strained bridges that incorporates effects of stress state, inelastic response, and so forth. For the present case of bridging by uncracked tensile ligaments along a bridging zone l (which is small compared with crack length) behind the crack tip (Fig. 20), the effect of the bridges can be modelled as a distributed force per unit thickness $dp(x)$ acting over $x = l$ along the crack faces (Fig. 21), analogous

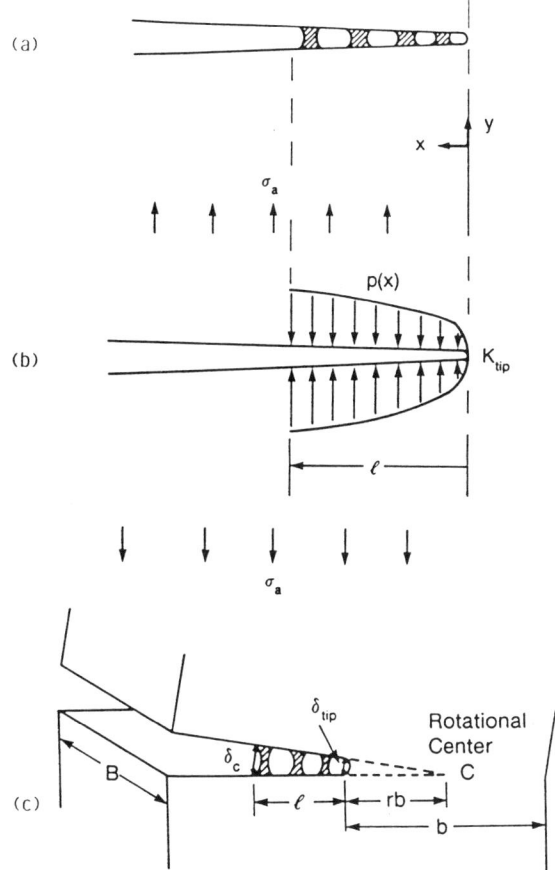

FIG. 20.   Schematic illustrations of the basis for models of (a) uncracked ligament bridging, with (b) a bridging zone stretching over distance l behind the tip of a crack subjected to an externally applied stress, σ_y, (c) the crack is assumed to open around a rotational center C located at distance rb ahead of the tip, where b is the uncracked specimen ligament and r is the rotational constant. Reprinted with permission from TMS-AIME, Jian Ku Shang and R. O. Ritchie, *Metall. Trans. A* **20A** (1989).

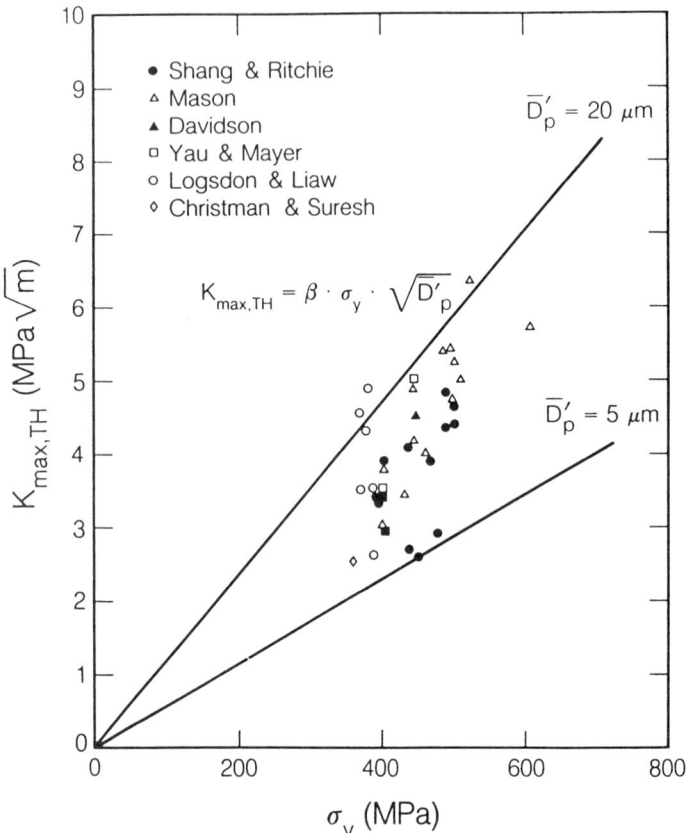

FIG. 21. Summary plot of the variation in fatigue threshold $K_{max,TH}$ values for SiC-reinforced aluminum alloys with yield strength σ_y, showing a comparison of experimentally measured threshold data [15–24] with the predicted trends from the crack-trapping model [Eq. (5)] for particle sizes $\bar{D}'_p$ of 5 and 20 μm.

to analyses developed for fiber bridging in ceramic composites and rubber toughening in polymers [e.g., 41–45]. For a semi-infinite crack, this force induces a shielding stress intensity [46],

$$K_s = (\sqrt{2/\pi}) \int_0^l dp(x)/\sqrt{x}, \qquad (8)$$

which, when superimposed upon the applied (far-field) stress intensity, K_a, yields an expression for the effective (near-tip) stress intensity

$$K_{tip} = K_a + K_s. \qquad (9)$$

Solutions to Eqs. (8, 9) involve the determination of the distributed force; in particle-reinforced metal matrix composites, this has been solved by two approaches, based either on a limiting crack-opening displacement or limiting tensile strain in the uncracked ligaments [20].

1. LIMITING CRACK-OPENING DISPLACEMENT APPROACH

The basis of this approach, which is more applicable to the case of coplanar bridges, lies with the assertion that the straining of the uncracked ligaments between two crack faces is limited by the critical crack-opening displacement δ_c that ligament material can withstand. By idealizing the fatigue crack as trapezoidal (Fig. 20c), the crack-opening displacement at any distance x along the crack, δ_x, can be related to the crack-tip opening displacement, δ_{tip} by [47]

$$\delta_x = \delta_{tip}\left(\frac{x + rb}{rb}\right), \tag{10}$$

where b is the specimen ligament, r is the rotational factor, which takes values of 0.195 (elastic deformation) to 0.470 (plastic deformation) [48], and δ_{tip} is given in terms of the near-tip stress intensity K_{tip} and yield stress σ_y by [33]

$$\delta_{tip} = \beta \frac{K_{tip}^2}{E'\sigma_y}, \tag{11}$$

where $E' = E$ in plane stress and $E/(1 - v^2)$ in plane strain, v is Poisson's ratio, and β is a constant varying between 0.3 and 1.0 depending upon the yield strain, work-hardening exponent, and whether plane-strain or plane-stress conditions apply. Since at $x = l$, $\delta_x = \delta_c$, the stress $\sigma(x)$ in the uncracked ligament becomes

$$\sigma(x) = -\frac{\pi}{2\sqrt{2}}\sqrt{\frac{\delta_c rbE'\sigma_y}{\beta}}\left[\frac{x^{1/2}}{(x + rb)^{3/2}}\right], \tag{12}$$

such that the degree of crack-tip shielding can be expressed in terms of the area fraction of ligaments, f_b, the applied stress intensity and l/rb [20]

$$K_s = -f_b K_a[(1 + l/rb)^{1/2} - 1]/[1 - f_b + f_b(1 + l/rb)^{1/2}]. \tag{13}$$

2. LIMITING STRAIN APPROACH

An alternative first-order approach, which is more applicable to the case of overlapping ligaments, represents the bridges as tensile ligaments and defines the extent of the bridging zone by the limiting strain in the ligament furthest away from the crack tip being equal to the fracture strain [20].

With reference to Fig. 20c, the strain $\varepsilon(x)$ in any ligament within the bridging zone can be estimated by assuming it to be equivalent to the strain in a bent beam, with rotational center at point C and neutral plane at the crack tip,

$$\varepsilon(x) = x/rb. \tag{14}$$

Converting to a true strain and substituting into a constitutive law, for the uncracked-ligament material, of the form

$$\sigma(x) = \sigma_y + k\varepsilon(x), \tag{15}$$

where σ_y is the (initial) yield stress and k is a constant, provides a second expression for the degree of crack-tip shielding due to uncracked-ligament bridging, in terms of the area fraction of ligaments, the ratio l/rb, and the (constrained) flow properties of the ligament [20],

$$K_s = f_b \sigma_y \frac{2\sqrt{2l}}{\pi} \left\{ 1 + \frac{k}{\sigma_y} \left[\ln\left(1 + \frac{l}{rb}\right) + 2\left(\sqrt{\frac{rb}{l}} \tan^{-1}\sqrt{\frac{l}{rb}} - 1\right)\right]\right\}. \tag{16}$$

Application of Eqs. (13) and (16) to predict uncracked-ligament bridging in a given system requires values for f_b, l, rb, K_a, σ_y, and k. This has been attempted for overaged $SiC_p/7091$ composites by serial sectioning through fatigue cracks loaded to a ΔK level corresponding to the intermediate range of growth rates [20]; at a ΔK of 8 MPa$\sqrt{m}$ ($K_a = 9$ MPa$\sqrt{m}$), $f_b \approx 27$ to 31%, $l \approx 300$ to 400 μm, rb is typically 1 mm for the tested DB(Mz) geometry, and σ_y and k are 405 and 2200 MPa, respectively, for reinforced bridges. For the fine SiC_p/Al composite where the uncracked ligaments were largely coplanar, the predicted shielding is minimal ($\sim 6\%$)—K_s is less than 0.5 MPa$\sqrt{m}$— consistent with the insignificant difference in growth rates between the reinforced and unreinforced alloys in this range (Fig. 16b). Conversely, in the coarser SiC_p/Al composites where overlapping ligaments are more prevalent, K_s is predicted to be ~ 2.4 MPa$\sqrt{m}$, which is consistent with the much larger shift in growth-rate curves between the reinforced and unreinforced alloys at $\Delta K = 8$ MPa$\sqrt{m}$ in Fig. 16a.

VII. Summary

The use of discontinuous reinforcements to improve the mechanical properties of metallic alloys has provided a relatively less expensive means of produce materials with improved stiffness, strength, wear, and high-temperature resistance for applications where the very highly directional properties of continuous fiber-reinforced composites are not required. In essence, for structural use, such metal matrix composites provide an opportunity to reduce both initial costs and life-cycle operating costs in many weight-critical applications [49]. For example, Lockheed has recently fabricated aircraft vertical tail structures entirely from a SiC-whisker-reinforced 2124 aluminum alloy [50]. However, although initial developments have been motivated by the aerospace industry, the greatest volume potential probably now lies in reciprocating or accelerating mass applications in the commercial/automotive industries or in the area of tribology and thermal management [51]. A notable example of this is the use of an aluminum matrix composite containing low-density chopped-alumina fibers for the Toyota diesel piston introduced in 1983; this, in fact, was the first celebrated commercial application of discontinuously-reinforced metal matrix composites [52].

Whether their primary use is with aerospace or automotive applications, the resistance of metal matrix composites to failure under cyclic or otherwise varying loads remains paramount. However, whereas the addition of such particulate or whisker reinforcements can lead to unacceptably low ductility and fracture-toughness properties, the fatigue properties of metal matrix composites are comparable and in many cases exceed that of their respective unreinforced matrices.

The present brief review has shown that improved fatigue-crack growth properties can be achieved in general with coarser particulate reinforcement-phase distributions, although details on the optimum size, shape, spacing, volume fraction, interface strength, and so forth are uncertain at this time and must await more specific models of their subcritical crack-growth behavior. At ultralow growth rates, however, where mechanistic interpretation suggests that the dominant mechanisms that impede crack advance involve crack deflection around the particles (to induce higher crack-closure levels from asperity wedging) and crack trapping by the particles, the coarser distributions appear to be the most effective. This is illustrated in Fig. 21, where all the fatigue-threshold data [15–22, 24] in the literature for SiC/Al alloys are plotted and compared with the predictions of the crack-trapping model (Eq. 5) for mean particle sizes of 5 and 20 μm. Similarly, at higher growth rates where particle fracture ahead of the crack tip leads to crack-bridging phenomena via uncracked ligaments, the alloys with coarser particle distributions again display the superior resistance to fatigue-crack growth.

Acknowlegments

This work was supported by the Air Force Office of Scientific Research, Contract No. F49620-87-C-0017 as part of the University Research Initiative program on High-Temperature Metal Matrix Composites at Carnegie Mellon University. Thanks are due to Dr. Alan

Rosenstein for his continued support, to L. Edelson, J. J. Mason, Drs. R. J. Bucci, W. H. Hunt, J. J. Lewandowski, T. G. Nieh, H. Rack, and A. W. Thompson for helpful discussions, and to Madeleine Penton for her assistance in preparing the manuscript.

References

1. A. K. Dhingra, *J. Met.*, **38**, 17 (1986).
2. R. C. Forney, *J. Met.*, **38**, 18 (1986).
3. A. P. Divecha, S. G. Fishman, and S. D. Karmarkar, *J. Met.*, **33**, 12 (1981).
4. R. J. Arsenault, *Mater. Sci. Eng.*, **64**, 171 (1984).
5. S. V. Nair, J. V. K. Tien, and R. C. Bates, *Int. Metall. Rev.*, **30**, 275 (1985).
6. D. L. McDanels, *Metall. Trans. A*, **16A**, 1105 (1985).
7. J. J. Lewandowski, C. Liu, and W. H. Hunt, in "Processing and Properties of Powder Metallurgy Composites" (P. Kumar, K. Vedula, and A. M. Ritter, eds.), p. 513. TMS-AIME, Warrendale, Pennsylvania, 1987.
8. C. P. You, A. W. Thompson, and I. M. Bernstein, *Scripta Metall.*, **21**, 181 (1987).
9. W. R. Mohn, *Res. and Devel.*, **54** (1987).
10. D. R. Williams and M. E. Fine, in "Proc. 5th Intl. Conf. on Composite Materials" (W. C. Harrigan, Jr., J. Strife, and A. K. Dhingra, eds.), p. 639, 1985.
11. N. J. Hurd, *Mater. Sci. Tech.* **4**, 513 (1988).
12. D. F. Hassen, C. R. Crowe, J. S. Ahearn, and D. C. Cooke, in "Failure Mechanisms in High Performance Materials" (J. G. Early, T. Robert Shives, and J. H. Smith, eds.), p. 147, 1984.
13. S. J. Harris, *Mater. Sci. Tech.*, **4**, 231 (1988).
14. D. R. Williams, Ph.D. Thesis, Northwestern University, 1985.
15. S. S. Yau and G. Mayer, *Mater. Sci. Eng.*, **82**, 45 (1986).
16. W. A. Logsdon and P. K. Liaw, *Eng. Fract. Mech.*, **24**, 737 (1986).
17. D. L. Davidson, *Metall. Trans. A*, **18A**, 2115 (1987).
18. J.-K. Shang, W. Yu, and R. O. Ritchie, *Mater. Sci. Eng.*, **102**, 181 (1988).
19. T. Christman and S. Suresh, *Mater. Sci. Eng.*, **102**, 211 (1988).
20. J.-K. Shang and R. O. Ritchie, *Metall. Trans. A*, **20A**, 897 (1989).
21. J.-K. Shang and R. O. Ritchie, *Acta Metall.*, **37**, 2267 (1989).
22. D. L. Davidson, "Micromechanisms of Fatigue Crack Growth and Fracture Toughness in Metal Matrix Composites," SWRI Report No. 06-8602/5, Southwest Research Institute, San Antonio, Texas, 1988.
23. S. Preston, A. Melander, H. L. Groth, and A. F. Blom, in "Thermal and Mechanical Behavior of Ceramic and Metal Matrix Composites," ASTM STP, in press. American Society for Testing and Materials, 1990.
24. J. J. Mason, M.S. Thesis, University of California, Berkeley, 1988.
25. P. C. Paris and F. Erdogan, *J. Basic Eng.*, **85D**, 528 (1963).
26. R. O. Ritchie, *Int. Metall. Rev.*, **20**, 205 (1979).
27. W. Elber, *Eng. Fract. Mech.*, **2**, 37 (1970).
28. S. Suresh and R. O. Ritchie, in "Fatigue Crack Growth Threshold Concepts" (D. L. Davidson and S. Suresh, eds.), p. 227. TMS–AIME, Warrendale, Pennsylvania, 1984.
29. R. O. Ritchie and W. Yu, in "Small Fatigue Cracks" (R. O. Ritchie and J. Lankford, eds.), p. 167. TMS–AIME, Warrendale, Pennsylvania, 1986.
30. N. Walker and C. J. Beevers, *Fat. Eng. Mat. Struct.*, **1**, 135 (1979).
31. K. Minakawa and A. J. McEvily, *Scripta Metall.*, **15**, 633 (1981).

32. S. Suresh and R. O. Ritchie, *Metall. Trans. A*, **13A**, 1627 (1982).
33. C. F. Shih, *J. Mech. Phys. Solids*, **29**, 305 (1981).
34. C. Atkinson, *Int. J. Engng. Sci.*, **10**, 45 (1972).
35. C. Atkinson, *Int. J. Engng. Sci.*, **10**, 127 (1972).
36. A. A. Rubinstein, *J. Appl. Mech.*, **53**, 505 (1986).
37. D. L. Davidson, *Acta Metall.*, **36**, 2275 (1988).
38. T. Lin, A. G. Evans, and R. O. Ritchie, *J. Mech. Phys. Solids*, **34**, 447 (1986).
39. T. Lin, A. G. Evans, and R. O. Ritchie, *Metall. Trans. A*, **18A**, 641 (1987).
40. A. G. Evans, in "Fracture Mechanics (20th Symp.)" (R. P. Wei and R. P. Gangloff, eds.), ASTM STP, 1020, p. 267. American Society for Testing and Materials, 1989.
41. S. Kunz-Douglass, P. W. R. Beaumont, and M. F. Ashby, *J. Mater. Sci.*, **15**, 1109 (1980).
42. D. B. Marshall, B. N. Cox, and A. G. Evans, *Acta Metall.*, **33**, 2013 (1985).
43. B. Budiansky, J. W. Hutchinson, and A. G. Evans, *J. Mech. Phys. Solids*, **34**, 167 (1986).
44. L. R. F. Rose, *Mech. Mater.*, **6**, 11 (1987).
45. Y. Mai and B. R. Lawn, *J. Am. Ceram. Soc.*, **70**, 289 (1987).
46. G. C. Sih, "Handbook of Stress Intensity Factors," Lehigh University Press, Bethlehem, Pennsylvania, 1972.
47. J.-L. Tzou, C. H. Hsueh, A. G. Evans, and R. O. Ritchie, *Acta Metall.*, **33**, 117 (1985).
48. C. C. Veerman and T. Muller, *Eng. Fract. Mech.*, **4**, 25 (1972).
49. P. Niskanen and W. R. Mohn, *Adv. Mater. Proc.*, **133**, 39 (1988).
50. Anon, *Res. and Devel.*, **83** (1988).
51. A. Mortensen, J. A. Cornie, and M. C. Flemings, *J. Met.*, **40**, 12 (1988).
52. T. Donomoto, N. Miura, K. Funatani, and N. Miyaka, "Ceramic Fiber Reinforced Piston for High Performance in Diesel Engines," SAE Technical Paper No. 83052, Detroit, Michigan, 1983.

Dynamic Mechanical Properties

A. WOLFENDEN

Mechanical Engineering Department and Amorphous Materials Research Group
Texas A&M University
College Station, Texas

J. M. WOLLA

Composites and Ceramics Branch
Naval Research Laboratory
Washington, D.C.

I. Introduction

Metal matrix composites (MMCs) are becoming a major component of advanced aerospace and hydrospace structures, and knowledge of their mechanical properties is critical. The current demands of dynamic structural applications depend upon high performance, structurally efficient materials for vibration and noise reduction capacity. The dynamic properties are, therefore, especially important. In particular, the mechanical damping and dynamic modulus of MMCs need to be analyzed, both theoretically and experimentally.

The mechanical damping of MMCs, or the ability of the MMC to dissipate vibrational energy, is increasingly of interest to the materials scientist and the materials designer. Damping measurements aid the researcher in basic studies such as crystal defect analyses. Knowledge of the mechanical damping is also critical to the designer seeking to reduce vibrations in rotating machinery and lightweight space structures, or for noise reduction in submarines. The ability to predict and measure the mechanical damping of MMCs accurately is essential. As materials and structures become more complex, reliable predictive capabilities will become even more important, since testing all possible materials configurations will be too cumbersome.

The dynamic modulus, or the stiffness of a material undergoing dynamic loading, is also of interest to both the researcher and the designer. For fundamental research, the dynamic modulus is of use in studies of, for example, interatomic potentials, creep behavior, thermal expansion, Debye temperature, and thermodynamics. On the other hand, design applications require information on the dynamic modulus when considering constitutive relations, fracture mechanics, buckling behavior, and so on. Both measurement and predictive capabilities are essential and become more important as the complexity of MMCs increases.

This chapter will address the analysis of dynamic properties in MMCs by reviewing some of the work in the field to date. The emphasis will be on damping, but dynamic modulus will be included, since the two properties are interrelated. First, measurement techniques for determining damping and modulus will be discussed. Second, theories concerning damping mechanisms and damping capacity will be presented along with some recent results for MMCs. Finally, the dynamic modulus of MMCs will be discussed briefly through theories and recent results. For a more general treatment of dynamic properties in composites, not just MMCs, the reader is directed to a series of extensive review articles [1–4].

II. Measurement Techniques

A. General Theory

The techniques for the measurement of damping and dynamic modulus span 17 orders of magnitude of frequency, from 10^{-6} Hz to 100 GHz. Although the details of the various apparatus are quite varied, there are four major categories: quasi-static methods, subresonance methods, resonance methods, and high-frequency wave propagation methods. Since the present concern is with dynamic techniques, only the last three categories are considered.

Most dynamic techniques have the capability of simultaneous measurement of dynamic modulus and damping, as is demonstrated by an examination of the primary response function for an anelastic solid subjected to dynamic loading [5]. For a periodic stress imposed on a specimen, the relations for stress σ and strain ε are given by

$$\sigma = \sigma_0 \exp(i\omega t), \tag{1}$$

$$\varepsilon = \varepsilon_0 \exp[i(\omega t - \phi)], \tag{2}$$

where σ_0 and ε_0 are the stress and strain amplitudes, $\omega = 2\pi f$ is the circular frequency (f is the vibrational frequency), t is the time, and ϕ is the loss angle by which the strain lags behind the stress. In an ideally elastic material, $\phi = 0$ and $\sigma/\varepsilon = E$, i.e., the elastic modulus. However, most materials are anelastic, so ϕ is not zero and the ratio σ/ε is complex. This complex modulus $E^*(\omega)$ is defined by

$$E^*(\omega) = \sigma/\varepsilon = |E|(\omega) \exp[i\phi(\omega)], \tag{3}$$

where $|E|(\omega)$ is the absolute dynamic modulus. The complex modulus may also be expressed as

$$E^*(\omega) = E'(\omega) + iE''(\omega), \tag{4}$$

where $E'(\omega)$ and $E''(\omega)$ are the real and imaginary parts of $E^*(\omega)$, respectively. From the appropriate vector diagram, it is easy to show that

$$|E|^2 = E'^2 + E''^2, \tag{5}$$

and

$$\tan \phi = E''/E'. \tag{6}$$

Alternatively, E' and E'' are referred to as the storage modulus and loss modulus, respectively, and they are related to the energy stored, W, and energy dissipated, ΔW, per cycle of vibration by

$$W = \int_{\omega t = 0}^{\omega t = \pi/2} \sigma \, d\varepsilon = \tfrac{1}{2}E'\varepsilon_0^2, \tag{7}$$

and

$$\Delta W = \oint \sigma \, d\varepsilon = \pi E''\varepsilon_0^2. \tag{8}$$

Hence, the specific damping capacity (Ψ), or SDC, is given by

$$\Psi = \Delta W/W = 2\pi(E''/E') = 2\pi \tan \phi. \tag{9}$$

For small values of ϕ ($\phi^2 \ll 1$), which is not unreasonable for MMCs, tan $\phi = \phi$ and $\Psi = 2\pi\phi$. The quantity ϕ is often termed internal friction. Thus, the general aim of dynamic measurements is to evaluate the real and imaginary parts of the complex modulus, namely, the dynamic modulus and the damping, respectively.

The dynamic response functions can be measured directly only in experiments conducted at frequencies well below the resonant frequency of the mechanical apparatus used. These experiments, known as subresonance methods, where the inertial force is negligible compared with the elastic force, are simple to do in principle. In practice, however, the measurement of the phase angle ϕ is difficult because the angle is small for most crystalline materials. Therefore, subresonance methods are not often used, and the crystalline materials are tested at frequencies where the inertia of the mechanical system is appreciable. Thus, techniques with resonant systems vibrating at a natural frequency (free decay or forced vibration), and wave propagation methods are considered.

B. Free Decay

A simple model of a resonant system with elastic and inertial components is shown in Fig. 1. This is a one degree of freedom system which can be set into either longitudinal or torsional vibration. Considering the longitudinal case, the periodic force F_a ($= F_0 e^{i\omega t}$) is applied to the inertial mass m, and the displacement of this inertial member is x, while the force acting on the

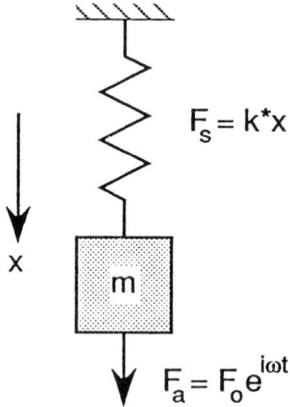

FIG. 1. Mechanical model of a single degree of freedom resonant system with an anelastic spring.

specimen is F_s. In general, F_s and x are given by

$$F_s = C_1 \sigma_{max} \tag{10}$$

and

$$x = C_2 \varepsilon_{max}, \tag{11}$$

where C_1 and C_2 are constants that depend on the dimensions of the specimen. For the case where the material is anelastic, the spring constant k in $F_s = kx$ is replaced by a complex spring constant k* such that

$$F_s = k^*x = k_1(1 + i \tan \phi)x, \tag{12}$$

where $k^* = (C_1/C_2)E^*$ and $E^* = \sigma/\varepsilon$.

In the decay method, the damping of free vibrations, shown in Fig. 2, is observed after the external exciting force is removed ($F_a = 0$). The equation of motion is

$$m\ddot{x} = F_a - F_s, \tag{13}$$

or

$$m\ddot{x} + k_1(1 + i \tan \phi)x = 0. \tag{14}$$

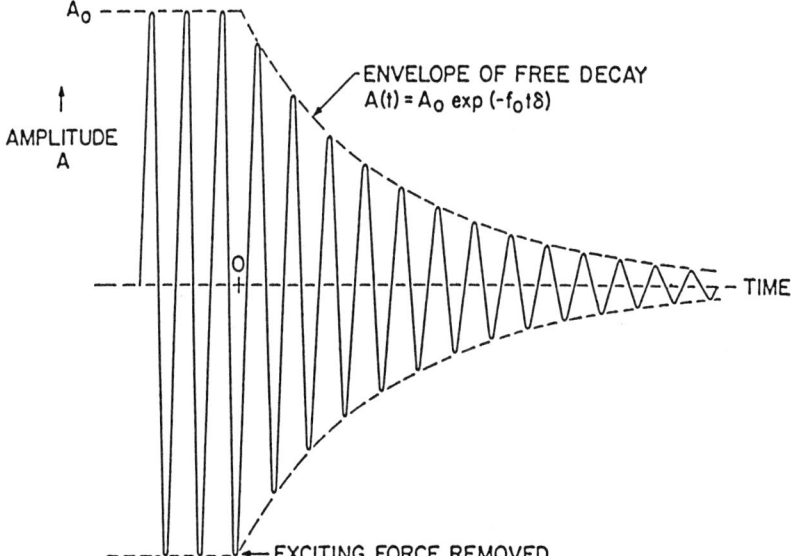

FIG. 2. The free decay or damping of natural vibrations of an anelastic solid (From Nowick and Berry [5]).

The solution of the equation is

$$x = x_0 \exp(-\delta f_0 t) \exp(i\omega_0 t) = A(t) \exp(i\omega_0 t), \tag{15}$$

where f_0 ($= \omega_0/2\pi$) is the natural frequency, δ is a constant, and $A(t)$ is the amplitude of the envelope of the free decay curve (see Fig. 2). If δ is small, the solution corresponds to exponentially damped vibrations. Further analysis yields

$$\omega_0^2 = k_1/\{m[1 - (\delta^2/4\pi^2)]\} = k_1/m \tag{16}$$

and

$$\delta = \pi\phi. \tag{17}$$

The term δ, called the logarithmic decrement, is given by

$$\delta = (1/q) \ln(A_n/A_{n+q}) \tag{18}$$

for vibrations that are q cycles apart, and is directly proportional to the loss angle.

In practical terms, the external exciting force F_a may be applied by eddy-current, electromagnetic, magnetic, mechanical, electrostatic, or piezo-electric transducers, whereas the displacement x may be detected by corresponding transducers or by optical means. An example of a well-known apparatus, the torsion pendulum, developed by Kê [6], is shown in Fig. 3. The typical ranges of strain and frequency explored by such apparatus are 10^{-6} to 10^{-4} and 10^{-5} to 100 Hz, respectively. The popular operating frequency is near 1 Hz. The limits on the temperature of the specimen during a test are often dictated by the difficulties of maintaining a uniform temperature along the relatively long wire specimen. A temperature range of 4 to 1000 K is reasonable.

C. Forced Vibrations

In experiments with forced vibrations, the periodic exciting force F_a is not zero. The equation of motion becomes

$$m\ddot{x} = F_0 \exp(i\omega t) - k^* x. \tag{19}$$

An expected solution of this equation is

$$x = x_0^* \exp(i\omega t) = x_0 \exp[i(\omega t - \theta)], \tag{20}$$

where θ is the angle by which x lags behind F_a. Then

$$x_0^2 = |x_0^*|^2 = (F_0/m)^2/[(\omega_r^2 - \omega^2)^2 + \omega_r^2 \tan^2 \phi], \tag{21}$$

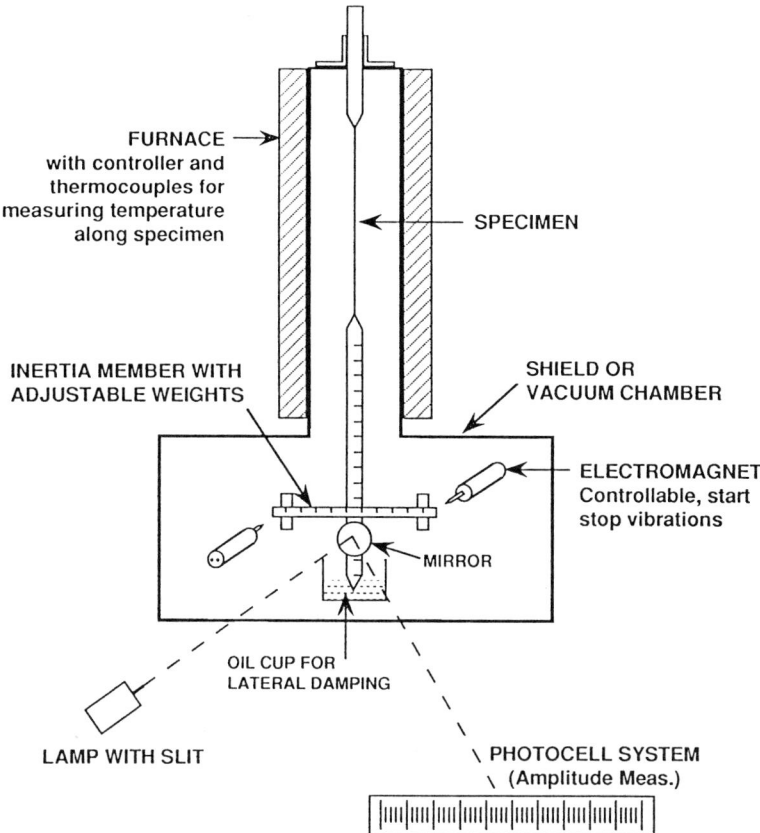

FIG. 3.   Schematic of a representative torsional pendulum apparatus. (Adapted from Nowick and Berry [5]).

where ω_r, the resonant circular frequency, is defined by

$$\omega_r = \sqrt{k_1/m}. \tag{22}$$

Equation (21) describes a resonant response where x_0 goes through a maximum at $\omega = \omega_r$ and decreases to small values at $\omega \gg \omega_r$ and $\omega \ll \omega_r$. The phase angle θ is given by

$$\tan \theta = \omega_r^2 \tan \phi/(\omega_r^2 - \omega^2). \tag{23}$$

Equation (23) shows that θ changes rapidly near ω_r, being equal to ϕ at $\omega \ll \omega_r$ and equal to π for $\omega \gg \omega_r$. For $\omega = \omega_r$, θ is $\pi/2$. Simplification of Eq. (21) is feasible when $\phi \ll 1$. Here the entire resonance peak is localized near $\omega = \omega_r$. Then $\omega_r + \omega = 2\omega_r$, $\omega_r^2 - \omega^2 = (\omega_r - \omega)2\omega_r$ and ϕ and k_1 are essentially constants. The solution for the displacement amplitude reduces to

$$x_0^2 = (F_0/m)^2/\{[4(\omega^2 - \omega_r^2)^2 + \omega_r^2\phi^2]\omega_r^2\}. \tag{24}$$

The term x_0^2 has a frequency dependence of the form

$$x_0^2 \propto (\omega - \omega_r)^2 + \omega_r^2\phi^2 - 1, \tag{25}$$

which is the Lorentzian curve. For this curve, it is well known that if ω_1 and ω_2 are the values of frequency for x_0^2 at half the maximum value (half-power points), then the damping is given by

$$\phi = Q^{-1} = (\omega_1 - \omega_2)/\omega_r = \delta/\pi = \Delta W/(2\pi W) = \Psi/2\pi, \tag{26}$$

where Q is the quality or amplification factor. The equivalent measures of damping, when damping is small, are also shown in Eq. (26). Thus, the determination of the width of the resonance peak at half-maximum in a plot of x_0^2 versus frequency gives the damping. The dynamic modulus can also be determined from the resonance peak via ω_r (see Eq. (22)).

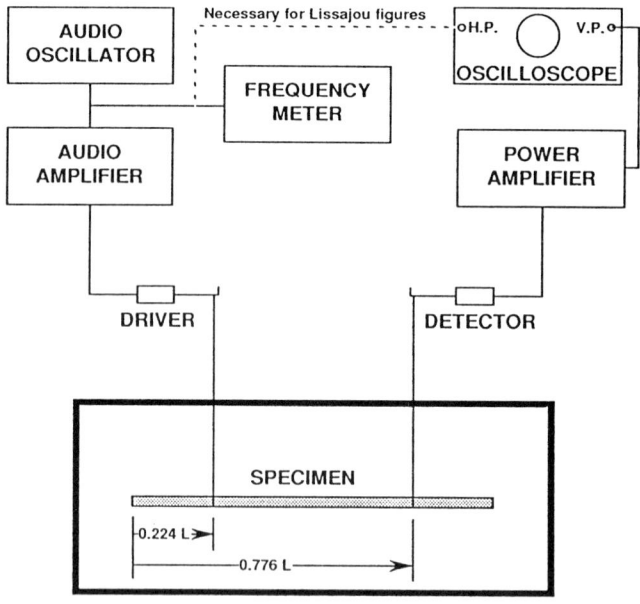

FIG. 4. Schematic diagram of instrumentation and specimen configuration for free–free beam test method. Note that the specimen is suspended at the nodal points.

1. FREE–FREE BEAM

Another example of a forced vibration method for measurement of dynamic modulus and damping is the free–free beam method. Whereas the details of the apparatus at various testing laboratories vary, the basic principle of operation is the same for all. To aid brevity, an apparatus will be described that may be regarded as typical for the technique.

The test method is patterned after the technique of Spinner and Tefft [7]. Figure 4 schematically shows the instrumentation used and the test configuration. The rectangular or cylindrical specimen of free length L is suspended near its nodal points by pure silk or cotton-covered polyester thread (for tests at room temperature). The nodal points for beams of uniform section in a free–free suspension are measured at distances from one end of approximately $0.224 L$ and $0.776 L$ [8]. The threads are suspended from the drive and detector transducers. Details of the string suspension are given elsewhere [9].

The sine wave signal from a function generator is fed through an amplifier to the drive transducer. The signal from the detector transducer is amplified and then sent to a suitable analyzer, which is configured in a peak-averaging mode using exponential averaging. The function generator is swept through the frequency range of interest while the output is examined for resonance peaks. The fundamental frequency is determined this way.

The equations used for calculating dynamic Young's modulus for cylindrical or rectangular specimens are as follows [8]

$$\text{Cylindrical:} \quad E = 64 \, \pi^2 f^2 L^4 \rho/(A^2 d^2), \quad (27)$$

$$\text{Rectangular:} \quad E = 48 \, \pi^2 f^2 L^4 \rho/(A^2 h^2), \quad (28)$$

where ρ is the density, d is the diameter of cylindrical specimens, h is the thickness of rectangular specimens, and A is a constant depending on the shape of the specimen.

The typical ranges of frequency and temperature used for these tests are 100 to 20,000 Hz and 78 to 1000 K. A measure of the damping is obtained by analysis of the Lissajou figure, a hysteresis loop for force and displacement, usually observed on an oscilloscope.

2. PUCOT

Also in the category of forced vibration techniques is the piezoelectric ultrasonic composite oscillator technique (PUCOT). The apparatus for the PUCOT, shown in Fig. 5, consists of a piezoelectric quartz drive (D) and gauge (G) crystal cemented together. This assembly is used to set up longitudinal or torsional ultrasonic (kHz) resonant stress waves in the test

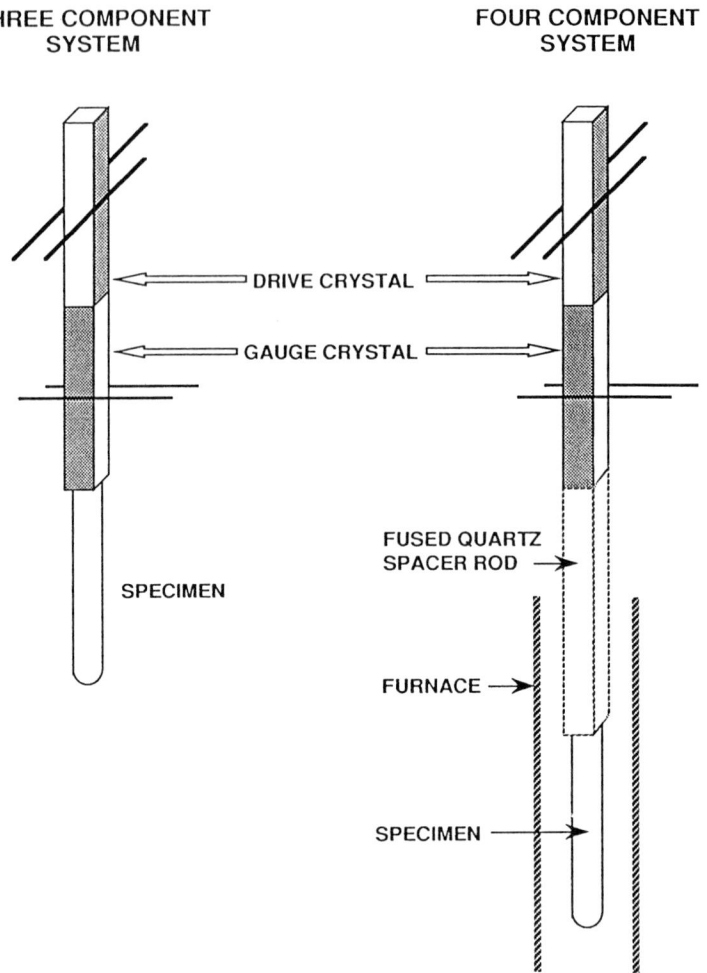

FIG. 5. Schematic of the PUCOT for measurements at room temperature (three-component system) and temperatures above room temperature (four-component system). Longitudinal crystals are shown.

specimen (S) of appropriate resonant length. The fused quartz spacer rod is used for tests above or below room temperature. The resonant system is driven by a closed-loop oscillator that maintains a constant, preselected gauge voltage and, hence, a constant maximum strain amplitude in the specimen. During a test, the drive and gauge voltages and the resonant period of the DGS system are recorded. Standard equations (Appendix 1) are used to calculate dynamic Young's modulus E and damping Q^{-1}. More details are given elsewhere [10–12]. The PUCOT is limited by quartz crystal considerations to frequencies between 20 and 200 kHz. Test specimens may be cylindrical, parallelpiped, or the cross-section may vary in size and shape. The ratio of specimen length to the largest dimension in the cross-section should exceed 5 to prevent dispersion of the ultrasonic waves. The strain amplitude is in the range of 10^{-8} to 10^{-4}, whereas the temperature range covered is 78 to 1400 K.

D. Wave Propagation Method

This technique is one of the better known methods for measuring dynamic moduli in materials. A brief account of the theory behind the technique and of the experimental arrangement is given.

All experimental methods of measuring dynamic modulus use, in some form, the basic wave equation for the propagation of an elastic wave in an elastic medium

$$E = \rho v^2, \tag{29}$$

where ρ is the mass density of the medium and v is the wave speed. The methods that measure transit time t of ultrasonic pulses over a known distance L in the material apply Eq. (29) directly with $v = L/t$, assuming that ρ is known or can be measured. For ultrasonic wavelengths of less than the dimensions of the specimen, two usual modes of wave propagation in isotropic media prevail: the longitudinal mode (with velocity V_L) and the shear mode (with velocity V_S). From these two wave speeds and the density, all the elastic parameters of the material can be calculated (Young's, bulk, and shear moduli, and Poisson's ratio).

The pulse-echo-overlap (PEO) technique is convenient for making transit time measurements with a single transducer [13, 14]. Figure 6 illustrates the principal components used in most tests. The transducer converts the excitation into a mechanical oscillation or sound wave that is coupled into the specimen to propagate at the sonic velocity. The signal is reflected back from the far end of the specimen and is received by the transducer. With this technique, at least two echoes are needed to provide an overlap of successive echoes on an oscilloscope by means of time-delaying circuitry. With the proper instrumentation, transit times are measured with adequate precision ($\pm 0.1\%$ or better). The usual frequencies for the tests are from 5 to 15 MHz.

To extract a measure of damping from high-frequency wave propagation

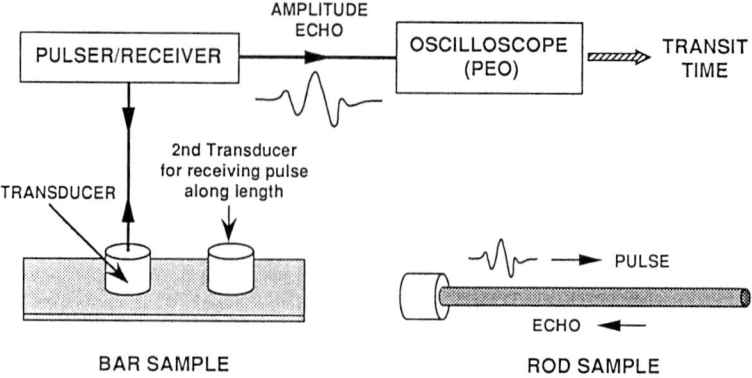

FIG. 6. Schematic of ultrasonic velocity measurement system with direct contact transducers. Transit time measured by pulse-echo-overlap (PEO) technique.

tests, it is necessary to consider the ultrasonic wave as a damped wave propagating in the $+x$ direction,

$$u = u_0 \exp(-Ax) \exp[i\omega(t - (x/v))], \tag{30}$$

where u is the particle displacement and A is the attenuation. For the decaying wave envelope

$$U(x) = u_0 \exp(-Ax), \tag{31}$$

and the amplitude of the envelope at positions x_1 and x_2 $(x_2 > x_1)$ are compared to obtain the attenuation as

$$A = \ln[U(x_2)/U(x_1)]/(x_2 - x_1). \tag{32}$$

Once v and A are known, the modulus and the damping can be obtained from Eq. (29) and

$$\phi = 2Av/\omega = \lambda A/\pi. \tag{33}$$

III. Damping Capacity

A. Mechanisms and Theory

The physical basis for internal friction or intrinsic damping in materials relies on internal structures, defects, or phenomena, called internal variables, which are influenced by the applied stress. The coupling between stress and strain for an anelastic material takes place via one or more of these internal variables. Some common internal variables that provide mechanisms for damping in crystalline materials are concentration of point defects (Snoek [15] and Zener [16] relaxations), dislocation lines with suitably impeded motion, grain boundaries, phase transformations (including ordering of atomic, magnetic, and dipolar types), and differential temperatures between the specimen and its surroundings or between one part of the specimen and another part (thermoelastic or Zener damping). Additional internal variables that provide damping mechanisms in MMCs are associated with the fiber–matrix (F–M) interface; these include Coulomb or sliding friction at imperfectly bonded interfaces, internal stresses generated by coefficient of thermal expansion (CTE) mismatch, and geometrical scattering. In this section, the damping mechanisms that will be applicable in most of the common MMCs, namely, point defect relaxations, dislocation damping, grain boundary damping, thermoelastic damping, and various forms of F–M interface damping will be addressed.

1. POINT DEFECT DAMPING

First consider the types of point defects that give rise to damping. For a crystal with a point defect (a vacancy or an extra atom, either of the crystal material or of an impurity atom), the symmetry of the perfect crystal has been altered. The defective crystal symmetry is termed the defect symmetry and is less than or equal to the perfect crystal symmetry (it cannot be greater). For the case where the defect symmetry is lower than that of the perfect crystal, the criterion for the existence of point defect damping is that there must be more than one distinguishable orientation of the defect.

An elastic distortion surrounds a point defect in a crystal, which causes the point defect to interact with the stress applied to the crystal. The defect behaves as an elastic dipole characterized by a second-rank tensor, λ. Nowick and Berry [5] show that this tensor determines the interaction of the point defect with the stress field. Two defects that differ in orientation but which give rise to the same λ tensor are indistinguishable in terms of their response to the stress. On the other hand, defects with different λ tensors will interact differently with a stress field, and due to one defect–stress field interaction being energetically favored over the others, a redistribution of the orientation of the defects occurs. This defect redistribution is the damping mechanism, and therefore, the criterion for the occurrence of damping due to point defects is that the number of independent λ tensors for a particular point defect must exceed 1. This criterion may be also expressed in the form that the defect symmetry must be lower than the crystal symmetry for damping to occur. For a material with a cubic crystal lattice, the following types of defects having nonsymmetrical strain fields can give rise to damping: interstitial impurities, vacancy–impurity pairs, and divacancies (vacancies with a spherical strain field do not give rise to damping). Important examples of damping arising from the first and second defect types are Snoek relaxation (e.g., interstitial C in Fe [17]) and Zener relaxation (e.g., Zn atom pairs in Ag [18]). Snoek relaxation is expected to contribute to damping in MMCs with *bcc* matrices or fibers, whereas Zener relaxation (solute pair reorientation) is expected to contribute to internal friction in alloys of all three common metallic structures (*fcc*, *bcc*, and *hcp*).

2. DISLOCATION DAMPING

Dislocation damping is expected to play an important role in all crystalline MMCs. This mechanism is based on dislocation movement lagging behind the applied stress. This condition occurs when the dislocations are impeded by obstacles such as point defects. Internal friction peaks in cold-worked Cu, Pb, Al, and Ag were observed at low temperatures (<150 K) by Bordoni [19, 20]. Other internal friction peaks at temperatures between room temperature and 150 K, known as Hasiguti peaks [21], have been studied in plastically deformed Cu and Au.

Most relevant to an understanding of the damping properties of MMCs are two particular mechanisms of dislocation damping: (a) relaxation or resonance absorption, and (b) hysteresis losses. Recognition by Read [22, 23] that damping in pure metals at ambient temperature was largely due to dislocation mechanisms led Koehler [24], and later, Granato and Lücke (G–L) [25, 26], to develop the vibrating string model (termed G–L model hereafter). The G–L theory of dislocation damping is particularly important because it can be used to calculate mobile dislocation densities and minor pinning lengths (spacing between impurity atoms on dislocation lines) from measurements of the strain amplitude dependence of damping.

The G–L model for dislocation damping considers a dislocation segment of length l pinned firmly at its ends and undergoing forced vibration caused by an applied alternating stress of circular frequency ω. The equation of motion examines the effects of an inertial term due to the effective mass of the dislocation, a damping term B, and a restoring force. Further analysis, with N_v dislocation segments per unit volume, dislocation density Λ, shear modulus G, and Burgers vector b, leads to this expression for the amplitude-independent damping (at strains usually below 10^{-6}),

$$\tan \phi = \Lambda B l^4 \omega / (36 G b^2). \tag{34}$$

Realistically, in materials there will be a distribution of loop lengths l, and some pinning points will be stronger than others. At sufficiently high applied stresses, the dislocation may break away from the minor (weak) pinning points while being restrained at the major pinning points, sweep out an increased area on the slip planes, and lead to a sharp increase in damping. For an exponential distribution of loop lengths, the amplitude-dependent damping ϕ_h (for strain amplitudes greater than the breakaway strain amplitude of about 10^{-5}) is predicted by the G–L model as

$$\phi_h = (C_1/\varepsilon_0) \exp(-C_2/\varepsilon_0), \tag{35}$$

where $C_1 = (\Omega\Delta_0/\pi^2)(\Lambda L_N^3/L_c)C_2$ and $C_2 = Ka\delta/L_c$. Here, K is a factor depending on the anisotropy of the elastic constants and the orientation of the specimen with respect to the applied stress, δ is the size factor of the pinning solute atom with respect to the solvent atom, a is the lattice parameter of the specimen, Ω is an orientation factor involved in summing the contributions from dislocations in each of the slip systems, and $\Delta_0 = 4(1 - v)/\pi^2$ with v being the Poisson ratio; L_c is the average minor pinning length and L_N is the major pinning or network length. Equation (35) indicates that a plot of $\log(\varepsilon_0\phi_h)$ versus ε_0^{-1} should be a straight line whose slope yields a measure of L_c and whose intercept at $\varepsilon_0^{-1} = 0$ gives a measure of $\Lambda L_N^3/L_c$ and hence Λ for known or estimated values of L_N (or, similarly, L_N for known or estimated values of Λ). Several damping results for MMCs have been analyzed in terms of the G–L theory and are presented later.

3. GRAIN-BOUNDARY DAMPING

Zener [27] predicted that grain-boundary relaxation should occur due to viscous sliding between adjacent grains in a material at appropriate temperatures. For a regular array of polyhedral grains, with certain simplifying assumptions, Kê [6] calculated that for a material with $v = 0.33$, the relaxation strength, Δ_G, for shear could be as high as 0.64. This high value relies on the assumption that grain-boundary sliding occurs across the entire boundary. The viscous slip model [5] predicts that the height of the internal friction peak, and thus the relaxation strength, is independent of grain size d, as long as d is less than the specimen diameter. The relaxation time τ, however, is predicted to be proportional to d. A satisfactory quantitative theory of grain-boundary relaxation is not yet available, in spite of much experimental research since 1947.

4. THERMOELASTIC DAMPING

Thermoelastic damping originates in the coupling that exists between the conjugate variables stress and strain, and the conjugate pair temperature and entropy. The best example of thermoelastic coupling is thermal expansion wherein the length of a material can be changed either by stretching it or by heating it. Thus the change in entropy with respect to stress (at constant temperature) equals the change in strain with respect to temperature (at constant stress) and is identical to the expansion coefficient α. In the transverse vibration of thin beams, plates, or reeds, thermoelastic relaxation occurs under conditions of inhomogeneous stress. The uniaxial strain induced by bending varies linearly with the distance from the neutral axis (for isotropic materials). Thus, an alternating temperature gradient is set up across the opposite faces of the vibrating beam. Relaxation occurs by heat flow from the hotter, compressed layers to the colder, stretched layers (transverse thermal currents). Analysis predicts the thermoelastic damping to be given by

$$\tan \phi = (E_U \alpha^2 T_0/c_\sigma)[\omega \tau/(1 + \omega^2 \tau^2)], \tag{36}$$

where $\tau = a^2/\pi^2 D_{th}$ for a beam of thickness a, and $\tau = d^2/13.55 D_{th}$ for a rod specimen of diameter d. Here, E_U is the unrelaxed Young's modulus, T_0 is the absolute temperature, c_σ is the specific heat per unit volume at constant stress, $D_{th} = K_{th}/c_\sigma$ is the thermal diffusivity, and K_{th} is the thermal conductivity. For longitudinal thermal currents induced in materials undergoing vibration, Nowick and Berry [5] show that this form of thermoelastic damping is negligibly small, compared with other sources of damping, at frequencies below 100 MHz.

5. FIBER–MATRIX (F–M) INTERFACE DAMPING

One of the attractive characteristics of MMCs is the possibility of introducing sources of damping by controlling the nature of the F–M interface. Poorly bonded interfaces could contribute to damping via a sliding friction mechanism (Coulomb friction). Nelson and Hancock [28] examined the effects of F–M interfacial slip on the damping of fiber-reinforced composites. They modeled a system with frictional energy loss at the interface and viscoelastic energy dissipation in the matrix while the material was subjected to cyclic tensile loading. Good agreement was obtained with results from a model composite consisting of discontinuous, aligned fibers that was loaded in the fiber direction. Kishore et al. [29, 30] also modeled a composite with F–M interfacial slip as a source for damping. In these studies, transverse loading of a linearly elastic matrix material with a rigid cylindrical reinforcement was modeled with regions of no slip, slip, and separation at the F–M interface. The bond between the matrix and the reinforcement was assumed to be purely mechanical, and therefore, only frictional loss was considered. The loss factor was found to be a function of the fiber volume fraction, coefficient of friction at the interface, load amplitude, and precompression present at the interface. Peak loss factors in the range of 1 to 8% were predicted. Unfortunately, gains in damping capacity through interfacial sliding will be accompanied by losses in the stiffness and strength of the composite.

Well-bonded interfaces could lead to increased damping via an increased dislocation density near the F–M interface. Whisker or particulate composites may have added potential for enhancement by this mechanism due to stress concentrations at the ends of the reinforcement that may lead to even greater increases in dislocation density at the F–M interface. This will be discussed in more detail in the following section. Of note here is the potential for developing composites with a combination of well-bonded and poorly-bonded regions to enhance damping while maintaining stiffness and strength.

Geometrical scattering of waves by F–M interfaces could also be considered as a source of internal friction in MMCs. Ledbetter and Datta [31] developed a model that predicts the internal friction for scattering of waves by elastic particles dispersed in an elastic matrix. The model predicts that the internal friction will increase with increased particle volume fractions, larger particle sizes, lower particle aspect ratios (approaching spherical), and increased differences between particle and matrix elastic stiffnesses.

There is further potential for exploiting the F–M interface for damping purposes. It is conceivable that fibers coated with ionic alloy-bonded material layers could result in increased damping due to charge-cloud damping associated with the electromechanical coupling of charged dislocation lines in the ionic material [32]. Interphases of the transient liquid-phase (TLP) type at and near the F–M interface could increase damping [33]. The relaxation effects at the F–M interfaces should be similar to those occurring at grain boundaries (details will be different because of the mismatch of crystal structure, CTE, etc., at F–M interfaces) and may provide additional damping. Rath et al. [34] have described a composite material design where damping is provided by collisions of freely vibrating particles with internal cavities of the matrix in which they are contained.

Hwang and Gibson [35] examined the effect of fiber interaction on damping in discontinuous fiber composites by a strain energy/finite element approach. They did not model the F–M interface explicitly, but considered the fiber-end gap size and fiber aspect ratio. Their analysis indicates that damping will be increased by increasing the fiber-end gap size (for a fixed fiber volume fraction) or decreasing the fiber aspect ratio.

6. COMBINATIONS OF DAMPING MECHANISMS

As with many physical and mechanical properties of mixtures, it is tempting to use the rule of mixtures (ROM) to predict the properties of the MMC from those of the components. Thus, in the case of damping, the general equation would be of the form

$$\Psi_{MMC} = V_m \Psi_m + V_f \Psi_f + V_i \Psi_i, \tag{37}$$

where each damping term could be comprised of several contributions from the various mechanisms that may be operative. For example, the matrix term could be taken as

$$\Psi_m = \Psi_{dislocations} + \Psi_{point\,defects} + \Psi_{grain\,boundaries} + \Psi_{other} \tag{38}$$

Thus, the full picture of damping could be extremely complicated. Equation (37) has been found to be oversimplistic and unable to predict damping in MMCs [36].

Hashin [37] has shown how the correspondence principle may be used to relate the effective viscoelastic functions for composites to the effective moduli. This method was used to derive expressions for effective complex moduli for particle- [38] and fiber- [39] reinforced composites. Application to the specific case of longitudinal damping of unidirectional, continuous fiber-reinforced composites (shown by Rawal et al. [40] and DiCarlo and Maisel [41]) results in the expression

$$\Psi_{MMC} = V_m(E_m/E_c)\Psi_m + V_f(E_f/E_c)\Psi_f, \tag{39}$$

where E_c is the longitudinal elastic modulus for the composite. Equation (39) and others derived by the same technique have been shown to be reasonable predictors of the damping of composites [40–42].

Fortunately, for many materials (including MMCs), one or two damping mechanisms are usually dominant for a combination of strain amplitude, temperature, and frequency, thus reducing the complexity of theoretical predictions. In particular, as will be shown in the following section, predictions of damping levels in MMCs from dislocation damping and thermoelastic damping have been found to be useful. However, for the prediction of damping based on the constituents of the MMC, the correspondence principle remains the most useful tool to date.

B. Results for Metal Matrix Composites

The progress in measuring the damping mechanisms in polymer matrix composites (PMCs), at least for frequencies in the audio range and for thin orthotropic sheets [43], seems to be more advanced that that for MMCs. However, outgassing and material degradation problems with PMCs reduce their effectiveness in extreme aerospace and hydrospace environments, and more resistant materials, such as MMCs, need to be examined for their damping capabilities. Results for some MMCs will be presented that document the dependence of damping on the primary variables that influence this property: strain amplitude, temperature, frequency, and microstructure of composite.

1. STRAIN-AMPLITUDE DEPENDENCE

Results for strain amplitude dependence of damping have been reported for several types of MMCs, and some representative studies will be discussed here. Comparison of the results from the different studies should be made with caution, since different techniques, frequencies, and vibrational modes were used. Strain distributions in the specimens are different for the various techniques, and how the strain is measured and reported varies. Trends with strain amplitude may be compared, but comparisons of damping behavior at exact strain levels are not recommended.

Recently, Wolfenden and Wolla [44] explored the damping at room temperature in several MMCs at 80 kHz and strain amplitudes of 10^{-8} to 10^{-4} with the PUCOT. Data for two MMCs (Al_2O_3/6061 Al, SiC/6061 Al) are shown in Fig. 7, along with the results for a pure Al specimen. It is evident that Al_2O_3/6061 Al and SiC/6061 Al, with either particle or continuous fiber reinforcement, exhibited essentially no dependence of damping

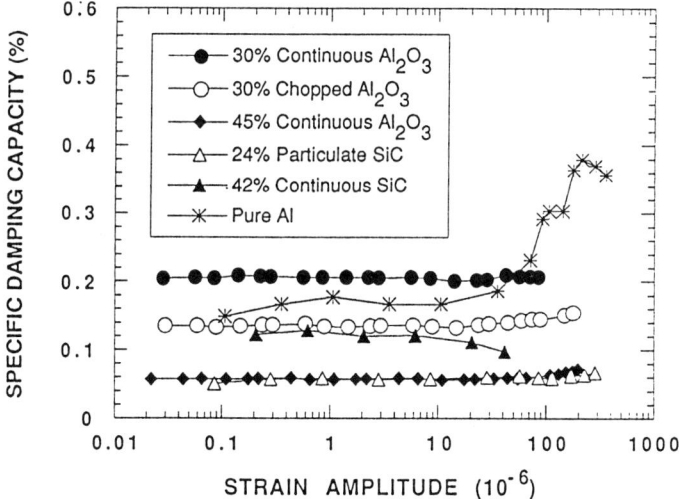

FIG. 7.   Strain-amplitude dependence of SDC in Al_2O_3/6061 Al and SiC/6061 Al MMCs at room temperature as measured with the PUCOT at 80 kHz and in the longitudinal vibration mode. The legend lists the percentage and type of fiber reinforcement. Data for pure Al with breakaway dislocation damping are included for comparison. (Adapted from Wolfenden and Wolla [44].)

on strain amplitude over the range of strains covered, whereas the aluminum sample exhibited breakaway damping typical of dislocation damping near a strain of 10^{-5}. Similar results were obtained by Kohyama et al. [45] in tests of SiC/A1050 Al at 1 Hz with a torsional vibration method. They found breakaway damping in pure Al near 10^{-7}, while the damping at room temperature in the MMC remained independent of strain amplitude until a strain near 5×10^{-5} was reached. It appears in all these cases that the damping arises from mechanisms in the aluminum matrix, speculated to be predominantly dislocation damping, and that the presence of reinforcement suppresses the breakaway behavior by strengthening the matrix through strong dislocation pinning.

Unidirectional W/6061 Al MMCs, also tested with the PUCOT at 80 kHz by Wolfenden and Wolla [44], showed amplitude-dependent damping for strains greater than about 5×10^{-6}, which was attributed to dislocation damping. The amplitude-dependent damping data for W/6061 Al (Fig. 8) have been analyzed in terms of the G–L theory to yield calculated values of the minor pinning length $L_c = 7 \times 10^{-8}$ m and dislocation density $\Lambda = 10^{12}$ m^{-2}. Thus, for the estimated network length (L_N) of 10×10^{-6} m, there were approximately 100 pinning points per network length, which is a reasonable result [25, 26]. The calculated dislocation density is also a reasonable number

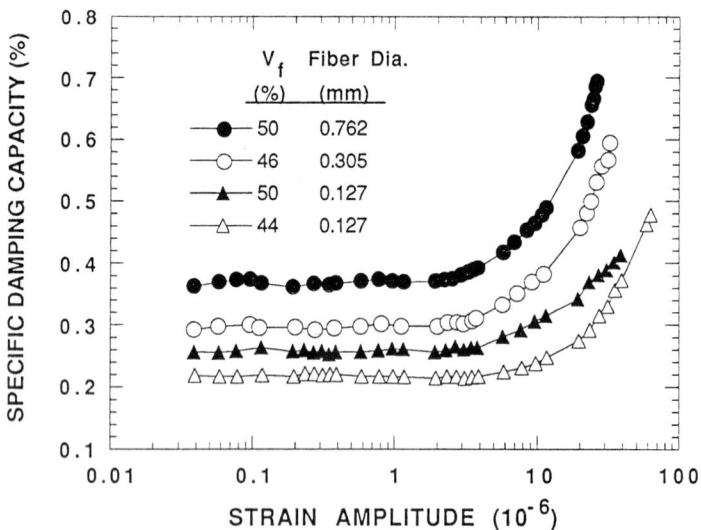

FIG. 8.   Strain-amplitude dependence of SDC at room temperature in unidirectional W/6061 Al MMCs as measured with the PUCOT at 80 kHz and in the longitudinal vibration mode. The effect of fiber diameter is shown. (Adapted from Wolfenden and Wolla [44].)

for an annealed material but is lower than the values calculated by Hartman *et al.* [46] and observed by Arsenault and Fisher [47] in ceramic fiber/metallic matrix (SiC/Al) composites. This discrepancy may be rationalized by considering that relaxation at the F–M interface or a less severe difference in elastic modulus between the metallic fiber and matrix, relative to the ceramic fiber case, may have led to reduced dislocation densities.

Strain-amplitude-dependent damping attributed to dislocation damping was also found by Misra [48] in tests of unidirectional P 55 C/6061 Al MMCs. Longitudinal specimens were tested in free–free flexure (half-power point method of damping determination) and were found to exhibit damping that increased with strain amplitude (strains kept below 10^{-4}). Transmission electron microscopy (TEM) analysis of C/Al interfaces [49] revealed high dislocation densities, good bonding, and three different precipitate structures. Analysis of the data by the G–L model for dislocation damping yielded dislocation densities that agreed with the TEM results. Therefore, it was suggested that dislocation damping within the matrix and at the interface was the operative damping mechanism in this strain-amplitude-dependent region.

Further testing of the damping at room temperature of unidirectional P 55 C/6061 Al was conducted by Rawal *et al.* [40]. They used both clamped-free cantilever beam and free-free beam techniques with a log decrement determination of the damping. Longitudinal and transverse specimens were tested over the strain range of 30×10^{-6} to 800×10^{-6}. Figure 9 shows some representative data from each study. Note that the damping was not significantly higher or lower than that for aluminum in any of the cases and there was no significant difference with fiber orientation. At the strain amplitudes tested, the free–free beam specimens exhibited nearly strain-amplitude-independent response. On the other hand, at the strain levels seen by the clamped-free specimens, strain-amplitude-dependent damping was observed. G–L analysis of these results yielded mobile dislocation densities of 10^{13} m^{-2} that corresponded well with the measured (from TEM images) average dislocation density of 10^{14} m^{-2} in the interfiber matrix regions. It was concluded that dislocation damping was the dominant mechanism in the strain-amplitude-dependent region.

In general, the results from these studies indicate that damping mechanisms within the matrix dominate the damping response to strain amplitude in MMCs. The strain-amplitude-independent response in MMCs is similar to that of the matrix and strain-amplitude-dependent behavior; when it is observed, is attributed to dislocation damping within the matrix. The presence of reinforcing fibers appears to lead to increased dislocation densities in the matrix at the F–M interface, but these dislocations are not always capable of providing additional damping. In view of the usual trend

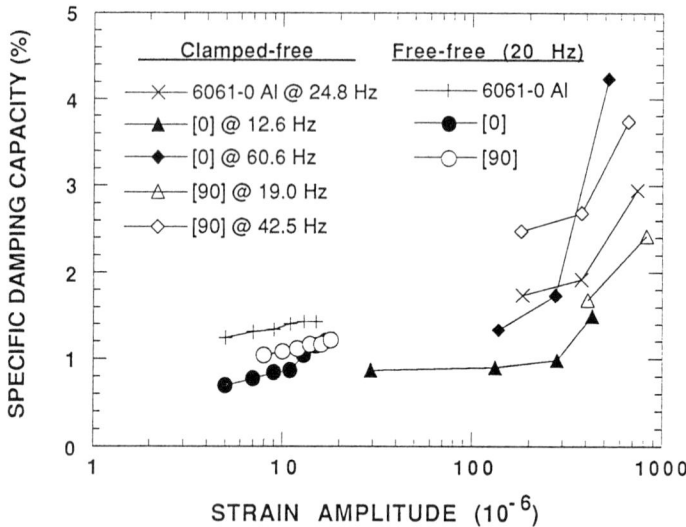

Fig. 9.   Strain-amplitude dependence of SDC at room temperature in unidirectional P 55 C/6061 Al as measured by free–free and clamped-free beam techniques in the flexural vibration mode. Both longitudinal [0] and transverse [90] specimens were tested over several frequencies. (Adapted from Rawal *et al.* [*40*].)

of lower damping in materials with greater stiffness, MMCs provide a benefit in that strength and stiffness are increased relative to the matrix material while damping levels are maintained.

2.   Temperature Dependence

Temperature is another variable that strongly influences damping. As advanced materials and structures will operate in service environments with temperatures much higher and lower than room temperature, it is necessary to examine the damping properties at these temperatures. For MMCs, this requires knowledge of the temperature-dependent properties of the constituents as well as the properties of the resulting combination.

In one such study of the temperature dependence of damping in MMCs, DiCarlo and Maisel [*41*] measured the damping of longitudinal and transverse B/Al specimens from $-200°C$ to $500°C$ with a free–free flexural technique (low strain, near 2000 Hz). The same technique was used to test longitudinal specimens of B(SiC)/Al, Al_2O_3/Al, and SiC/Ti from room temperature to $590°C$ [*42*]. They found that, as a function of temperature, the highly anelastic B fibers dominated the axial damping behavior in longitudinal specimens of B/Al composites (Fig. 10) while the matrix

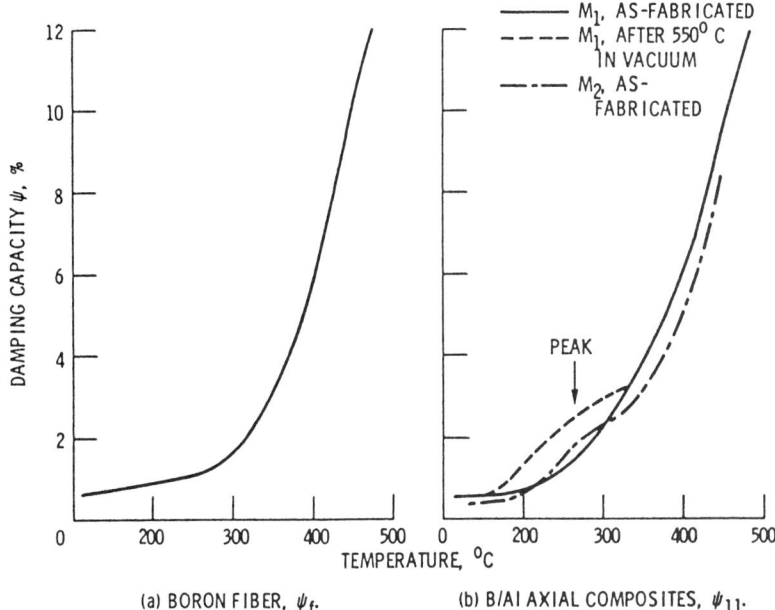

FIG. 10.   Temperature dependence of SDC at ≈ 2000 Hz in (a) B fibers and (b) longitudinal (axial) B/Al MMCs. Note how the axial damping in the MMC, Ψ_{11}, closely follows that of the fiber, Ψ_f. M_1 and M_2 refer to specimens cut from different composite panels. (From DiCarlo and Maisel [41]. Copyright ASTM. Reprinted with permission.)

dominated the damping in transverse specimens (Fig. 11). In fact, the B and B(SiC) fibers improved the damping of the MMCs relative to that of the matrix material, with the increase attributed to intrinsic fiber damping, not F–M interfaces. On the other hand, the use of elastic SiC and Al_2O_3 reinforcement in these studies resulted in decreased axial damping relative to that of the matrix.

K ohyama et al. [45] studied the effect of temperature on the damping of unidirectional SiC/A1050 Al MMCs with a torsional vibration setup. Over the range of 20°C to 400°C, pure aluminum exhibited a significant grain-boundary damping peak. Addition of 30 vol-% SiC fibers resulted in complete suppression of this peak, and an annealing treatment recovered only a small portion of the peak. This suppression phenomena was attributed to the constraint effect of the fibers on the matrix, differences in grain-boundary microstructure (between pure Al and A1050 Al), and differences in the thermal and mechanical histories.

Effects at the F–M interface were speculated to be a strong influence on the damping behavior in SiC/A357 Al with whisker SiC reinforcement (10,

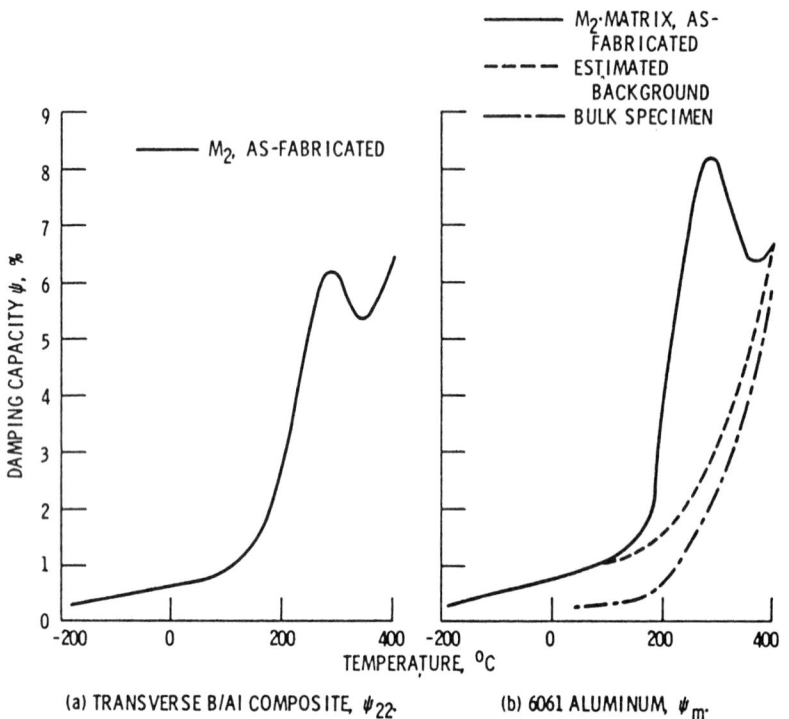

FIG. 11.   Temperature dependence of SDC at ≈ 2000 Hz in (a) transverse B/Al MMCs and (b) 6061 Al. Note how the transverse damping in the MMC, Ψ_{22}, closely follows that of the matrix, Ψ_m. M_2 is a panel designation. (From DiCarlo and Maisel [41]. Copyright ASTM. Reprinted with permission.)

15, and 20 vol-%), which was tested with the PUCOT at 100 kHz and at temperatures from 20 to 200°C [50]. It was found that at 100°C, the SDC increased from 0.3% to 0.9% as the whisker volume percentage increased from 10 to 20%. However, at 200°C the trend partially reversed, with the SDC becoming 1.4% at 10% whisker, 0.5% at 15% whisker, and 0.8% at 20% whisker. It was postulated that at 100°C, high dislocation densities existed in the matrix due to the presence of the whiskers and that the volume of matrix with high dislocation densities, and the associated dislocation damping, became greater with increased whisker volume fractions. At 200°C, annealing of dislocations occurred, and the damping due to dislocations was reduced.

The damping of longitudinal and transverse C/Cu MMCs was examined over the temperature range of 25°C to 875°C with the PUCOT at 150 kHz by Wickstrom and Wolfenden [51]. As shown in Fig. 12, they found a

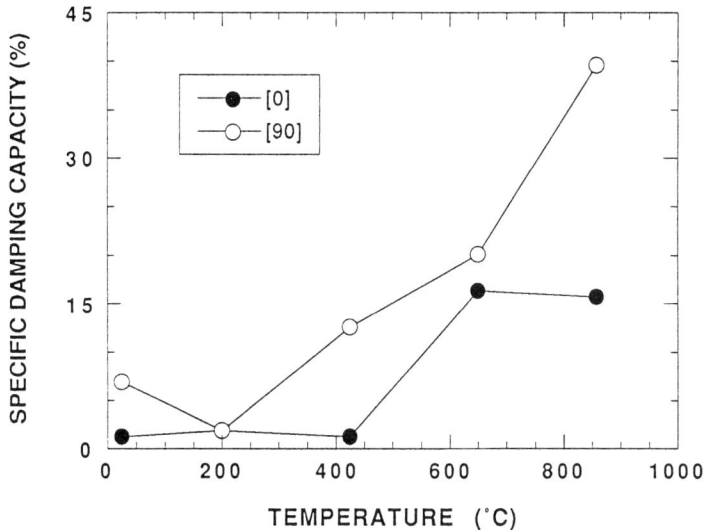

FIG. 12. Temperature dependence of SDC at 150 kHz in unidirectional C/Cu MMCs as measured by the PUCOT in the longitudinal vibration mode. Both longitudinal [0] and transverse [90] specimens were tested. (Adapted from Wickstrom and Wolfenden [51]).

large increase in damping at the higher temperatures (650°C and 875°C) that was speculated to be due to degradation of the F–M interface. The damping in the transverse specimen showed a greater increase due to the additional effect of the temperature-dependent damping of the copper matrix.

As with the strain-amplitude studies, studies of the temperature dependence of damping in MMCs have suggested that the matrix and F–M interface play a dominant role in the damping behavior. Except in the case where the fiber was highly anelastic [41], the temperature-dependent damping behavior of the MMCs was dominated by the matrix and proposed F–M interface effects. The matrix effects were, as expected, more prevalent in transverse specimens. More detailed studies are needed to identify and verify the proposed F–M interface effects on damping in MMCs.

3. FREQUENCY DEPENDENCE

As with strain amplitude and temperature, the frequency of vibration will affect which damping mechanisms are operative in the MMC. For practical applications, noise or vibrations need to be reduced over a particular range of frequencies. Therefore, knowledge of the frequency response of damping for materials and the ability to design materials that damp well over a specified frequency range are needed. Some studies that examined the dependence of damping in MMCs on frequency will be discussed here.

Wren and Kinra [52] measured the damping as a function of frequency (0.2 to 50 Hz) in unidirectional P 55 C/6061 Al composites. A clamped-free beam setup was used with a log-decrement determination of the damping. In Fig. 13, the damping for flexural vibrations is plotted as a function of frequency for single longitudinal and transverse plies, with the calculated thermoelastic damping for aluminum also shown. Damping increased with frequency over this range, and a majority of the damping was due to mechanisms other than thermoelastic damping. Also, transverse damping was slightly higher than axial damping. In a similar study, clamped-free flexure tests of C/Al over a frequency range of 10 to 113 Hz were reported by Rawal *et al.* [40]. Their results, some of which are shown in Fig. 9, show that at most strain amplitudes in the 30×10^{-6} to 1000×10^{-6} range, the damping increased with frequency over this limited range. This was true for both longitudinal and transverse specimens.

Higher frequencies were used in a recent study by Bhagat *et al.* [53], where the damping of C/6061 Al MMCs was measured by a clamped-free beam and log-decrement method for frequencies up to 12 kHz. As shown in Fig. 14, they found damping levels significantly greater than those of the fiber or

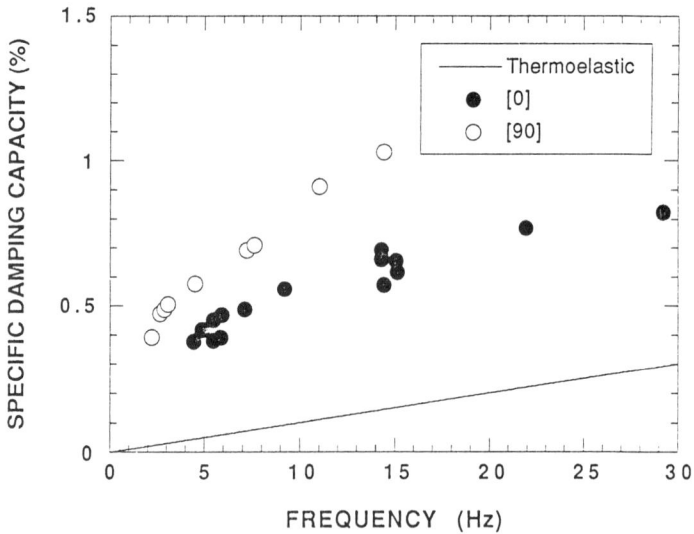

FIG. 13.   Frequency dependence of SDC at room temperature in unidirectional single plies of P 55 C/6061 Al MMCs as measured by a clamped-free beam technique. Both longitudinal and transverse specimens were tested and peak strains were 55×10^{-6}. Prediction of thermoelastic damping for an equivalent thickness of 6061 Al is also shown. (Adapted from Wren and Kinra [52].)

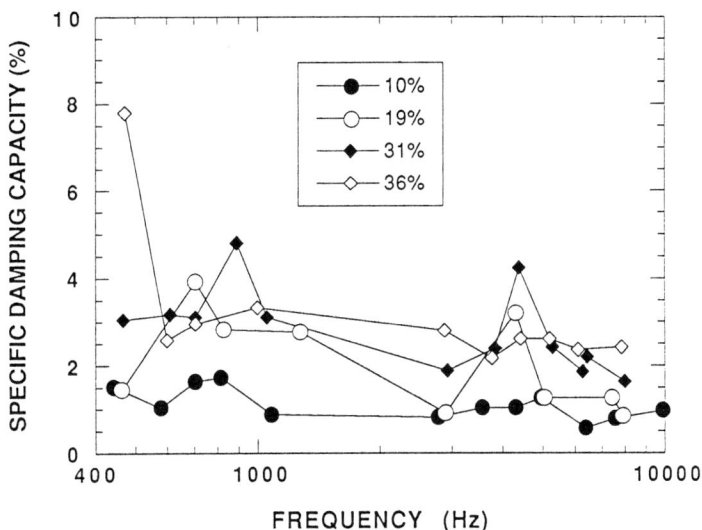

FIG. 14.   Frequency dependence of SDC at room temperature in unidirectional C/Al MMCs as measured by a clamped-free beam technique. Four different fiber volume fractions were tested and damping peaks with frequency were noted. (Adapted from Bhagat *et al.* [*53*]. Copyright ASTM. Reprinted with permission.)

matrix and damping peaks near resonant frequencies in the first and second modes of flexural vibration. The enhanced damping in the MMC relative to that of the constituents was attributed to microplasticity in the matrix and dislocation breakaway damping at the F–M interface.

Timmerman [*54*] used clamped-free beams in both a resonant-dwell and a log-decrement mode to test several MMCs over the range of 4 to 10,000 Hz. Aluminum and magnesium reinforced with continuous carbon fibers exhibited the highest specific damping capacities (0.7 to 2.3%) and little frequency dependence, but the damping values were not measurably greater than those obtainable from the matrix materials alone. Aluminum reinforced with SiC whiskers or particles had the lowest specific damping capacities measured (0.2 to 1.3%). As a function of frequency, the SiC/Al MMCs exhibited minima, near 2000 Hz for a 45% particulate composite and near 3500 Hz for a 20% whisker-reinforced composite.

Steckel and Nelson [*55*] tested a variety of carbon fiber-reinforced mag-nesium composites over the range of 12 to 350 Hz with a clamped-free beam setup. Selected data are shown in Fig. 15. In all cases (varying fiber modulus, fiber volume fraction, matrix alloy, surface foils, and F–M interface quality), a mild frequency dependence was found, with damping increasing with frequency. Ross and Rubin [*56*] used a similar setup to test graphite and

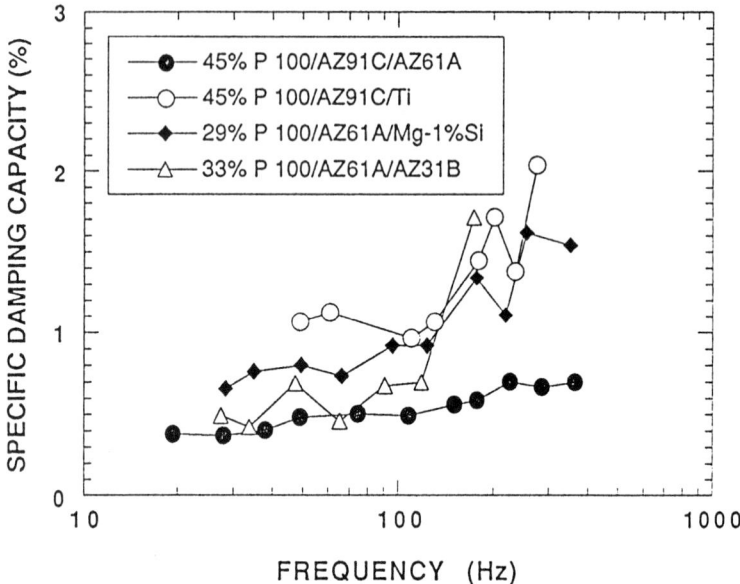

FIG. 15. Frequency dependence of SDC at room temperature in unidirectional C/Mg MMCs as measured by a clamped-free beam technique. Longitudinal specimens were used and the maximum strain in the specimens was 2×10^{-6}. The legend shows the fiber volume percentage followed by fiber/matrix/surface foil description for each specimen. (Adapted from Steckel and Nelson [55].)

silicon carbide (particle or whisker) reinforced aluminum composites. They detected a small drop in damping from 53 to 111 Hz for these materials. They also observed that, relative to aluminum, graphite reinforcement reduced the damping and SiC reinforcement increased the damping.

Crawley and van Schoor [57] used a form of free–free flexural vibration (beam specimen in free fall) to study damping in C/Al and C/Mg MMCs. Results from the MMCs were compared to predictions of thermoelastic damping as a function of frequency (340 to 1985 Hz) for an identical size sample of the matrix alloy. In the comparisons, it was assumed that the carbon fibers were lossless (contribute no intrinsic damping), and therefore, damping in these MMCs should come only from the matrix or the F–M interface. The thermoelastic prediction (matrix only) provided a lower bound for damping in C/Al composites, with the average measured value slightly higher than the prediction, indicating that the matrix was the dominant source for damping in this MMC. On the other hand, for C/Mg, the thermoelastic prediction provided a lower bound with the average measured value significantly higher than the prediction. It was inferred that a significant loss mechanism other than the matrix or the fiber was active in C/Mg at the test frequencies, with poorly bonded F–M interfaces considered a leading possibility.

The study of damping in SiC/Al from 0.001 to 10 Hz with a torsion device by Kohyama *et al.* [45] found a damping peak near 0.07 Hz. The peak disappeared after heat treatment. They attributed the peak in the as-received material to relaxation at the F–M interface.

An investigation of damping in the kilohertz range and over four orders of strain amplitude in an Al_2O_3/Pb MMC with the PUCOT failed to show any significant frequency dependence [58]. Moreover, the damping in the MMC (45 vol-% Al_2O_3) was independent of strain amplitude over the range 10^{-7} to 4×10^{-4}. The restricted range in frequencies used prevented a definitive exploration of the frequency dependence. Although the G–L theory of amplitude-independent damping predicts that damping is proportional to frequency [see Eq. (34)], many experimental measurements of damping in metals, alloys, and ceramics [36] in the frequency range 30 to 200 kHz have failed to support this prediction.

As these results indicate, frequency-dependent damping was usually observed at the lower frequencies where thermoelastic damping is operative. When distinct damping peaks were found, the behavior was attributed to mechanisms at the F–M interface. There are differences among studies as to the effects of the fiber and the F–M interface on damping in MMCs relative to the matrix material. For similar MMCs, some authors found damping levels greater than those found in the matrix and speculated on damping mechanisms at the F–M interface, while others found no evidence of interface effects and measured similar or lower damping levels in the MMC relative to that in the matrix. Differences in measurement techniques could account for many of the apparent differences, but an even larger effect may be differences in the actual microstructure of the MMCs used in each study due to variations in material fabrication techniques.

4. DEPENDENCE ON COMPOSITE MICROSTRUCTURE

As alluded to in the previous section, the microstructure of MMCs can be quite complex and may vary from specimen to specimen or even within the specimen itself. Variations include fiber volume fraction, fiber form (continuous, whisker, or particle, for example), fiber orientation, F–M interface area, quality of F–M bond, and matrix microstructure, on each of which damping in the MMC may be dependent. The degree to which these characteristics of the MMC can be controlled depends largely on the material synthesis process. In practice, therefore, it is often difficult to isolate and examine the dependence of damping on a single microstructure variable. With this in mind, some studies that attempted to examine the effect of some of these variables on damping in MMCs will be reviewed.

Damping as the fiber volume fraction, and thus the F–M interface area, varied was examined for three MMCs [44]. For Al_2O_3/Al, the trend (Fig. 7 shows a portion of the data) was for decreased damping as the F–M interface area increased. Therefore, increasing the F–M interface area did not increase the level of damping, even though it was hoped that increasing the volume of material where mechanisms such as dislocation damping could operate would increase the level of damping. It is possible that the material volume with dislocations available for damping was increased but the dislocations were strongly pinned and unable to provide damping. In addition, it is possible that the higher volume percentage of a low-damping fiber resulted in the lower strain-amplitude-independent damping for the Al_2O_3/Al MMC. For SiC/Al, there was no clear trend for damping as a function of fiber volume fraction. The W/Al MMC results (Fig. 8) showed that there was a trend toward lower damping as the fiber diameter decreased, and therefore, the F–M interface area increased (at constant volume fraction). In this case, the W fiber itself appeared to control the damping behavior, not the matrix or F–M interface.

Another study discussed previously also examined the effect of fiber volume fraction on damping in MMCs, whisker-reinforced SiC/A357 Al in this case [50]. At 100°C, the SDC was observed to increase threefold as the whisker volume fraction increased from 10 to 20%. The increase was postulated to be due to the increased F–M interface area at the higher volume fraction and an associated increase in operative dislocation damping at these interfaces.

Bhagat et al. [53] also examined the effect of fiber volume fraction and the F–M interface area on damping in MMCs. In flexural tests of C/6061 Al with 10 to 36 vol-% reinforcement, they found that peak damping (damping peaks as a function of frequency as shown in Fig. 14) increased with fiber volume fraction up to a volume fraction near 30% and then decreased with higher volume fractions. They attributed this behavior to there being an optimum fiber volume fraction where the buildup of elastic energy equaled the cumulative loss of energy through matrix plasticity and dislocation breakaway damping at the F–M interface. On the other hand, off-peak damping increased with fiber volume fraction, with no peak as a function of volume fraction observed. It was speculated that different damping mechanisms were dominant during off-peak damping.

The F–M interface was the focus of attention in tests by Misra and LaGreca [59] on C/Al and C/Mg MMCs with continuous fiber reinforcement and SiC/Al with discontinuous reinforcement. In cantilever beam tests, they found that the damping from the discontinuous SiC/Al specimens was similar to that of the matrix alone, even as the fiber volume fraction varied from 17 to 30%. Examination of fracture surfaces in these specimens indicated that the F–M interface strength was greater than that of the matrix, which led to the conclusion that all the elastic and anelastic strains were concentrated in the matrix. This, in turn, led to matrix domination of the damping behavior. On the other hand, the MMCs with unidirectional continuous fiber reinforcement exhibited higher damping than the matrix material for both longitudinal and transverse fiber orientations. In addition, the damping increased as fiber volume fraction increased. Examination of the microstructure of these MMCs revealed that deformation was concentrated at the diffusion-bonded interfaces between precursor wires (not the actual F–M interface). It was concluded that deformation and, therefore, vibrational energy dissipation was concentrated at these interfaces. Further study of unidirectional C/Al MMCs by Rawal and Misra [60], where well- and imperfectly bonded composites were tested, also indicated that interfaces provided significant damping mechanisms that led to enhanced damping levels for the MMC relative to the matrix material. The highest damping levels were measured for the imperfectly bonded composites. Dislocation damping and frictional damping at the interfaces were suggested as possible mechanisms for the increased damping. A subsequent report by Rawal et al. [40] did not confirm that imperfectly bonded composites provided enhanced damping relative to well-bonded composites.

A detailed study of the effect of ply angle on flexural damping in C/Al MMCs was performed by Wren and Kinra [52]. They measured the flexural damping at 35 Hz for cantilever-beam specimens of symmetric 4-ply C/Al. Figure 16 shows the results as the ply angle varied from 0 to 90° along with a calculated curve from laminate theory. It was concluded that laminate theory provided an adequate prediction for the damping of the MMC laminate.

It is apparent from the studies discussed in this and previous sections that the ability to incorporate significant damping at the F–M interface in MMCs, a desirable goal, has yet to be proved. Some studies [48, 50, 51, 53, 57, 59, 60] attributed a portion of measured damping to the F–M interface, whereas other studies [40–42, 44, 54, 55], some of which specifically examined the damping in MMCs with good and poor bonding where the poorly bonded composites were presumed to yield higher damping, found essentially no enhancement of damping in MMCs by interfaces. The fibers and matrix materials used in MMCs are usually selected to meet other property goals (e.g., stiffness, strength, corrosion resistance), leaving the F–M interface as the remaining site for potential enhancement of damping in MMCs, but an effective means by which to accomplish this enhancement has yet to be found.

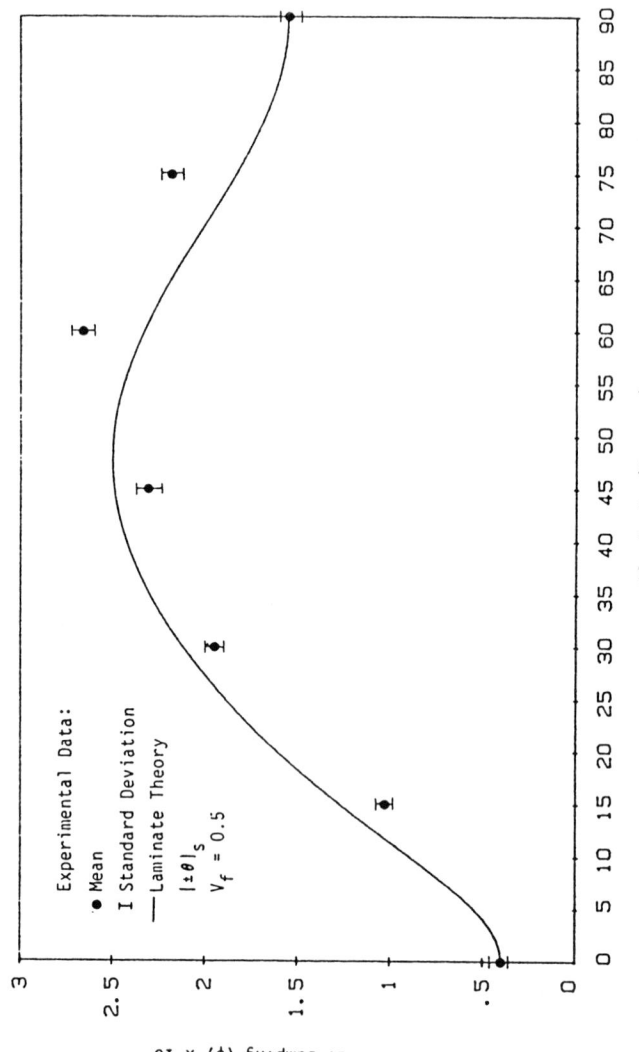

FIG. 16. Ply-angle dependence of flexural SDC at room temperature in symmetric 4-ply P 55 C/6061 Al MMCs and comparison to laminate theory prediction. Measurements made with clamped-free beam apparatus at 35 Hz and a strain of 55×10^{-6}. (From Wren and Kinra [52]. Copyright ASTM. Reprinted with permission.).

IV. Dynamic Modulus

As described previously, damping and dynamic modulus are interrelated, and a discussion of some dynamic modulus studies applicable to MMCs will be beneficial. Theories for predicting modulus in composites have been well documented and will be referenced, followed by some representative results from studies of MMCs. Also of interest is a recent round-robin study of dynamic modulus measurements on metallic materials undertaken by a group of ASTM members [61].

A. Theory

In order to simplify the analysis of dynamic modulus in composites, it will be assumed that the wavelength is large compared with the characteristic dimensions of the composite. Examples of typical characteristic dimensions in MMCs are matrix grain sizes and particle or fiber diameters. With this assumption, which is valid for all the studies considered in this review, it becomes possible to apply theories based on the analysis of static modulus to the analysis of dynamic modulus [62]. Rather than going into details of the many theories that may be used to predict dynamic modulus of MMCs from the properties of the constituents, the reader is directed to an extensive review of analysis techniques by Hashin [63]. In this survey of techniques, the theoretical approaches are grouped into three main categories: (1) direct approaches for exact solutions, (2) variational bounding techniques, and (3) approximations such as self-consistent schemes. Christensen [64] also provides a review of the methods for predicting properties of composites in terms of the constituent properties. Exact solutions are available for special cases of composite microstructure but are usually unavailable or become too cumbersome as the complexity of the composites and their constituents increases. As a result, bounding techniques and approximations are frequently used in comparisons with experimental results.

Specific mention of one recent model by Ledbetter and Datta [65] will be made, since it considers the unique characteristics of particulate MMCs when predicting effective wave speeds and the associated dynamic moduli in these materials. The model allows the particles, modeled as prolate spheroids, to be distributed randomly, both in position and orientation. This allows wave-speed anisotropy and nonhomogeneous particle distributions to be considered, which provides a more accurate representation of the actual material structure.

B. Results for Metal Matrix Composites

An initial study of interest is that of Ledbetter [66], where the dynamic modulus of several composite materials (MMCs and PMCs) was measured by a technique similar to the PUCOT and compared with modulus values obtained by static techniques. It was observed that the modulus values were measured more accurately by dynamic means than by static means and that the static moduli scattered about the dynamic values. This is an advantage, since in addition to better accuracy, dynamic measurements are often easier to perform than static measurements and can accommodate a wider variety of specimen shapes and sizes.

1. DEPENDENCE ON COMPOSITE MICROSTRUCTURE

The dependence of dynamic modulus on composite microstructure in MMCs will be examined in terms of the effects of fiber volume fraction, fiber form, fiber orientation, fiber distribution, and internal defects on the measured and predicted dynamic moduli.

In MMCs with continuous, unidirectional reinforcement, the dynamic modulus as a function of fiber volume fraction is expected to follow the ROM. For the case of W/Al, tested at room temperature with the PUCOT at 80 kHz [36], the agreement between measured values and ROM predictions was excellent. Fiber volume fractions from 36 to 55% were tested, and the data fit very well to a straight line ($E = 68.2 + 3.36V_f$, $R = 0.999$). The extrapolations to 0% fiber (6061 Al) and 100% fiber (W) yielded values of modulus of 68.2 and 404.2 GPa respectively, both in excellent agreement with handbook values of moduli for these materials.

The room-temperature dynamic moduli for both fibrous and particulate SiC/Al MMCs were measured by Wolla and Wolfenden [67] (PUCOT at 80 kHz), with the data shown in Fig. 17. For the fibrous or continuous fiber case, most of the data agreed well with the ROM prediction. The points that fell away from the straight line were from specimens that were found to have significant void volume fractions and fiber damage as a result of the synthesis and machining processes. The measured dynamic moduli for these specimens should have been lower than the ideal ROM prediction due to these defects. Accounting for these defects, it was found that the dynamic modulus measurements were accurate. Therefore, deviation of the dynamic modulus measurements from ROM predictions served as an indicator of microstructural defects. The particulate SiC/Al composites are a case where the use of predicted bounds is necessary because an exact solution does not exist. The Hashin–Shtrikman bounds [68] are used in Fig. 17, and the measured

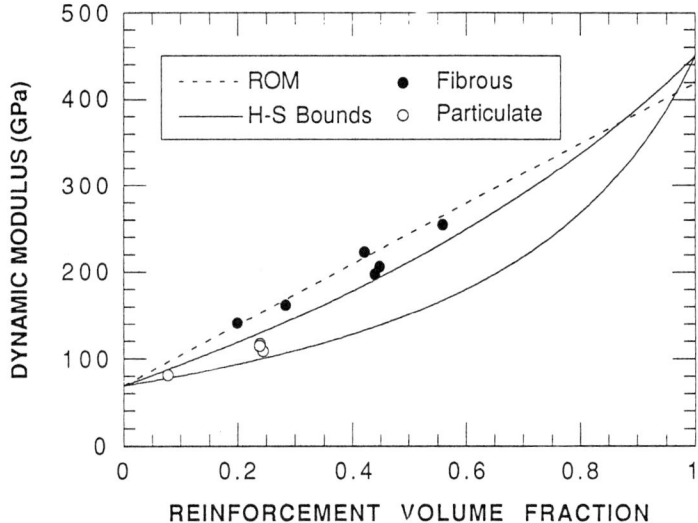

FIG. 17. Dynamic modulus as a function of reinforcement volume fraction for SiC/Al MMCs. Data are shown for fibrous MMCs with corresponding ROM prediction and for particulate MMCs with corresponding Hashin–Shtrikman (H–S) bounds. (Adapted from Wolla and Wolfenden [67]. Copyright ASTM. Reprinted with permission.)

dynamic moduli for the particulate SiC/Al fell within the bounds as they must do to be valid.

Particulate SiC/Al was also tested by Ledbetter and Datta [69] with a setup similar to the PUCOT. In this study, the dynamic modulus was tested in the three principal directions of a rolled-plate specimen of 30 vol-% SiC/Al. The rolling induced anisotropy in the specimen that was measurable. Their results measured values (at 90 kHz and room temperature) that agreed well with the values predicted by the model that allowed anisotropy in the MMC [65].

2. TEMPERATURE DEPENDENCE

The studies discussed in the previous section all dealt with room-temperature testing. Now, the effects of temperature on the dynamic modulus of MMCs will be examined along with comparisons to predictions of the temperature-dependent behavior.

The temperature dependence of dynamic elastic modulus for two different aluminum matrix MMCs was measured with the PUCOT and was found to be a strongly matrix-dominated property [58]. The modulus as a function of temperature up to about 260°C for two MMCs is shown in Fig. 18. Both materials were reinforced with 15 vol-% SiC particles. One MMC had a

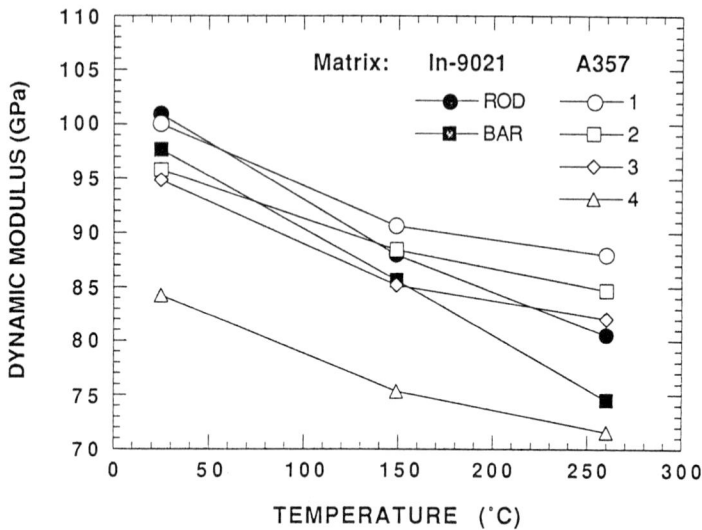

FIG. 18.   Temperature dependence of dynamic elastic modulus in 15% particulate SiC/Al MMCs as measured by the PUCOT in the kHz range. A rod and bar specimen of In-9021 matrix MMC was tested and four different specimens of A357 matrix MMCs. (Adapted from Cribb *et al.* [*58*]).

matrix of In-9021 and was in the form of a rod extrusion or a bar extrusion. The second MMC had a matrix of A357 casting alloy. The temperature dependencies were linear with values of R over 0.96 for all specimens. The slopes dE/dT were in the range of -0.088 to -0.099 GPa/K for the In-9021 matrix specimens, and -0.049 to -0.058 GPa/K for the Al A357 matrix specimens. The values of $-(1/E)(dE/dT)$ for all the specimens fell in the range of 5×10^{-4} to 10×10^{-4} K^{-1}, which is the range documented by Friedel [*70*] for many common metals at 297 K. Therefore, it seems likely that the decrease in modulus as testing temperature T_t increases for these Al-based MMCs can be attributed mostly to the matrix. This is logical in two senses: First, the MMC is 85 vol-% Al; and second, there is a correlation between the melting point T_m, the modulus, and the interatomic forces in materials, making the decrease in modulus high for Al ($T_t/T_m \approx 0.6$) and low for SiC ($T_t/T_m \approx 0.2$). Similar behavior was observed in $Al_2O_3/6061$ Al by Wolfenden and Wolla [*71*].

The DiCarlo and Maisel studies [*41, 42*] also examined the temperature dependence of dynamic modulus in MMCs. A clamped-free beam setup measured the flexural dynamic modulus at temperatures up to 500°C of aluminum matrix MMCs with continuous B, B(SiC), SiC, or Al_2O_3 fibers. SiC/Ti–6Al–4V and SiC/Ti were also tested. Predictions of the temperature dependence of longitudinal and transverse modulus were made by using the ROM for the longitudinal modulus and the Halpin–Tsai equation for the transverse modulus. These predictive models require knowledge of the temperature-dependence of the moduli of the constituents. They found that the B/Al exhibited a predictable nonlinear dependence of flexural dynamic modulus on temperature caused by the anelastic behavior of the constituents (both B and Al) over this temperature range. They also observed that the MMCs with SiC and Al_2O_3 fibers exhibited an essentially linear loss of longitudinal modulus with temperature that agreed with the results described earlier [58, 71]. They attributed this linear behavior to the elastic deformation of the fibers over this temperature range (fibers dominated axial deformation). The combination of being able to measure and predict accurately the temperature dependence of dynamic modulus (or damping) led to the ability to use temperature dependence tests to monitor the effect of the environment on composite microstructure. Deviations of the measured behavior from the predicted behavior were traced to specific changes in the microstructure. In a similar manner, Wickstrom and Wolfenden [51] used temperature-dependent modulus and damping data to postulate the occurrence of F–M interface degradation at high temperatures in C/Cu MMCs.

As the B/Al results showed, the temperature dependence of elastic modulus for MMCs is not always a monotonic linear function. Another example of nonlinear dependence is shown in Fig. 19 for unidirectional P 55 C/6061 Al [44]. Two specimens were tested: Specimen A was tested from 25 to 150°C, then retested over the range of 50 to 100°C; specimen B was tested once over the range of 25 to 150°C. In each of the three tests, as the temperature increased, the modulus decreased first to a minimum value and then began to increase. The minimum modulus was 150 to 160 GPa in the temperature range 50 to 75°C. A hysteresis effect with respect to thermal cycling was noted as successive tests with the specimen A resulted in different responses. Liu [72] found similar effects for C/Al in which Young's modulus decreased with decreasing temperature over the range of 80 to −60°C. These results can be attributed to the strong dependence of the modulus of the fiber on the residual stresses that exist at the F–M interface as a result of prior thermal history and the mismatch in CTEs between the fiber and the matrix. It is possible to alleviate the residual stresses by successive thermal cycling runs and therefore to reduce the extent of the initial modulus decrease.

The above studies of room-temperature and temperature-dependent damping indicate that dynamic modulus measurements are accurate, predictable with relatively simple theories, and sensitive to the composite microstructure. The effects of internal defects, material anisotropy, and microstructural changes are measurable and can be identified through dynamic modulus measurements.

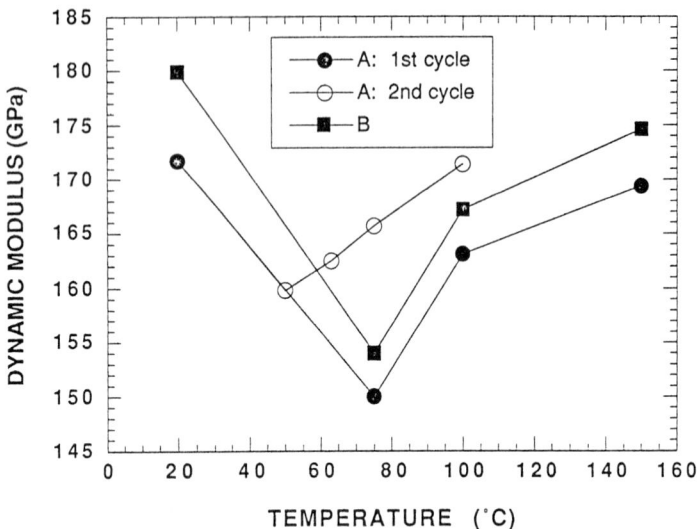

FIG. 19. Temperature dependence of longitudinal dynamic modulus in two unidirectional C/Al specimens (A and B) as measured by the PUCOT at 80 kHz. Specimen A was cycled through the temperature range twice. (Adapted from Wolfenden and Wolla [*44*]).

V. Summary

An examination of the current state of damping studies in MMCs shows that much work needs to be done. As for predictive capabilities, the correspondence principle shows the most promise but is severely limited due to a lack of exact solutions for mechanical properties that can be carried over into the complex damping domain. As for measurement of damping in MMCs, each technique has its advantages and disadvantages. Unfortunately, the many factors that influence damping, including but not limited to, frequency, strain amplitude, temperature, vibration mode, composite micro-structure, and material fabrication processes, render the results presented in previous sections difficult, if not impossible, to compare in order to draw definitive conclusions. These complications arise in testing any material but are specially detrimental to analysis of MMCs due the complex micro-structures and interactions between constituents. For practical applications, damping tests and analyses must be limited to the range of conditions the MMC will expect to see in service, and even this is not always possible with existing techniques.

Beyond being able to predict and measure the damping of MMCs is the need to enhance the damping in these materials. Attempts to accomplish this have been largely unsuccessful because the measured values are usually below levels of practical engineering significance. The fact that the MMCs often provide damping similar to that of the matrix material while greatly improving other properties is positive, but the goal is to improve damping in conjunction with the other property gains in order to reduce or eliminate the need for external damping treatments. Taking advantage of the hetero-geneous nature of MMCs, F–M interfaces in particular, appears to be a promising method of increasing the damping, but studies have not yet been able to show definitively this to be effective.

The simpler case of dynamic modulus analysis and measurement has met with much more positive results. Though not discussed in detail here, predictive capabilities are much more advanced, and excellent correlation with experimental results has been obtained. Studies have shown that dynamic modulus can be measured accurately and that these measurements can be used to monitor other phenomena within MMCs. Measurement of dynamic modulus in MMCs often provides a more flexible and accurate alternative to standard static testing technqiues.

Appendix
PUCOT Equations for the three-component system.

$$E = \text{Young's modulus} = 4\rho L^2/\tau_S^2,$$

$$\varepsilon_{11} = \text{longitudinal strain amplitude} = \{(\sqrt{2}\pi C_m)/N\lambda\}V_G,$$

$$Q^{-1} = \text{damping} = \{(2/(m_S C_m))(N\tau_s/\pi)^2\}V_D/V_G,$$

where

$$\tau_S = \sqrt{m_S \tau_{DG}\tau_{DGS}/A}$$

$$A = \sqrt{\tau_{DG}^2 m_{DGS} - \tau_{DGS}^2 m_{DG}},$$

and

$D = \text{drive crystal}$

$G = \text{gauge crystal}$

$S = \text{specimen}$

$L = \text{specimen length}$

$m = \text{mass}$

$\rho = \text{density}$

$f = \text{frequency}$

$$\lambda = \text{wavelength} = 2L$$
$$\tau = \text{resonant period}$$
$$C_m = \text{capacitance of gauge circuit}$$
$$N = \text{transformer ratio for quartz}$$
$$V = \text{crystal voltage}$$

References

1. R. F. Gibson and R. Plunkett, *Shock and Vib. Dig.*, **9**, 9 (1977).
2. R. F. Gibson and D. G. Wilson, *Shock and Vib. Dig.*, **11**, 3 (1979).
3. R. F. Gibson, *Shock and Vib. Dig.*, **15**, 3 (1983).
4. R. F. Gibson, *Shock and Vib. Dig.*, **19**, 13 (1987).
5. A. S. Nowick and B. S. Berry, "Anelastic Relaxation in Crystalline Solids." Academic Press, New York, 1972.
6. T. S. Kê, *Phys. Rev.*, **71**, 533 (1947).
7. S. Spinner and W. E. Tefft, *Proc. ASTM*, **61**, 1221 (1961).
8. C. M. Harris and C. E. Crede, in "Shock and Vibration Handbook" (C. M. Harris and C. E. Crede, eds.), Chapter 1. McGraw-Hill, New York, 1976.
9. R. V. Dickson and S. Spinner, *J. Mater.*, **3**, 716 (1968).
10. J. Marx, *Rev. Sci. Instrum.* **22**, 503 (1951).
11. W. H. Robinson and A. Edgar, *IEEE Trans. Sonics and Ultrasonics*, **SU-21**, 98 (1974).
12. M. R. Harmouche and A. Wolfenden, *J. Test. Eval.*, **13**, 424 (1985).
13. E. P. Papadakis, in "Physical Acoustics: Principles and Methods" (W. P. Mason and R. N. Thurston, eds.), Vol. XII, Chapter 5. Academic Press, New York, 1976.
14. E. P. Papadakis, *IEEE Trans. Sonics and Ultrasonics*, **SU-16**, 210 (1969).
15. J. L. Snoek, *Physica*, **6**, 591 (1939).
16. C. Zener, *Trans. AIME*, **152**, 122 (1943).
17. J. C. Swartz, J. W. Shilling, and A. J. Schwoeble, *Acta Metall.*, **16**, 1359 (1968).
18. D. P. Seraphim and A. S. Nowick, *Acta Metall.*, **9**, 85 (1961).
19. P. G. Bordoni, *Ricerra Sci.*, **19**, 851 (1949).
20. P. G. Bordoni, *J. Acous. Soc. Am.*, **26**, 495 (1954).
21. R. R. Hasiguti, N. Igata, and G. Kamoshita, *Acta Metall.*, **10**, 442 (1962).
22. T. A. Read, *Phys. Rev.*, **58**, 371 (1940).
23. T. A. Read, *Trans. AIME*, **143**, 30 (1941).
24. J. S. Koehler, in "Imperfections in Nearly Perfect Crystals" (W. Shockley, J. H. Hollomon, R. Maurer, and F. Seitz, eds.), Chapter 7. Wiley, New York, 1958.
25. A. Granato and K. Lücke, *J. Appl. Phys.*, **27**, 583 (1956).
26. A. Granato and K. Lücke, *J. Appl. Phys.*, **27**, 789 (1956).
27. C. Zener, *Phys. Rev.*, **60**, 906 (1941).
28. D. J. Nelson and J. W. Hancock, *J. Mater. Sci.*, **13**, 2429 (1978).
29. N. N. Kishore, A. Ghosh, and B. D. Agarwal, *J. Reinf. Plas. Comp.*, **1**, 40 (1982).
30. N. N. Kishore, A. Ghosh, and B. D. Agarwal, *J. Reinf. Plas. Comp.*, **1**, 64 (1982).
31. H. M. Ledbetter and S. K. Datta, in "Vibration Damping Workshop 1984," AFWAL-TR-84-3064, pp. W1–W18. AFWAL, Wright–Patterson AFB, Ohio, 1984.
32. W. H. Robinson, A. J. Glover, and A. Wolfenden, *Physica Status Solidi*, **48**, 155 (1978).

33. A. Wolfenden, T. G. Aldridge, Jr., E. W. Davis, Jr., V. K. Kinra, G. G. Wren, and J. M. Wolla, in "Proceedings of Damping '89," WRDC-TR-89-3116, Vol. 3, (L. Rogers, ed.), pp. JDC1–JDC18. WRDC, Wright–Patterson AFB, Ohio, 1989.

34. B. B. Rath, M. A. Imam, and N. P. Louat, in "Fifth European Conference on Internal Friction and Ultrasonic Attenuation in Solids" (R. DeBatist and J. van Humbeeck, eds.), *Journal de Physique*, **48**, Colloque C8, 365 (1987).

35. S. J. Hwang and R. F. Gibson, *J. Eng. Mater. and Tech.*, **109**, 47 (1987).

36. A. Wolfenden and J. M. Wolla, in "Proc. 19th Int. SAMPE Tech. Conf.," pp. 37–46. SAMPE. Corina, California, 1987.

37. Z. Hashin, *Trans. ASME*, **32**, 630 (1965).

38. Z. Hashin, *Int. J. Solids Structures*, **6**, 539 (1970)

39. Z. Hashin, *Int. J. Solids Structures*, **6**, 797 (1970).

40. S. P. Rawal, J. H. Armstrong, and M. S. Misra, "Interfaces and Damping in Metal Matrix Composites," Report MCR-86-684. Martin Marietta Denver Aerospace, Denver, Colorado, 1986.

41. J. A. DiCarlo and J. E. Maisel, in "Composite Materials: Testing and Design (Fifth Conference), ASTM STP 674" (S. W. Tsai, ed.), pp. 201–227. American Society for Testing and Materials, Philadelphia, Pennsylvania, 1979.

42. J. A. DiCarlo and J. E. Maisel, in "Advanced Fibers and Composites for Elevated Temperatures" (I. Ahmad and B. R. Noton, eds.), pp. 55–79. The Metallurgical Society of AIME, Warrendale, Pennsylvania, 1980.

43. M. E. McIntyre and J. Woodhouse, *Acta Metall.*, **36**, 1397 (1988).

44. A. Wolfenden and J. M. Wolla, in "Mechanical and Physical Behaviour of Metallic and Ceramic Composites: Proc. 9th Int. Symp. on Metall. and Mat. Sci." (S. I. Andersen, H. Lilholt, and O. B. Pedersen, eds.), pp. 511–516. RISØ National Laboratory, Roskilde, Denmark, 1988.

45. A. Kohyama, S. Saito, H. Tezuka, and S. Sato, in "Interfaces in Polymer, Ceramic, and Metal Matrix Composites" (H. Ishida, ed.), pp. 259–267. Elsevier Science Publishers, New York, 1988.

46. J. T. Hartman, Jr., K. H. Keene, R. J. Armstrong, and A. Wolfenden, *J. Metals*, **38**, 33 (1986).

47. R. J. Arsenault and R. M. Fisher, *Scripta Metall.*, **17**, 67 (1983).

48. M. S. Misra, "Metallurgical Characterization of the Interfaces and the Damping Mechanisms in Metal Matrix Composites," Progress Report, MCR-85-605, Issue 4, Martin Marietta Denver Aerospace, Denver, Colorado, 1986.

49. L. F. Allard, S. P. Rawal, and M. S. Misra, *J. Metals*, **38**, 40 (1986).

50. A. Wolfenden, C. K. Frisby, K. J. Heritage, S. S. Vinson, and R. C. Knight, in "Fifth European Conference on Internal Friction and Ultrasonic Attenuation in Solids" (R. DeBatist and J. van Humbeeck, eds.), *Journal de Physique*, **48**, Colloque C8, 377 (1987).

51. S. N. Wickstrom and A. Wolfenden, *Scripta Metall.*, **23**, 839 (1989).

52. G. G. Wren and V. K. Kinra, in "Dynamic Elastic Modulus Measurements in Materials, ASTM STP 1045," (A. Wolfenden, ed.), pp. 58–74. ASTM, Philadelphia, Pennsylvania, 1990. Also in "Damping in Metal-Matrix Composites: Theory and Experiment," Department of Aerospace Engineering, Texas A & M University, College Station, Texas, 1990.

53. R. B. Bhagat, M. F. Amateau, and E. C. Smith, *J. Comp. Tech. Res.*, **11**, 113 (1989).

54. N. S. Timmerman, "Damping Characteristics of Metal Matrix Composites," Army Materials and Mechanics Research Center Report, AMMRC TR 82-19, Watertown, Massachusetts, 1982.

55. G. L. Steckel and B. A. Nelson, "Mechanical Damping Behavior of Graphite–Magnesium," Report TOR-0086(6726-01)-01, Aerospace Corporation, El Segundo, California, 1985.

56. F. D. Ross and L. Rubin, "Acoustic Attenuation of Metal Matrix Composites," Report SD-TR-82-62, Aerospace Corporation, El Segundo, California, 1982.

57. E. F. Crawley and M. C. van Schoor, *J. Comp. Mater.*, **21**, 553 (1987).
58. V. N. Cribb, A. Wolfenden, R. C. Knight, and M. A. Boyle, in "Mechanical and Physical Behaviour of Metallic and Ceramic Composites: Proc. 9th Int. Symp. on Metall. and Mat. Sci." (S. I. Andersen, H. Lilholt, and O. B. Pedersen, eds.), pp. 321–326. RISØ National Laboratory, Roskilde, Denmark, 1988.
59. M. S. Misra and P. D. LaGreca, in "Vibration Damping Workshop Proceedings" (L. Rogers, ed.), pp. U1–U13, AFWAL TR-84-3064, AFWAL, Wright–Patterson AFB, Ohio, 1984.
60. S. P. Rawal, and M. S. Misra, in "Roles of Interfaces on Material Damping" (B. B. Rath and M. S. Misra, eds.), pp. 43–49, ASM International, Metals Park, Ohio, 1986.
61. A. Wolfenden, M. R. Harmouche, G. V. Blessing, Y. T. Chen, P. Terranova, V. Dayal, V. K. Kinra, J. W. Lemmens, R. R. Phillips, J. S. Smith, P. Mahmoodi, and R. J. Wann, *J. Test. Eval.*, **17**, 2 (1989).
62. V. K. Kinra and V. Dayal, in "Experimental Methods for Mechanical Testing" (R. L. Pendleton and M. E. Tuttle, eds.), in press. Society for Experimental Mechanics, Bethel, Connecticut.
63. Z. Hashin," *J. Appl. Mech.*, **50**, 481 (1983).
64. R. M. Christensen, in "Mechanics of Composite Materials: Recent Advances" (Z. Hashin and C. T. Herakovich, eds.), pp. 1–16, Pergamon Press, New York, 1982.
65. H. M. Ledbetter and S. K. Datta, *J. Acoust. Soc. Am.*, **79**, 239 (1986).
66. H. M. Ledbetter, in "Nonmetallic Materials and Composites at Low Temperatures" (A. F. Clark, R. P. Reed, and G. Hartwig, eds.), pp. 267–281, Plenum Press, New York, 1979.
67. J. M. Wolla and A. Wolfenden, in "Dynamic Elastic Modulus Measurements in Materials, ASTM STP 1045" (A. Wolfenden, ed.), pp. 110–119. ASTM, Philadelphia, Pennsylvania, 1990.
68. Z. Hashin and S. Shtrikman, *J. Mech. Phys. Solids*, **11**, 127 (1963).
69. H. M. Ledbetter and S. K. Datta, *Mater. Sci. Eng.*, **67**, 25 (1984).
70. J. Friedel, "Dislocations," pp. 454–457, Pergamon Press, New York, 1964.
71. A. Wolfenden and J. M. Wolla, *J. Mater. Sci.*, **24**, 3205 (1989).
72. J. M. Liu, *Appl. Phys. Lett.*, **48**, 469 (1986).

Thermal Expansion of Metal Matrix Composites

THOMAS A. HAHN

Naval Research Laboratory
Washington, D.C.

I. Introduction

Thermal expansion characteristics of metal matrix composites can be analyzed by using many existing theories developed in the analysis of the expansivity of metallurgical alloys, glasses, ceramics, plastics, and composites of these materials. Much of the early analysis considered elastic interactions between isotropic materials with either cylindrical fiber or spherical particle reinforcements. A review of the expansivity of composites based on polymer matrices reinforced with particles or fibers was giving by Holliday and Robinson [1], and a bibliography of thermophysical properties of composites by Chamis and Sendeckyj [2]. A review of thermal expansion measurement techniques has been given in [3]. Recent advances, such as the development of high-modulus graphite fibers with transversely isotropic properties, fiber-coating technology, and short-fiber reinforcements, have necessitated the development of additional analysis for the prediction of the expansion of composites. Earlier analyses had treated only isotropic fibers. The use of

Metal Matrix Composites:
Processing and Interfaces
ISBN 0-12-341833-X

fiber reinforcement in high-temperature composites, with the greater possibility for chemical interaction between the materials, has resulted in the need to first coat the fibers with barrier materials. A new mathematical technique has been developed to predict the internal stresses and expansion of multilayer coated continuous fiber composites. With the mismatch in the expansion between the coatings, matrices, and reinforcements of current interest, large thermal stresses are developed between the materials that are of sufficient magnitude to cause yielding in the constituents. Therefore, it has become necessary to extend the analysis of thermally generated stresses and for the expansion to include the nonelastic response of materials. In addition, the developing use of discontinuous reinforcements in metals, particularly SiC whiskers, has also spurred the analysis of the effects on the expansion of aligned or randomly aligned short-fiber-reinforced composites.

In the present chapter, both elastic and plastic analysis for particulate and continuously reinforced composites will be reviewed. Extensions of the analysis will be presented that include multiple-layered composites and plastic deformation of the outer layer of a coaxial cylinder model that includes linear work hardening of the outer cylinder to better approximate the stress–strain curve of the matrix material.

II. Elastic Interaction

A. Spherical-Particulate-Reinforced Composites

When there are no elastic interactions between the constituents of a composite or when the elastic properties of the constituents of the composite are equal, the simplest description of the linear coefficient of thermal expansion (herein referred to as the expansivity) of the composite (α_c) is the volume fraction rule of mixtures. For a two-component composite, the volume fraction rule of mixtures states

$$\alpha_c = V_1 \alpha_1 + V_2 \alpha_2, \tag{1}$$

where V_i is the volume fraction, and α_i the expansivity of the ith component. Early development on the expansivity of metallic and plastic composites was carried out by Turner [4]. By considering that each component of the composite was constrained to change dimensions with temperature equal to the aggregate dimensional change with temperature and using a force-balance criteria, Turner gave the expansivity of a two-component composite as

$$\alpha_c = \frac{V_1 K_1 \alpha_1 + V_2 K_2 \alpha_2}{V_1 K_1 + V_2 K_2}, \tag{2}$$

where K_i is the bulk modulus of the ith component of the composite. The bulk modulus is related to the Young modulus E of isotropic materials by $K = E/3(1 - 2v)$, where v is Poisson's ratio. Turner [4] obtained good agreement between the predictions of this model with measured expansivity values supplied by Hidnert for lead-antimony mixtures, beryllium–aluminum alloys, and some commercial wood products. Another early contribution for predicting the expansivity of composites was derived by Kerner [5] specifically for packed spherical particles. Using an averaging process for finding the bulk modulus of the composite (K_c), the expansivity of a two-component composite was found to be

$$\alpha_c = V_1 \alpha_1 + V_2 \alpha_2 + \frac{4G_2}{K_c} \frac{(K_c - K_1)(\alpha_2 - \alpha_1)V_1}{4G_2 + 3K_1}, \tag{3}$$

where

$$K_c = \frac{\dfrac{V_1 K_1}{3K_1 + 4G_2} + \dfrac{V_2 K_2}{3K_2 + 4G_2}}{\dfrac{V_1}{3K_1 + 4G_2} + \dfrac{V_2}{3K_2 + 4G_2}}, \tag{4}$$

with K_1 and K_2 being the bulk modulus of the spherical reinforcement and matrix, respectively, and G_2 the shear modulus of the matrix equal to $E_2/2(1 + v_2)$ for isotropic phases. Predictions of the expansivity based upon Eqs. (1)–(3) and using the material properties of 6061 Al alloy and SiC are shown in Fig. 1. As can be seen, the elastic interactions introduced in Eqs. (2) and (3) reduce the predicted expansivity values below the volume fraction rule of mixtures. Predictions of the expansivity from other models, which will be described later, are also shown in Fig. 1.

Predictions from a two-concentric-spheres model were considered in a number of different studies. It is interesting to note that the expansivity values calculated from Eq. (3) are in agreement with this simple model. For example, Wang and Kwei [6] presented a concentric-spheres model from which they derived the expansivity using the boundary conditions of the equality of strain and radial stress at the interface between the two spheres and zero radial stress at the outer surface. Good agreement was obtained between the predictions of the model and expansivity data on two different titanium-dioxide-filled plastics taken from unpublished data of Thomas. By using the

concentric-spheres model, Hahn and Armstrong [7, 8] obtained good agreement with the low-temperature expansivity of Al–Si eutectic alloys, and on
particulate SiC-reinforced 6061 Al alloy composites. Fahmy and Ragai [9]
derived the expansivity using the same model and obtained good agreement
between their predictions and expansivity data for systems with various
volume fractions of lead-fused silica and aluminum-silicon. Feltham et al.
[10] made an extensive study on the expansivity of epoxy resin reinforced
with silica flour, glass micro-spheres, and copper powder. The size of the
various reinforcements was between 22 and 237 μm. The size of the reinforcement, for the same volume fraction of reinforcement, to the limit of their
very sensitive interferometric apparatus, did not change the expansivity of
the composites. Tummala and Friedberg [11] used the concentric spheres
model, but with an infinite matrix, and obtained good agreement between
predictions from their model and expansivity measurements on spherical
particles of yttria-stabilized zirconia in a glass matrix. Predictions of the
expansivity from this model are greater than values calculated from the other
concentric spheres models and even greater than the rule-of-mixtures calculation using (1).

B.   Continuous-Fiber-Reinforced Composites

In the prediction of the thermal expansion of continuous-fiber composites,
the simplest geometrical model is that of two coaxial cylinders. An early
analysis using this model was giving by Laszlo [12] who also considered the
pressure developed in concentric spheres and the expansivity of capped
cylinders to approximate the expansivity of two concentric spheres. With
some changes in notation, more consistent with the current notation, the
composite longitudinal expansivity (α_L) and transverse expansivity (α_T) were
given as

$$\alpha_L = \alpha_2 + \frac{\sigma_{z,2}}{E_2 \Delta T} - \frac{(\sigma_{t,1} + P)v_2}{E_2 \Delta T}, \tag{5}$$

$$\alpha_T = \alpha_2 + \frac{\sigma_{t,2}}{E_1 \Delta T} - \frac{\sigma_{z,2} v_2}{E_2 \Delta T}, \tag{6}$$

with

$$\sigma_{z,2} = -\frac{V_1}{V_2}\left(\frac{\beta_1}{\beta_2}\frac{\beta_2 + \beta_3}{\beta_1\beta_3 - 2\beta_2^2} - \frac{1}{\beta_2}\right)(\alpha_2 - \alpha_1)\Delta T, \tag{7}$$

$$P = \left(\frac{\beta_2 + \beta_3}{\beta_1\beta_3 - 2\beta_2^2}\right)(\alpha_2 - \alpha_1)\Delta T, \tag{8}$$

$$\sigma_{t,1} = -\frac{1 + V_1}{V_2}P, \tag{9}$$

$$\sigma_{t,2} = -\frac{2PV_1}{V_2}, \tag{10}$$

$$\beta_1 = \frac{1 - \nu_1}{E_1} + \frac{\nu_2}{E_2} + \frac{1 + V_2}{V_1 E_2},$$

$$\beta_2 = \frac{\nu_1}{E_1} + \frac{V_1\nu_2}{V_2 E_2}, \tag{11}$$

$$\beta_3 = \frac{1}{E_1} + \frac{V_1}{V_2 E_2},$$

where P is the radial stress at the interface between the two cylinders, $\sigma_{z,2}$ is the axial stress in the outer cylinder, $\sigma_{t,1}$ is the tangential stress in the inner cylinder, and $\sigma_{t,2}$ is the tangential stress at the outer boundary.

As with the many different representations of the concentric-spheres model, Eqs. (5) and (6) of the coaxial-cylinders model are open to different representations, many of which give the same predicted values of the longitudinal and transverse expansivities. For example, Craft and Christensen [13] also derived the expansivity for the coaxial cylinders in considering the expansivity of a random array of fibers. Whereas the exact mathematical equivalence was not checked, predictions of the longitudinal and transverse expansivity from the analytical equations in [8], [13], and [14] agree with Eqs. (5) and (6) to better than one part in 10^8. Predictions from Eqs. (5) and (6) are compared in Fig. 1. Hahn and Armstrong [8] found that the thermal expansivity of aligned whisker-SiC-reinforced 6061 Al, as might be expected, was greater than the expansivity as predicted by the continuous-fiber coaxial-cylinders model. Hsueh and Becher [14], with a correction in [15], also derived the expansivity for two coaxial cylinders that included allowances for differences in axial and transverse expansivities of the inner cylinder. Predictions from the model were compared with the measured expansivities of a series of different volume fractions of glass-fiber-reinforced epoxy resins.

Hsueh and Becher also provide a critique of predictions of the expansivity by using the bounds on the elastic constants of a composite developed by Levin [16]. For example, Levin's effective elastic-constants solution for the maximum axial expansion agrees with the coaxial-cylinder model. If the matrix (i.e., the outer cylinder) is the material with the smaller modulus, the

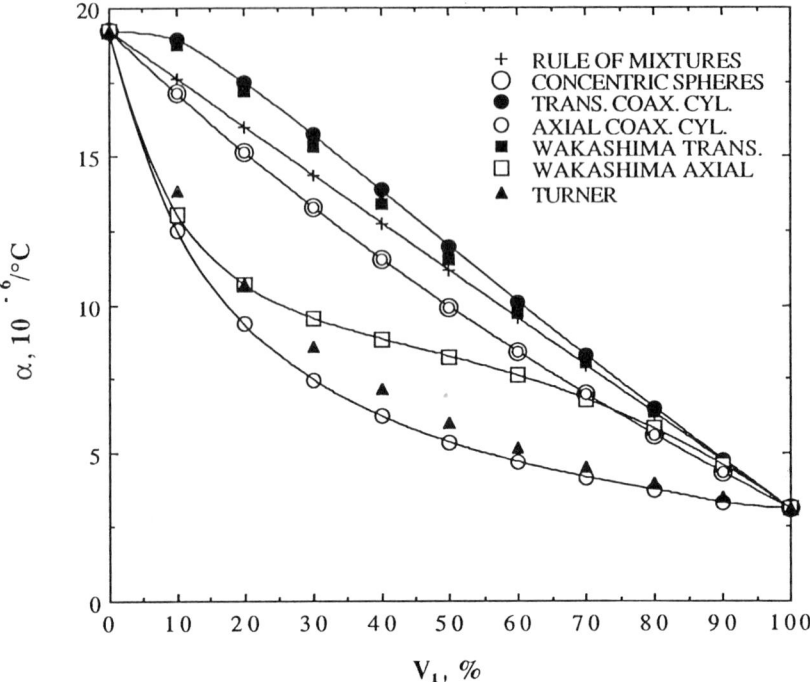

FIG. 1.   Comparison of predicted thermal expansivities from various analytical models for SiC (V1) reinforced 6061 aluminum alloy.

coaxial-cylinder model agrees with the upper bound of the transverse expansivity and the lower bound of the axial expansivity. However, if the outer cylinder is the high-modulus material, the solution from the coaxial-cylinder model agrees with the lower bound of the transverse expansivity and the upper bound of the axial expansivity.

In a similar fashion, Schapery [17] has also performed detailed analysis of the effective modulus technique using energy principles to show the applicability of many of the early predictions of the expansivity. The upper bound on the composite expansivity was found to agree with Eq. (3), while the lower bound using the current notation was given as

$$\alpha = \bar{\alpha} - \frac{\dfrac{1}{K_L} - \dfrac{1}{\hat{K}}}{\dfrac{1}{K_2} - \dfrac{1}{K_1}}(\alpha_2 - \alpha_1), \tag{12}$$

where $\bar{\alpha} = \alpha_1 V_1 + \alpha_2 V_2$ and

$$\frac{1}{K_L} = \frac{V_1}{K_1} + \frac{V_2}{K_2}. \tag{13}$$

$$\hat{K} = K_1 + \frac{V_2}{\dfrac{1}{K_2 - K_1} + \dfrac{V_1}{K_1 + 4G_1/3}}. \tag{14}$$

In addition, Schapery gives an approximation for the transverse expansivity (α_T) of a unidirectionally reinforced composite with isotropic fiber and matrix properties as

$$\alpha_T = (1 + v_1)\alpha_1 V_1 + (1 + v_2)\alpha_2 V_2 - \alpha_L \bar{v}, \tag{15}$$

where (α_L) is the longitudinal expansivity of the composite and $\bar{v} = V_1 v_1 + V_2 v_2$. Predictions from Eqs. (12) and (15), using Eq. (12) as α_L, are shown in Fig. 1. The lower bound given by Eq. (12) is greater than the longitudinal expansivity of the coaxial-cylinder model, whereas that given by Eq. (15) is greater than the transverse expansivity of the coaxial-cylinder model. In the case of transversely isotropic fibers, Rojstaczer et al. [18] propose replacing the first term in Eq. (12) containing the fiber properties with $\bar{\alpha} = \alpha_{1,T} V_1 + \alpha_{1,L} v_{1,L} V_1$, where $\alpha_{1,L}$ and $\alpha_{1,T}$ are the longitudinal and transverse expansivities of the fiber, and $v_{1,L}$ is the longitudinal Poisson's ratio of the fiber, which is also used in the averaged Poisson's ratio equation. Good agreement with measurements on Kevlar fiber-reinforced epoxy was obtained by using this new equation. Klemens [19], also using energy principles, derived the expansivity for both continuous fiber- and spherical-particle-reinforced composites. Ishikawa et al. [20] studied the effects of hexagonal and square arrays of unidirectional fiber-reinforced composites for both isotropic (glass) and transversely isotropic (graphite) fibers in an epoxy matrix. There were only small differences between the predicted expansivities for the hexagonal and square arrays. For the isotropic fibers, the predicted values were in good agreement with Turner's predictions from Eq. (2). In the case of the transverse values, the predictions were less than the upper bounds given by Schapery's prediction in Eq. (11). Predictions of the expansivity for both the glass- and graphite-fiber-reinforced composites were in good agreement with measurements on a series of samples cut at various angles to the fiber direction. Other research dealing with hexagonal and square arrays of fibers includes that of Rogers et al. [21] and Schneider [22].

In order to account for both the axial and transverse mechanical properties and expansivities of a fiber with coatings in a unidirectionally reinforced composite, as well as more than two coaxial cylinders, Mikata and Taya [23] developed a solution technique for the thermomechanical stresses using four coaxial cylinders. The outer cylinder has the volume-averaged properties of the three inner cylinders and so approximates the properties of the surrounding composite. Their model can easily be extended to include five coaxial cylinders along with calculations of the expansivity. The large number of zones is necessary for modeling high-temperature, intermetallic matrix composites with diffusion barriers on the fibers and reaction zones formed during high-temperature processing. Isotropic or transversely isotropic materials can be considered in each zone; and by assigning equal properties to contiguous zones, it is possible to calculate the stresses and strains for a four, three, or two-coaxial cylinders composite. For the two-zone composite, predictions from this model are in agreement with Eqs. (4) and (5). The five-coaxial-cylinders ensemble, consisting of a fiber, two barrier coatings, the matrix, and the surrounding composite, is shown in Fig. 2 along with

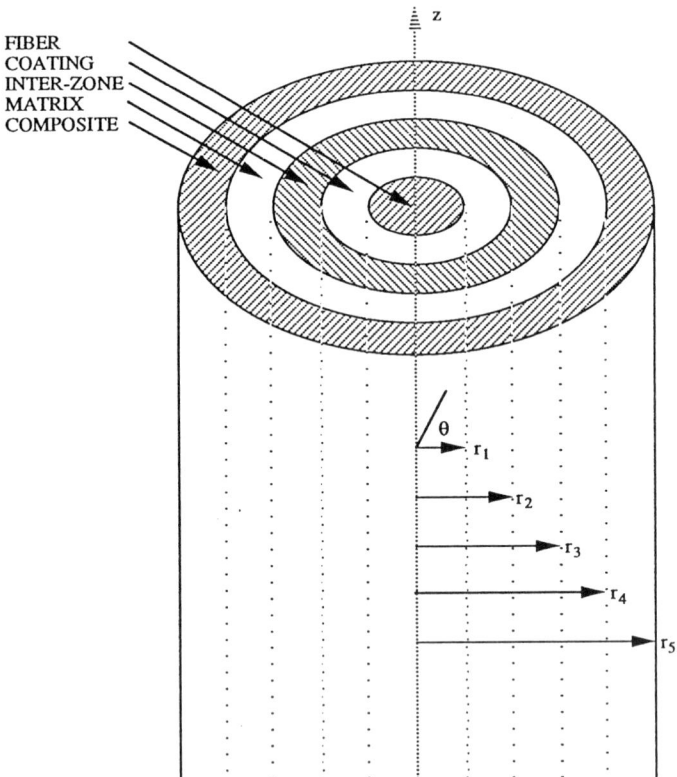

FIG. 2.   Representation of the five-coaxial-cylinders model.

the various radii used in the analysis. End effects are neglected and isothermal temperature changes are assumed with no temperature gradients. Mikata and Taya allowed for a radial temperature gradient. The ensemble may also have applied stresses in the axial or radial directions. Following this development for cylindrical symmetry, the stress–strain equations are

$$\sigma_{rr}^{(n)} = C_{11}^{(n)} e_{rr}^{(n)} + C_{12}^{(n)} e_{\theta\theta}^{(n)} + C_{13}^{(n)} e_{zz}^{(n)} - \beta_1^{(n)} T_n$$

$$\sigma_{\theta\theta}^{(n)} = C_{12}^{(n)} e_{rr}^{(n)} + C_{11}^{(n)} e_{\theta\theta}^{(n)} + C_{13}^{(n)} e_{zz}^{(n)} - \beta_1^{(n)} T_n \qquad (16)$$

$$\sigma_{zz}^{(n)} = C_{13}^{(n)} e_{rr}^{(n)} + C_{13}^{(n)} e_{\theta\theta}^{(n)} + C_{33}^{(n)} e_{zz}^{(n)} - \beta_3^{(n)} T_n$$

where C_{ij} are the elastic constants, the superscript (n) refer to cylinders 1–5, and

$$\beta_1^{(n)} = (C_{11}^{(n)} + C_{12}^{(n)})\alpha_1^{(n)} + C_{13}^{(n)}\alpha_3^{(n)},$$

$$\beta_3^{(n)} = 2C_{12}^{(n)}\alpha_1^{(n)} + C_{33}^{(n)}\alpha_3^{(n)}, \qquad (17)$$

where $\alpha_1^{(n)}$ and $\alpha_3^{(n)}$ are the expansivities of cylinder n in the transverse (1, 2) and longitudinal (3) directions, and T_n is the temperature change of cylinder n. Due to the axisymmetry, the strains in terms of the radial displacement, u, and axial displacement, w, are given by

$$e_{rr}^{(n)} = \frac{\partial u_n}{\partial r},$$

$$e_{\theta\theta}^{(n)} = \frac{u_n}{r}, \qquad (18)$$

$$e_{zz}^{(n)} = \frac{\partial w_n}{\partial z},$$

where u is a function of r, and w is a function of z.

The stress displacement equations are giving by

$$\sigma_{rr}^{(n)} = C_{11}^{(n)} \frac{\partial u_n}{\partial r} + C_{12}^{(n)} \frac{u_n}{r} + C_{13}^{(n)} \frac{\partial w_n}{\partial z} - \beta_1^{(n)} T_n,$$

$$\sigma_{\theta\theta}^{(n)} = C_{12}^{(n)} \frac{\partial u_n}{\partial r} + C_{11}^{(n)} \frac{u_n}{r} + C_{13}^{(n)} \frac{\partial w_n}{\partial w} - \beta_1^{(n)} T_n, \qquad (19)$$

$$\sigma_{zz}^{(n)} = C_{13}^{(n)} \frac{\partial u_n}{\partial r} + C_{13}^{(n)} \frac{u_n}{r} + C_{33}^{(n)} \frac{\partial w_n}{\partial z} - \beta_3^{(n)} T_n,$$

In case the materials are isotropic, the stiffness constants reduce to

$$C_{11}^{(n)} = C_{33}^{(n)} = 2\mu_n + \lambda_n = \frac{(1 - v_n)E_n}{(1 + v_n)(1 - 2v_n)},$$

$$C_{12}^{(n)} = C_{13}^{(n)} = \lambda_n = \frac{v_n E_n}{(1 + v_n)(1 - 2v_n)}, \tag{20}$$

$$\beta_1^{(n)} = \beta_3^{(n)} = (2\mu_n + 3\lambda_n)\alpha_1^{(n)} = \frac{E_n \alpha_1^{(n)}}{(1 - 2v_n)}, \tag{21}$$

where, for material in cylinder $n = 1$–5, μ_n and λ_n are the Lamé constants, v_n is Poisson's ratio, and E_n is Young's modulus.

For isothermal temperatures, equilibrium conditions require the displacements to satisfy

$$\frac{d^2 u_n}{dr^2} + \frac{1}{r}\frac{du_n}{dr} - \frac{u_n}{r^2} = 0,$$

$$\frac{d^2 w_n}{dz^2} = 0, \tag{22}$$

which have the general solutions

$$u_n(r) = A_n r + \frac{B_n}{r},$$

$$w_n(z) = G_n z + H_n. \tag{23}$$

From the equality of axial displacement

$$G_1 = \cdots = G_5 = G,$$

$$H_1 = \cdots = H_5 = H, \tag{24}$$

H can be seen to be just a rigid body displacement in the axial direction, which can be set equal to zero. Substitution into the stress displacement equations gives

$$\sigma_{rr}^{(n)} = C_{11}^{(n)}\left[A_n - \frac{B_n}{r^2}\right] + C_{12}^{(n)}\left[A_n - \frac{B_n}{r^2}\right] + C_{13}^{(n)}G - \beta_1^{(n)}T_n,$$

$$\sigma_{\theta\theta}^{(n)} = C_{12}^{(n)}\left[A_n - \frac{B_n}{r^2}\right] + C_{11}^{(n)}\left[A_n - \frac{B_n}{r^2}\right] + C_{13}^{(n)}G - \beta_1^{(n)}T_n, \tag{25}$$

$$\sigma_{zz}^{(n)} = 2C_{13}^{(n)}A_n + C_{33}^{(n)}G - \beta_3^{(n)}T_n,$$

where the constants A_n, B_n, and G will be determined by the boundary conditions. For the five-cylinders model, these boundary conditions of continuity of stresses and displacements are

$$u_1 = u_2, \quad w_1 = w_2, \quad \sigma_{rr}^{(1)} = \sigma_{rr}^{(2)} \quad \text{at} \quad r = r_1,$$

$$u_2 = u_3, \quad w_2 = w_3, \quad \sigma_{rr}^{(2)} = \sigma_{rr}^{(3)} \quad \text{at} \quad r = r_2,$$

$$u_3 = u_4, \quad w_3 = w_4, \quad \sigma_{rr}^{(3)} = \sigma_{rr}^{(4)} \quad \text{at} \quad r = r_3, \qquad (26)$$

$$u_4 = u_5, \quad w_4 = w_5, \quad \sigma_{rr}^{(4)} = \sigma_{rr}^{(5)} \quad \text{at} \quad r = r_4,$$

$$\sigma_{rr}^{(4)} = \sigma_{0r} \quad \text{at} \quad r = r_5.$$

In addition, the balance of axial forces gives

$$\int_0^{r_1} \sigma_{zz}^{(1)} r \, dr + \cdots + \int_{r_4}^{r_5} \sigma_{zz}^{(5)} r \, dr = \int_0^{r_5} \sigma_{0z} r \, dr, \qquad (27)$$

where σ_{0z} is the applied stress in the z-direction, and σ_{0r} is the applied stress in the radial direction.

Substitution of the stress–strain equations in the boundary-condition equations and the balance-of-forces equation gives a set of linear equations for the unknown parameters A_n, B_n, and G. These equations can be represented by the following matrix equation,

$$
\begin{bmatrix}
a_{11} & a_{12} & a_{13} & 0 & 0 & 0 & 0 & 0 & 0 & 0 \\
0 & a_{22} & a_{23} & a_{24} & a_{25} & 0 & 0 & 0 & 0 & 0 \\
0 & 0 & 0 & a_{34} & a_{35} & a_{36} & a_{37} & 0 & 0 & 0 \\
0 & 0 & 0 & 0 & 0 & a_{46} & a_{47} & a_{48} & a_{49} & 0 \\
a_{51} & a_{52} & a_{53} & 0 & 0 & 0 & 0 & 0 & 0 & a_{510} \\
0 & a_{62} & a_{63} & a_{64} & a_{65} & 0 & 0 & 0 & 0 & a_{610} \\
0 & 0 & 0 & a_{74} & a_{75} & a_{76} & a_{77} & 0 & 0 & a_{710} \\
0 & 0 & 0 & 0 & 0 & a_{86} & a_{87} & a_{88} & a_{89} & a_{810} \\
0 & 0 & 0 & 0 & 0 & 0 & 0 & a_{98} & a_{99} & a_{910} \\
a_{101} & a_{102} & 0 & a_{104} & 0 & a_{106} & 0 & a_{108} & 0 & a_{1010}
\end{bmatrix}
\begin{bmatrix}
A_1 \\ A_2 \\ B_2 \\ A_3 \\ B_3 \\ A_4 \\ B_4 \\ A_5 \\ B_5 \\ G
\end{bmatrix}
$$

$$
=
\begin{bmatrix}
0 \\ 0 \\ 0 \\ 0 \\ c_5 \\ c_6 \\ c_7 \\ c_8 \\ c_9 \\ c_{10}
\end{bmatrix}. \qquad (28)
$$

The individual a_{ij} and c_j are listed in Appendix I. A_n, B_n, and G are dependent on the various radii, which in turn are dependent on the temperature change. Thus, A_n, B_n, and G will also be dependent on the temperature change. This dependency is small and can be adjusted for by correcting the individual r_n when the temperature change is large. After solving for the unknown constants, the expansion can be calculated from Eq. (18). Dividing the expansion by the temperature change then gives the expansivity.

C. Short-Fiber-Reinforced Composites

The expansivity of oriented and randomly oriented short-fiber composites has been considered by a number of authors using widely different approaches. Craft and Christensen [13] based their analyses on the coaxial-cylinder model, and Halpin [24] starts with laminate theory, which is then generalized to the expansivity of randomly oriented short-fiber composites. In the analysis of Craft and Christensen, as mentioned earlier, they derived properties for two coaxial cylinders including $(E_{11}, E_{22}, K_{23}, \alpha_L, \alpha_T, \text{ and } v_{12})$, which are used to find the expansivity of composites with either two-dimensional or three-dimensional random arrays of fibers. These are

$$\alpha_{2D} = \frac{E_{11}\alpha_L + E_{22}(\alpha_L + \alpha_T)v_{12} + E_{22}\alpha_T}{E_{11} + E_{22}(1 + 2v_{12})}, \tag{29}$$

and

$$\alpha_{3D} = \frac{[E_{11} + 4v_{12}(1 + v_{12})K_{23}]\alpha_L + 4v_{12}(1 + v_{12})K_{23}\alpha_T}{E_{11} + 4(1 + v_{12})^2 K_{23}}. \tag{30}$$

These expressions were also used to find asymptotic expressions for the expansivity when the properties of the fiber and matrix satisfy $E_2/V_1E_1 \ll 1$ and $K_2/K_1 \ll 1$, and terms that are of the order of the squares of these terms can be neglected and the Poisson ratios of both the fiber and matrix were equal to $1/4$.

As mentioned, Halpin also considered the expansivity of oriented short-fiber composites using an estimate given by Schapery [17] when $v_1 = v_2$. The expansivity of the composite in the longitudinal direction is

$$\alpha_L = \bar{\alpha} + \left(\frac{\overline{E\alpha}}{\overline{E}} - \bar{\alpha}\right) \frac{\dfrac{1}{E_L} - \dfrac{1}{E_{11}}}{\dfrac{1}{E_L} - \dfrac{1}{E_U}}, \tag{31}$$

where

$$\frac{1}{E_L} = \frac{V_1}{E_1} + \frac{V_2}{E_2}, \tag{32}$$

$$E_U = V_1 E_1 + V_2 E_1, \tag{33}$$

and the quantities with the bar over them denote volume averages. An estimate for E_{11} from Halpin [25] was used in the analysis and is given by

$$E_{11} = E_2 \frac{(1 + \xi \eta V_1)}{(1 - \eta V_1)}, \tag{34}$$

with

$$\eta = \frac{E_1/E_2 - 1}{E_1/E_2 + \xi}, \tag{35}$$

and

$$\xi = 2\frac{L}{D} \tag{36}$$

where L/D is the length-to-diameter ratio of the reinforcement. In [24], the ξ is omitted in Eq. (34). When the Poisson ratios are not equal, the values of the bulk moduli are used in expressions for the volume average of $E\alpha$ and E and Eqs. (31)–(34). Halpin and Pagano [26] extended this analysis to randomly oriented short fibers. In this case, the composite expansivity is

$$\alpha = \tfrac{1}{2}(\alpha_L + \alpha_T) + \frac{1}{2}\frac{E_{11} - E_{22}}{E_{11} + (1 + 2v_{12})E_{22}}(\alpha_L - \alpha_T), \tag{37}$$

where E_{11} is estimated using Eqs. (34)–(36). The same equations are used to find E_{22} but with $L/D = 1$, and v_{12} is set equal to the volume average of the fiber and matrix Poisson ratios. Predictions from Eqs. (29), (30), (31), and (37) are shown in Fig. 3. Maron and Weinberg [27] also consider the expansivity of short-fiber composites but modify Turner's expression [Eq. (2)] by making Poisson's ratio of the reinforcement and the matrix equal and by including an efficiency factor that depends on the length of the fiber. A systematic comparison between these analytical expressions and experimental data has not been carried out; however, Maron and Weinberg present data on the longitudinal and transverse expansivities of epoxy resin–glass fiber composites. The data exhibit a large scatter and only general agreement with the predicted expansivities.

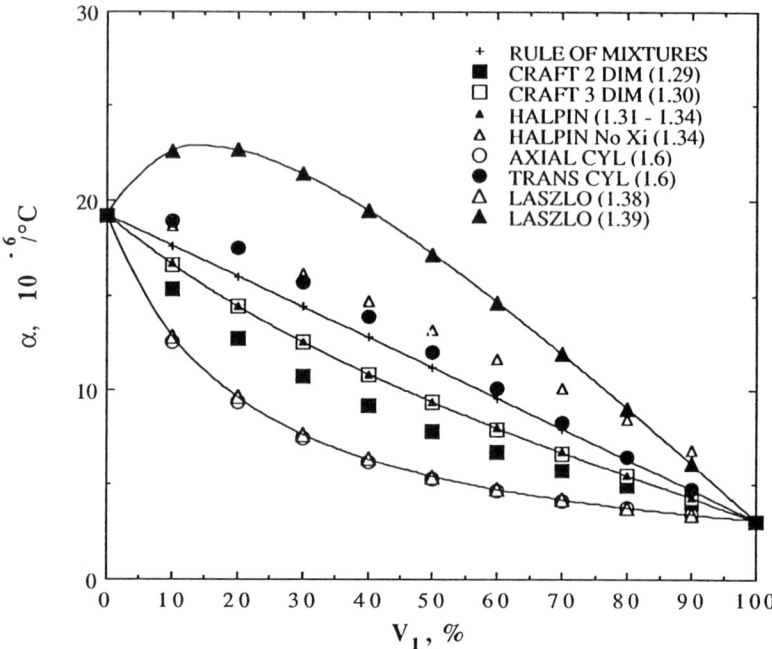

FIG. 3.   Comparison of predicted expansivities for short-fiber reinforcements and laminated composites.

D.   Laminated Composites

In addition to the analysis on spheres and cylinders, Laszlo [12] also gives the expansivity parallel and perpendicular to the planes of a lamellar structure. These are giving by

$$\alpha_{\parallel} = \alpha_1 + \frac{(1 - v_1)f_1}{E_1 \Delta T} - \frac{qv_1}{E_1 \Delta T} = \alpha_2 + \frac{(1 - v_2 f_2)}{E_2 \Delta T} - \frac{qv_2}{E_2 \Delta T}, \qquad (38)$$

$$\alpha_{\perp} = V_1 \alpha_1 + V_2 \alpha_2 + \left(\frac{V_1}{E_1} + \frac{V_2}{E_2}\right)q - \frac{2V_1 f_1 v_1}{E_1 \Delta T} - \frac{2V_2 f_2 v_2}{E_2 \Delta T}, \qquad (39)$$

with $q = -2(V_1 f_1 + V_2 f_2)$, where f_1 and f_2 are the stresses in the lamellar layers given by

$$f_1 = \frac{\varepsilon_1 \varepsilon_2 V_2}{\varepsilon_1 V_1 + \varepsilon_2 V_2} + (\alpha_1 - \alpha_2)\Delta T, \qquad (40)$$

$$f_2 = \frac{\varepsilon_1 \varepsilon_2 V_1}{\varepsilon_1 V_1 + \varepsilon_2 V_2} + (\alpha_1 - \alpha_2)\Delta T, \qquad (41)$$

with $\varepsilon_i = E_i/(1 - v_i)$. Gulati and Plummer [28] also derived $\alpha_{\parallel}$ and compared the predictions with measurements of the expansivity of glass-ribbon-reinforced epoxy and a thermoplastic matrix material. Good agreement between the predictions and the measurements was obtained if the glass ribbon was not damaged during consolidation. Predictions from [12] and [28] are numerically the same.

E. Equivalent-Inclusion Analysis

A number of authors have used the equivalent-inclusion analysis introduced by Eshelby [29] and additionally developed by Mura [30] in the analysis of thermal stress and expansivity for elliptical inclusions, which, depending upon the excentricity, can be used for spherical and cylindrical inclusions. Using these considerations, Wakashima et al. [31] derived the expansivity for spherical, disc-shaped, and continuous-fiber-reinforced composites. The solution for the spherical inclusion agreed with the solution previously given in Eq. (3) as derived by Kerner. For the disc-shaped inclusion, with the limiting equation

$$\frac{x_1^2 + x_1^2}{a^2} + \frac{x_3^2}{c^2} = 1 \quad \text{with} \quad \frac{c}{a} \to 0, \tag{42}$$

the expansivity of the composite perpendicular to the x_3-axis of the disc is

$$\alpha_{\perp} = V_1\alpha_1 + V_2\alpha_2 + V_1V_2(\alpha_1 - \alpha_2)\left[\frac{(1 - v_2)E_1 - (1 - v_1)E_2}{V_2(1 - v_1)E_2 + V_1(1 - v_2)E_1}\right]. \tag{43}$$

Parallel to the x_3-axis,

$$\alpha_{\parallel} = V_1\alpha_1 + V_2\alpha_2 + V_1V_2(\alpha_1 - \alpha_2)\left[\frac{(1 - v_2)E_1 - (1 - v_1)E_2}{V_2(1 - v_1)E_2 + V_1(1 - v_2)E_1}\right]. \tag{44}$$

For fiber-reinforced composites, Eq. (42) with the limiting value $c/a \to \infty$ was used to derive the expansivity of composites with cylindrical symmetry. The transverse expansivity perpendicular to the x_3-axis is

$$\alpha_{\perp} = V_1\alpha_1 + V_2\alpha_2 + V_1V_2(\alpha_1 - \alpha_2)\left[\frac{A - B}{AC - BD}E_1 - 1\right], \tag{45}$$

and the axial expansivity parallel to the x_3-axis is

$$\alpha_{\parallel} = V_1\alpha_1 + V_2\alpha_2 + V_1V_2(\alpha_1 - \alpha_2)\left[\frac{C - D}{AC - BD}E_1 - 1\right], \tag{46}$$

where

$$A = V_2 \frac{2v_2 v_1 G_1 + 2(1 - 2v_1)G_2}{1 - v_2} + 2V_1(1 - v_1)G_1, \qquad (47)$$

$$B = V_2 \frac{v_2 G_1 + (1 - 2v_1)v_2 G_2}{1 - v_2} + 2V_1 v_1 G_1, \qquad (48)$$

$$C = V_2 \frac{G_1 + (1 - 2v_1)G_2}{1 - v_2} + 2V_1 G_1, \qquad (49)$$

$$D = V_2 \frac{2v_1 G_1 + 2(1 - 2v_1)v_2 G_2}{1 - v_2} + 4V_1 v_1 G_1. \qquad (50)$$

The two limiting values parallel and perpendicular to the axis of the fiber are shown in Fig. 1 as a function of the fiber volume fraction, V_1. Plastic deformation of the matrix was also considered by Wakashima and coworkers and will be considered later.

Takao and Taya [32], also using Eshelby's equivalent-inclusion method, developed a solution for the expansivity and thermal stresses in aligned anisotropic short-fiber-reinforced composites with a distribution of fiber lengths. Although it is difficult to get numerical predictions from the equations, the analysis will be outlined here. The composite expansivity is given by the simple tensor equations

$$\alpha = \alpha_2 + V_1(e^* + \alpha^*)/\Delta T, \qquad (51)$$

with

$$\alpha^* = V_1(\alpha_1 - \alpha_2)\Delta T \qquad (52)$$

where α, α_1, α_2, e^*, and α^* are tensors, and e^* is the "transformation strain" found from the solution of

$$\bar{e} + V_1(S - I)(\alpha^* + e^*) = 0. \qquad (53)$$

S is the Eshelby tensor and I is the identity matrix. The solution of the strain tensor $\bar{e}$ is found from

$$[I - V_1(S - I)\{(C_1 - C_2)(S - I) + C_1\}^{-1}(C_1 - C_2)]\bar{e}$$
$$= -V_1(S - I)\{(C_1 - C_2)(S - I) + C_1\}^{-1}C_1\alpha^* \qquad (54)$$

where C_1 and C_2 are the elastic stiffness constants of the fiber and the matrix, respectively. Various solutions of these equations were presented graphically by Takao and Taya for carbon fiber-reinforced epoxy and aluminum composites. For the case of anisotropic fibers with variable aspect ratios, the tensorial inversion equations are summed over either a discrete distribution of fiber aspect ratios or integrated over a continuous distribution of aspect ratios. Solutions in this case were presented for SiC whisker-reinforced aluminum alloy composites.

III. Plastic-Deformation Analysis

Thermal expansion measurements provide a testing procedure for determining the plastic deformation that may result from thermally induced stresses caused by differences in the thermal expansivity between metal matrices and reinforcements used in metal matrix composites. Indications of plastic deformation from an experimental expansion-versus-temperature curve or a plot of the expansivity versus temperature are (1) changes in slope of the expansion curve with either continuous or discontinuous changes in expansivity, (2) differences between the heating and cooling expansion values with different expansivity values at the same temperature, (3) permanent residual length changes after thermal cycling, or (4) changes in length at constant temperature. These considerations are applicable if there are no phase changes in the materials or stress-induced microcracking when brittle phases are present in the composite.

Kreider and Paterini [33] considered a force balance between the matrix and fiber, analogous to the analysis of Turner; however, after yielding of the matrix, the force balance is

$$A_1 \sigma_1 = A_2 \sigma_y, \tag{55}$$

where A_1 and A_2 are the areas of the fiber and the matrix, respectively, σ_1 is the stress in the fiber, and σ_y is the yield stress of the matrix, which is taken as the limiting stress that the matrix can sustain. The composite axial expansivity derived from these considerations is

$$\alpha_c = \alpha_1 + \frac{1}{\Delta T} \frac{A_2}{A_1} \frac{\sigma_y}{E_1}. \tag{56}$$

The predictions from this formulation were in good agreement with the axial measurements over a temperature range that induced plastic deformation of the matrix and were well below the predictions for elastic interaction between the fibers and the matrix. Three different volume fractions of 1100 Al alloy reinforced with SiC-coated boron fibers were included in the study. Measurements of the transverse expansion were in good agreement with predictions from Schapery's transverse formula [Eq. (15)]. Good agreement was also obtained between measurements on a 2024 Al alloy reinforced with

SiC-coated boron fibers on samples cut with various fiber orientations between $0°$ and $90°$ and the predictions given by Schapery [17] for the angular dependence of the expansivity.

Garmong [34] has developed a model for predicting the axial thermal expansion of two-component composites that was applied to the analysis of the measured thermal expansion of directionally solidified Al–Al$_3$Ni eutectic alloy. Excellent agreement between the measured and predicted expansivities was obtained on both heating and cooling when the matrix deforms by either elastic or plastic processes, and with predictions of the transition temperature between the two different types of behavior. This agreement occurred even without taking into account the change in expansivity and modulus with temperature and a questionable temperature dependence of the yield stress of the Al matrix as pointed out by Tyson [35]. Tyson also suggests an alternate solution technique, but the original analysis of Garmong will be outlined here. The difference in expansion between the two constituents is viewed as the driving mechanism for the resultant deformation of the composite. The matrix stress is used as the parameter that determines the elastic limit and subsequent plastic flow of the matrix. Added features of the analysis were the inclusion of creep relaxation of the stress at high temperatures, and work hardening of the matrix. The overall governing equation used to calculate the stress in the matrix (σ_2), which is subsequently used in calculating the expansivity of the composite, is

$$\int_{T_1}^{T_2} (\alpha_2 - \alpha_1)\, d\text{T} = \frac{\sigma_a - \sigma_2 V_2}{V_1 E_1} - \left[\frac{\sigma_2}{E_2} + \frac{\sigma_2 - \sigma_y}{K^*}\right]^{1/n}$$
$$\pm \int_{T_1}^{T_2} A\left(\frac{|\sigma_2|}{G}\right)^s \frac{Gb}{RT} D_0 e^{-Q/RT} \left(\frac{dt}{d\text{T}}\right) d\text{T}, \qquad (57)$$

where σ_a is the applied stress, σ_1 is the stress in the reinforcement, σ_2 the stress in the matrix, σ_y is the matrix yield strength, K^* and n are parameters in Ludwik's stress–strain equation for a yielding matrix, and the last integral is the Dorn formulation [36] for stage-II dislocation-controlled, steady-state creep with $(dt/d\text{T})$ as the reciprocal rate of change of temperature. The sign is chosen in order to decrease the strain with the plus sign for tensile matrix stress and the negative sign for compressive matrix stress.

Larsson [37] presents a theoretical analysis and experimental results on the thermal expansion of a tungsten-wire uniaxial reinforced stainless-steel composite. Two methods were used in calculating the stress and strain of the composites. The first used the plastic deformation relation of Ludwik and the creep relaxation of Dorn, as was used in the analysis of Garmong, which resulted in good agreement between the matrix stress calculated from expansion measurements and the analytical models. However, better agreement between the experimental measurements and predictions was obtained by using a first-order differential equation, employing a hyperbolic sine strain-rate model for the matrix behavior developed by Miller [38].

Min and Crossman [39] and Kural and Min [40] present theoretical and experimental results on 6061 Al alloy and A291C Mg alloy reinforced with graphite fiber plus built-up laminate structures. They assume a Prandtl–Reuss flow rule after the matrix has yielded as predicted by the von-Mises yield criteria. In one study [39], perfect plastic deformation, with no work hardening, was assumed, whereas in the other study [40], the complete stress–strain curve, with plastic deformations, was approximated in the analysis by using a piecewise linear relationship. With either approximation, they were able to predict a hysteresis in the thermal expansion that agreed reasonably well with the measured hysteresis.

In their study of the expansion of Cu–W composites mentioned earlier, Wakashima $et\ al.$ [31] also included the effects of plastic deformation in the matrix by using an energy-balance criterion. Their formulation predicted a hysteresis in expansion between the heating and cooling values that was in good agreement with the measurements. Although solutions were given both for disc-shaped and for continuous-fiber reinforcements, only the fiber equations will be presented here. These expansivities, parallel and perpendicular to the reinforcements, are

$$\alpha_{c\|} = \alpha_{\|} + \frac{\varepsilon_p}{\Delta T}\left[1 - \frac{\nu_1 G_1(1 - \nu_1)(2C + D)}{AC - BD}\right] \tag{58}$$

and

$$\alpha_{c\perp} = \alpha_{\perp} + \frac{1}{2}\frac{\varepsilon_p}{\Delta T}\left[1 - \frac{\nu_1 2G_1(1 - \nu_1)(A + 2B)}{AC - BD}\right], \tag{59}$$

where

$$\varepsilon_p = \frac{2(1 + \nu_2)E_1(\alpha_1 - \alpha_2)\Delta T}{2(1 + \nu_2)E_1 + 3V_2(1 - 2\nu_1)E_2 + 3V_1(1 - 2\nu_2)E_1}$$

$$\pm \frac{(1 - \nu_1)(AC - BD)\sigma_y}{\nu_1(1 - 2\nu_1)G_1G_1[2(1 + \nu_2)E_1 + 3V_2(1 - 2\nu_1)E_2 + 3V_1(1 - 2\nu_2)E_1]}, \tag{60}$$

with A, B, C, and D being the same as in Eqs. (47)–(50). The (+) sign is used when $(\alpha_1 - \alpha_2)\Delta T < 0$ and the (−) sign when $(\alpha_1 - \alpha_2)\Delta T > 0$.

One method of including plastic deformation in the analysis of composite behavior is to use a bilinear relationship to approximate the actual stress–strain response of the composite matrix material. When the stresses are below the yield point of the matrix, the matrix modulus is one constant; above the yield point, the response is plastic with linear work hardening so that the stress increases linearly with strain but with a different slope as compared with the elastic response. In an analysis of the plastic relaxation of stress around a spherical precipitate in an infinite matrix, Earmme et al. [41] employed this bilinear approximation for the actual stress–strain behavior of the matrix material. This bilinear stress–strain analysis was used by Hahn [42] for calculating the expansivity of two concentric spheres with finite radius in order to take into account the volume-fraction effect on the expansivity. Further development of the analysis, reported by Hahn and Armstrong [8], resulted in a more compact representation for the expansivity of the two concentric spheres. For the concentric spheres with a totally plastic matrix,

$$
\alpha_c = \alpha_2 - (\alpha_2 - \alpha_1)\mathbf{V}_1 \left[\frac{3\left(\dfrac{f}{g}\right)\dfrac{1}{2\mathbf{E}_2}}{\dfrac{1}{g}\left(\dfrac{2\mathbf{m}}{1-\mathbf{m}}\right)\dfrac{1}{3}\left(\dfrac{1}{\mathbf{K}_1} + \left\{\dfrac{1}{\mathbf{K}_2} - \dfrac{1}{\mathbf{K}_1}\right\}\mathbf{V}_1\right) + \dfrac{1}{2\mathbf{E}_2}} \right]
$$

$$
+ \left[\frac{\dfrac{2}{3}\left(\dfrac{1}{g}\right)\left(\dfrac{\Delta\sigma_y}{\Delta\mathbf{T}}\right)\left(\dfrac{1}{\mathbf{K}_2} - \dfrac{1}{\mathbf{K}_1}\right)\mathbf{V}_1\ln\left(\dfrac{1}{\mathbf{V}_1}\right)}{\dfrac{1}{g}\left(\dfrac{2\mathbf{m}}{1-\mathbf{m}}\right)\dfrac{1}{3}\left(\dfrac{1}{\mathbf{K}_1} + \left\{\dfrac{1}{\mathbf{K}_2} - \dfrac{1}{\mathbf{K}_1}\right\}\mathbf{V}_1\right) + \dfrac{1}{2\mathbf{E}_2}} \right], \qquad (61)
$$

where $\Delta\sigma_y$ is the change in yield stress accompanying the temperature change $\Delta\mathrm{T}$, and f and g are constants related to the linear work hardening of the plastic stress–strain curve of slope $m\mathrm{E}$ in accordance with the equations $f = 1 + [2m/(1-m)][1-v_2]$ and $g = 3 + [2m/(1-m)][1+v_2]$.

In Eq. (61), with no work hardening, $m = 0 = \Delta\sigma_y/\Delta\mathrm{T}$, the rule-of-mixtures equation for α_b is obtained. For $m = 1$, the system is elastic so that the last term in Eq. (61) is omitted, and the remainder of the equation agrees with the elastic concentric spheres analysis [Eq. (3)]. Equation (61) has been applied to compute composite expansivity by using the material properties for 6061 Al and SiC for the cases of an elastic matrix and a totally plastic matrix. The results are shown in Figs. 4 and 5 for the 5 and 20 v/o SiC composites, respectively.

In the case of a continuous-fiber-reinforced composite, a model similar to that of the concentric spheres has been utilized for two coaxial cylinders representing the inner fiber and outer matrix [42]. Complicated, closed-form mathematical solutions were obtained for the axial and transverse expansivities and internal stresses. An alternate numerical method of analysis was used similar to the analysis of the five coaxial cylinders. For cylindrical

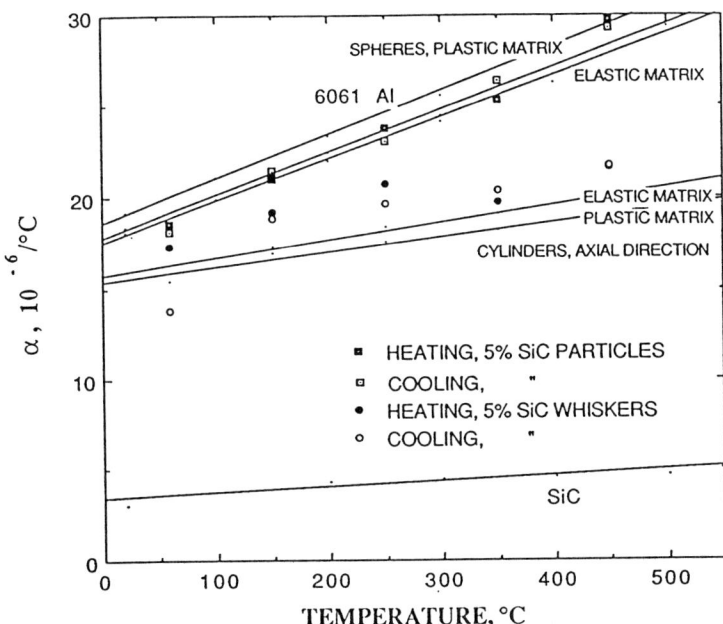

FIG. 4. Thermal expansivities of 6061 Al–0.05 SiC particles or whiskers, and predictions from the sphere and cylinder models.

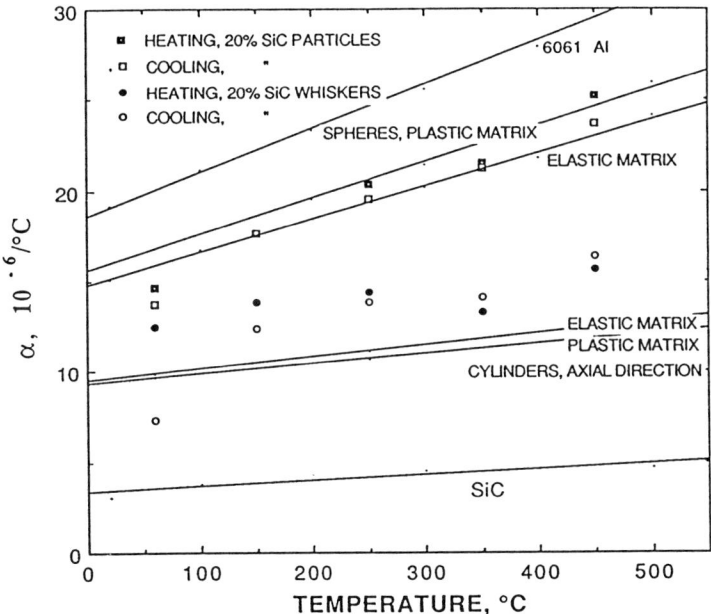

FIG. 5. Thermal expansivities of 6061 Al–0.20 SiC particles or whiskers, and predictions from the sphere and cylinder models.

349

symmetry, the strains are

$$\varepsilon_r = \frac{du}{dr} = \frac{1}{E}[\sigma_r - \nu(\sigma_\theta + \sigma_z)] + \varepsilon_r^p + \alpha\Delta T,$$

$$\varepsilon_\theta = \frac{u}{r} = \frac{1}{E}[\sigma_\theta - \nu(\sigma_r + \sigma_z)] + \varepsilon_\theta^p + \alpha\Delta T, \qquad (62)$$

$$\varepsilon_z = \frac{dw}{dz} = \frac{1}{E}[\sigma_z - \nu(\sigma_r + \sigma_\theta)] + \varepsilon_z^p + \alpha\Delta T,$$

where u and w are the radial and axial displacements respectively, σ_i are the stresses, and ε_i^p are the plastic strains in the three principle directions. The solution for elastic interaction between the cylinders gives σ_z as the intermediate stress so that, according to the Tresca yield criteria, plastic flow will occur in the r–θ plane. With $\varepsilon_z^p = 0$, the assumption of constant volume requires that $\varepsilon_\theta^p + \varepsilon_r^p = 0$. Considering the equilibrium of stresses [41], the plastic strain is

$$\varepsilon_r^p = \left(\frac{1 - m}{m}\right)\frac{(\sigma_\theta - \sigma_r) - \sigma_y}{E_2}. \qquad (63)$$

The differential equation for the radial displacement is

$$\frac{d^2u}{dr^2} + \frac{1}{r}\frac{du}{dr} - \frac{u}{r^2} = \frac{2\sigma_y}{hE_2'r}, \qquad (64)$$

where $h = 1 + \phi(1 - \nu_2^2)$, $\phi = m/1 - m$, and $E_2' = E_2/(1 + \nu_2)(1 - 2\nu_2)$. The general solution is

$$U = \frac{\sigma_y}{hE_2'}R\ln(R) + AR + \frac{B}{R}, \qquad (65)$$

where $U = u/r_1$, $R = r/r_1$, and A and B are constants that will be determined by the boundary conditions. The axial displacement is assumed to be $w = C_z$ and is the same in both cylinders. The stresses in the outer cylinder, with $g = 2 + \phi(1 + \nu_2)$, are

$$\sigma_{rr}^{(2)} = \frac{\sigma_y}{g}\ln R + E_2'A_2 - \frac{\phi E B_2}{gr^2} + \nu_2 E_2'C - 3K_2\alpha_2\Delta T,$$

$$\sigma_{\theta\theta}^{(2)} = \frac{\sigma_y}{g}[\ln R + 1] + E_2'A_2 + \frac{\phi E B_2}{gr^2} + \nu_2 E_2'C - 3K_2\alpha_2\Delta T, \qquad (66)$$

$$\sigma_{zz}^{(2)} = \frac{2\nu_2\sigma_y}{h}\left[\ln R + \frac{1}{2}\right] + 2\nu_2 E_2 A_2 + (1 - \nu_2)E_2'C - 3K_2\alpha_2\Delta T.$$

The inner cylinder (representing the fiber) is assumed to remain elastic, thus requiring that the radial displacement be finite. Therefore, $u_1 = A_1 r$, where A_1 is a constant, and the stresses are

$$\sigma_{rr}^{(1)} = A_1 E_1' + v_1 E_1' C - 3K_1 \alpha_1 \Delta T,$$

$$\sigma_{\theta\theta}^{(1)} = A_1 E_2' + v_1 E_1' C - 3K_1 \alpha_1 \Delta T, \tag{67}$$

$$\sigma_{zz}^{(1)} = 2v_1 A_1 E_1' + (1 - v_1)E_1' C - 3K_1 \alpha_1 \Delta T.$$

The boundary conditions are $\sigma_{rr}^{(2)} = 0$ at $r = b$, $u_1 = u_2$, $w_1 = w_2$, and $\sigma_{rr}^{(1)} = \sigma_{rr}^{(2)}$ at $r = a$. The balance of forces requires that

$$\int_0^a \sigma_{zz}^{(1)} dA_1 + \int_a^b \sigma_{zz}^{(2)} dA_2 = 0. \tag{68}$$

With these equations and boundary conditions, a linear set of equations for the four dependent variables A_1, A_2, B_2, and C can be solved in a manner similar to Eq. (28). The matrix elements for the linear-equations solution are given in Appendix 2. The solution for the constants can be employed to calculate stresses and displacements, which in turn can be used to calculate the axial and transverse expansivities. Separate cases for elastic deformation of the matrix and of plastic deformation of the matrix with linear strain hardening have been considered. The axial expansivities for the 5 and 20 vol-% SiC-reinforced 6061 Al composites are shown in Figs. 4 and 5, respectively.

Figure 4 compares the computed expansivities from this model with those values determined from the expansion data given in Reference 8 for the 5% SiC composites. There is very good agreement between the model-computed expansivities and the experimental results obtained for the 5% SiC particle composite. However, large deviations do occur at the lowest and highest temperatures. The deviation between measurements obtained from the heating and cooling cycles approximately matches the shift between the fully elastic matrix and the perfectly plastic matrix. Less agreement between the model calculations and the experimental results is shown for the 5% SiC whisker system. Two reasons for the difference are that the model calculations are for an infinite, continuous fiber, leading to an underestimation of expansivity, and the first occurrence here of significant permanent deformation effects. Figure 5 shows reasonable agreement between the model and experimental expansivities for both the 20% SiC particle and the whisker composites. Better agreement occurs for the particle composite, as expected. Again, the whisker composite model calculations for the continuous fiber fall below the measurements. The largest range in measured expansivities occur for the low-temperature determinations, as indicated in the expansion figures themselves.

For the concentric-spheres case, as the slope of the stress–strain behavior in the plastic region decreased from that of an elastic matrix to zero slope as appropriate for a "perfectly plastic" behavior, the expansivity increased from that of the elastic concentric spheres [Eq. (63)] to that of the rule-of-mixtures value [Eq. (61)].

IV. Summary

A review has been presented of many of the models for predicting the thermal expansivity of both continuous-fiber- and spherical-particle-reinforced metal matrix composites. Many of the recently derived models concerned with elastic interaction between the constituents have been shown to be equivalent to earlier analysis. A new mathematical technique has been presented for calculating the expansivity of multiple cylinders. Various models that incorporate the effects of plastic deformation of the matrix material have also been presented along with predictions using the two-concentric-spheres and the two-coaxial-cylinders models for the case of two-component composites. For temperature changes producing stresses that give either elastic or plastic deformation of the matrix, good agreement between predictions and measurements can generally be obtained by selecting the proper model for the composite material.

Appendix I

$$a_{11} = r_1 \qquad a_{35} = \frac{1}{r_3} \qquad a_{51} = C_{11}^{(1)} + C_{12}^{(1)} \qquad a_{62} = C_{11}^{(2)} + C_{12}^{(2)} \qquad a_{74} = C_{11}^{(3)} + C_{12}^{(3)}$$

$$a_{12} = -r_1$$
$$a_{36} = -r_3 \qquad a_{52} = -C_{11}^{(2)} - C_{12}^{(2)} \qquad a_{63} = \frac{C_{12}^{(2)} - C_{11}^{(2)}}{r_2^2} \qquad a_{75} = \frac{C_{12}^{(3)} - C_{11}^{(3)}}{r_3^2}$$

$$a_{13} = -\frac{1}{r_1} \qquad a_{37} = -\frac{1}{r_3} \qquad a_{53} = \frac{C_{11}^{(1)} - C_{12}^{(1)}}{r_1^2} \qquad a_{64} = -C_{11}^{(3)} - C_{12}^{(3)} \qquad a_{76} = -C_{11}^{(4)} - C_{12}^{(4)}$$

$$a_{22} = r_2$$
$$a_{46} = r_4 \qquad a_{510} = C_{13}^{(1)} - C_{13}^{(2)} \qquad a_{65} = \frac{C_{11}^{(3)} - C_{12}^{(3)}}{r_2^2} \qquad a_{77} = \frac{C_{11}^{(4)} - C_{12}^{(4)}}{r_3^2}$$

$$a_{23} = \frac{1}{r_2} \qquad a_{47} = \frac{1}{r_4}$$

$$a_{24} = -r_2 \qquad a_{610} = C_{13}^{(2)} - C_{13}^{(3)} \qquad a_{710} = C_{13}^{(3)} - C_{13}^{(4)}$$
$$a_{48} = -r_4$$

$$a_{25} = -\frac{1}{r_2} \qquad a_{49} = -\frac{1}{r_4}$$

$$a_{34} = r_3$$

$$a_{86} = C_{11}^{(4)} + C_{12}^{(4)}$$

$$a_{87} = \frac{C_{12}^{(4)} - C_{11}^{(4)}}{r_4^2}$$

$$a_{88} = -C_{11}^{(5)} - C_{12}^{(5)}$$

$$a_{89} = \frac{C_{11}^{(5)} - C_{12}^{(5)}}{r_4^2}$$

$$a_{810} = C_{13}^{(4)} - C_{13}^{(5)}$$

$$a_{98} = C_{11}^{(5)} + C_{12}^{(5)}$$

$$a_{99} = \frac{C_{12}^{(5)} - C_{11}^{(5)}}{r_5^2}$$

$$a_{910} = C_{13}^{(5)}$$

$$a_{101} = 2C_{13}^{(1)}r_1^2$$

$$a_{102} = 2C_{13}^{(2)}(r_2^2 - r_1^2)$$

$$a_{104} = 2C_{13}^{(3)}(r_3^2 - r_2^2)$$

$$a_{106} = 2C_{13}^{(4)}(r_4^2 - r_3^2)$$

$$a_{108} = 2C_{13}^{(5)}(r_5^2 - r_4^2)$$

$$a_{1010} = C_{33}^{(1)}r_1^2 + C_{33}^{(2)}(r_2^2 - r_1^2) + C_{33}^{(3)}(r_3^2 - r_2^2)$$
$$+ C_{33}^{(4)}(r_4^2 - r_3^2) + C_{33}^{(5)}(r_5^2 - r_4^2)$$

$$c_5 = (\beta_1^{(1)} - \beta_1^{(2)})\Delta T$$

$$c_6 = (\beta_1^{(2)} - \beta_1^{(3)})\Delta T$$

$$c_7 = (\beta_1^{(3)} - \beta_1^{(4)})\Delta T$$

$$c_8 = (\beta_1^{(4)} - \beta_1^{(5)})\Delta T$$

$$c_9 = \sigma_{0r} + \beta_1^{(5)}\Delta T$$

$$c_{10} = \beta_3^{(1)}r_1^2 + \beta_3^{(2)}\Delta T(r_2^2 - r_1^2) + \beta_3^{(3)}\Delta T(r_3^2 - r_2^2)$$
$$+ \beta_3^{(4)}\Delta T(r_4^2 - r_3^2) + \beta_3^{(5)}\Delta T(r_5^2 - r_4^2)$$

Appendix II

$$[A] \times [B] = [C]$$

$$[A] = \begin{bmatrix} r_1 & -R_1 & -\dfrac{1}{R_1} & 0 \\[2ex] E_1' & -E_2' & \dfrac{\phi E_2}{gR_1^2} & v_1 E_1' - v_2 E_2' \\[2ex] 0 & E_2' & -\dfrac{\phi E_2}{gR_2^2} & v_2 E_2' \\[2ex] 2v_1 E_1' r_1^2 & 2v_2 E_2'(r_2^2 - r_1^2) & 0 & (1 - v_1)E_1' r_1^2 + (1 - v_2)E_2'(r_2^2 - r_1^2) \end{bmatrix}$$

$$[B] = \begin{bmatrix} A_1 \\ A_2 \\ B_2 \\ C \end{bmatrix}$$

$$[C] = \begin{bmatrix} \dfrac{\sigma_y}{hE'_2} R_1 \ln R_1 \\[2ex] \dfrac{\sigma_y}{g} \ln(R_1) + 3K_1\alpha_1\Delta T - 3K_2\alpha_2\Delta T \\[2ex] -\dfrac{\sigma_y}{g} \ln(R_2) + 3K_2\alpha_2\Delta T \\[2ex] 3K_1\alpha_1\Delta T r_1^2 + 3K_2\alpha_2\Delta T(r_2^2 - r_1^2) - \dfrac{2v_2\sigma_y}{h} r_1^2 \ln R_2 \end{bmatrix}$$

References

1. L. Holliday and J. Robinson, *J. Mater. Sci.*, **8**, 301 (1973).
2. C. C. Chamis and G. P. Sendeckyj, *J. Comp. Mater.*, **2**, 332 (1968).
3. Y. S. Touloukiasn, R. K. Kirby, R. E. Taylor, and P. D. Desai, "Thermal Expansion—Metallic Elements and Alloys," p. 17a. IFI/Plenum, New York, 1975.
4. P. S. Turner, *J. Res. NBS*, **37**, 239 (1946).
5. E. H. Kerner, *Proc. Phys. Soc. London*, **B69**, 808 (1956).
6. T. T. Wang and T. K. Kwei, *J. Polymer Science*, **7**, 889 (1969).
7. T. A. Hahn and R. W. Armstrong, *Int. J. Thermophys.*, **9**, 179 (1988).
8. T. A. Hahn and R. W. Armstrong, *Int. J. Thermophys.*, **9**, 861 (1988).
9. A. A. Fahmy and A. N. Ragai, *J. Appl. Phys.*, **41**, 5108 (1970).
10. S. J. Feltham, B. Yates, and R. J. Martin, *J. Mater. Sci.*, **17**, 2309 (1982).
11. R. R. Tummala and A. L. Friedburg, *J. Am. Ceram. Soc.*, **53**, 376 (1970).
12. F. Laszlo, *J. Iron Steel Inst.*, **147**, 173 (1943); *ibid.*, **148**, 137 (1943).
13. W. J. Craft and R. M. Christensen, *J. Comp. Mater.*, **15**, 2 (1981).
14. Chun-Hway Hsueh and Paul F. Becher, *J. Am. Ceram. Soc.*, **71**, C438 (1988).
15. *Ibid.*, **72**, C359 (1989).
16. V. M. Levin, *Mekh. Tverd. Tela.*, **1**, 88 (1967).
17. R. A. Schapery, *J. Comp. Mater.*, **2**, 380 (1968).
18. S. A. Rojstaczer, D. Cohn, and G. Marom, *J. Mater. Sci.*, **4**, 1233 (1985).
19. P. Klemens, *Int. J. Thermophys.*, **7**, 197 (1986).
20. T. Ishikawa, K. Koyana, and S. H. Kobayashi, *J. Comp. Mater.*, **12**, 153 (1978).
21. K. F. Rogers, L. N. Phillips, D. M. K. Lee, B. Yates, M. Overy, J. P. Sargent, and B. A. McCalla, *J. Mater. Sc.* **12**, 718 (1977).
22. W. Schneider, *Kunstoffe*, **61**, 23 (1971).
23. Y. Mikata and M. Taya, *J. Comp. Mater.*, **19**, 554 (1985).
24. J. C. Halpin, *J. Comp. Mat.*, **3**, 732 (1969).
25. J. C. Halpin, "Primer on Composite Materials: Analysis." Technomic Publishing Co., Lancaster, Pennsylvania, 1984.
26. J. C. Halpin and N. J. Pagano, *J. Comp. Mat.*, **3**, 72 (1962).
27. G. Maron and A. Weinberg, *J. Mater. Sci.*, **10**, 1005 (1975).
28. S. T. Gulati and W. A. Plummer, in "Thermal Expansion–1971" (M. G. Graham and H. E. Hagy, eds.), p. 257. American Institute of Physics, New York, 1972.

29. J. D. Eshelby, *Proc. Roy. Soc. Lond.*, **A241**, 376 (1957).
30. T. Mura, "Micromechanics of Defects in Solids." Martinus-Nijhoff, The Hague, 1982.
31. K. Wakashima, M. Otsuka, and S. Umekawa, *J. Comp. Mater.*, **8**, 391 (1974).
32. Y. Takao and M. Taya, *J. Comp. Mat.*, **21**, 140 (1987).
33. K. G. Kreider and V. M. Patarini, *Metall. Trans.*, **1**, 3431 (1970).
34. G. Garmong, *Metall. Trans.*, **5**, 2183 (1974).
35. W. R. Tyson, *Metall. Trans.*, **6A**, 1674 (1975).
36. S. K. Makerjee, J. E. Bird, and J. E. Dorn, *Trans. ASM*, **62**, 155 (1969).
37. L. O. K. Larsson, in "Advances in Composite Materials; ICMM–3" (A. R. Bunsell, C. Bathias, A. Martenchar, D. Menkes, and G. Verchery, eds.), p. 649. Pergamon Press, Oxford, 1980.
38. A. Miller, *J. Eng. Mater. Tech.*, April, 1976.
39. B. K. Min and F. W. Crossman, in "Thermal Expansion–8" (T. A. Hahn, ed.), p. 175. Plenum Press, New York, 1984.
40. M. H. Kural and B. K. Min, *J. Comp. Mat.*, **18**, 519 (1984).
41. Y. Y. Earmme, W. C. Johnson, and J. K. Lee, *Metall. Trans.*, **12A**, 1521 (1981).
42. T. A. Hahn, Ph.D. dissertation, University of Maryland, College Park, 1986.

Electrical Conductivity in Continuous-Fiber Composites

JACQUES E. SCHOUTENS

Kaman Sciences Corporation
Santa Barbara, California

I. Introduction[1]

This chapter presents a discussion of the electrical conductivity properties of continuous-fiber-reinforced metal-matrix *composite* materials. Electrical conductivity can be separated into two types: transverse to the fiber direction and along the fiber direction. The transverse conductivity is a complex function of the geometry that electrons must follow in their motion along the electric field and of the distortions in the electric field in the matrix due to the presence of the fibers. The modeling discussed is applicable to a wide range of temperatures, from liquid-hydrogen temperatures to that near the solidus of the matrix. At low temperatures, there exists a complex interaction between electrons, matrix, and the fiber surfaces. There are no known models or theories describing these phenomena. At temperatures approaching the solidus of the matrix, the validity of the modeling describing transverse conductivity is not known.

The longitudinal electrical conductivity is described fairly well by a simple model involving the volume fraction of fibers in the matrix when the temperature exceeds that of liquid hydrogen. At temperatures below approximately 10 K, the composite electrical conductivity exhibits a drastic increase in its conductivity with decreases in temperatures, which is due to the interaction of electrons with fiber surfaces.

The above-mentioned phenomena greatly increases in complexity when a magnetic field is superimposed on the electric field, and to date only one model describing this phenomena has been published, which we shall discuss in the following.

II. Longitudinal Conductivity[2]

No theoretical analyses that predict the electrical resistivity of continuous-fiber-reinforced metal matrix composites (MMCs) at cryogenic and higher temperatures have been found. However, a number of experimental studies on the electrical conductivity of MMCs and in situ composites have been reported [1–6]. In these studies, experimental results are discussed and related to conduction models that are either modified forms of the rule of mixture or simple equations based on Dingle's [7] asymptotic solution of the Boltzmann equation for electrical conduction in thin wires or thin films when the reinforcing filaments are either much larger or much smaller than the bulk-electron mean free path.

As the temperature of an MMC decreases from room temperature, the composite resistivity is expected to decrease steadily until it rises sharply at cryogenic temperature. The steady decrease is due to thermal effects in the matrix that can be modeled as $\rho(T) \approx T^n$, where $\rho(T)$ is the temperature-dependent resistivity, T is the absolute temperature, and n is an exponent between 1 and 2 [8]. The sharp increase in resistivity at cryogenic temperature has been referred to by Dingle [7] and [9–11] as the size effect. In the analysis of MMC resistivity, the fiber resistivity is generally considered to be orders of magnitude greater than that of the matrix material, since these fibers consist of ceramic materials or boron (B) or carbon (C).

[2] Adapted from the following sources:

F. S. Roid and J. E. Schoutens, "Theory of Longitudinal Electrical Resistivity of Metal Matrix Composites at Cryogenic Temperatures," *Proceedings ICCM-VI/ECCM-2*, Volume 4, p. 4.311, Elsevier Applied Science Publishers, Ltd.

F. S. Roig and J. E. Schoutens, *J. Mater. Sci.*, **22**, (Chapman & Hall, 1987), 3749–3754.

F. S. Roig and J. E. Schoutens, *J. Mater. Sci.*, **22** (Chapman & Hall, 1987), 4002–4010.

In this chapter, a theory is developed to describe the electron-scattering process from fiber surfaces using a classical theory based on the solution of Boltzmann's equation. In a sense, the theory discussed is the inverse problem solved by Dingle [7] for very thin wires. The resulting triple integral is integrated numerically, and the results are applied to continuous-fiber-reinforced MMCs. The present theory is applicable only to conduction along the fiber direction in the absence of a magnetic field. The case of a magnetic field is presented in two papers [12, 13].

A. Theory

The theory considers a single nonconducting fiber (B, C, or ceramic) in a metallic matrix with the fiber along the z-direction. A small electric field E is applied in the z-direction. Let $\mathbf{v}$ be the electron velocity at point $\mathbf{r}$ in the metal, where $\mathbf{r}$ originates at the fiber axis. Then, the Boltzmann equation in cylindrical coordinates is

$$v_r \frac{\partial F^1}{\partial r} + \frac{v_\theta^2}{r} \frac{\partial F^1}{\partial v_r} - \frac{v_r v_\theta}{r} \frac{\partial F^1}{\partial v_\theta} + \frac{v_\theta}{r} \frac{\partial F^1}{\partial \theta} + \frac{F^1}{\tau} = \frac{eE}{m^*} \frac{\partial F^0}{\partial v_z}, \tag{1}$$

where e is the absolute value of the electronic charge; m^* is its effective mass; τ is the relaxation time for scattering; v_r, v_θ, and v_z are the radial, angular, and axial electron velocities, respectively; E is the electric field, and

$$F^1(\mathbf{v}, \mathbf{r}) = F(\mathbf{v}, \mathbf{r}) - F^0(\mathbf{v}) \tag{2}$$

represents the deviation of the electron distribution function F from the equilibrium distribution F^0. Equation 1 can be solved directly by using general methods from the theory of quasilinear first-order partial differential equations [14]. The general solution is [15, 16]

$$F^1 = \frac{eE\tau}{m^*} \frac{\partial F^0}{\partial v_r} \left\{ 1 - f(rv_\theta, v_r^2 + v_\theta^2, \theta + \varphi) \exp\left[-\frac{1}{\tau} \left(\frac{rv_r}{v_r^2 + v_\theta^2} \right) \right] \right\}, \tag{3}$$

which is the same solution found by Dingle [7] by using an indirect method.

The function f is determined from the boundary conditions of the problem. It is an arbitrary function that must be even in the variable rv_θ, since F^1 has to be an even function of v_θ from the symmetry of the problem and is determined by the fiber surface. When $v_r > 0$ (electron motion away from the fiber), f_+ is used; for $v_r < 0$ (electron motion toward fiber), f_- is used. If p is the probability for elastic scattering at the fiber surface, then the boundary condition at $r = a$ is [16]

$$F(v_r, v_\theta; r = a) = pF(-v_r, v_\theta; r = a) + g, \tag{4}$$

where g is the distribution function for diffusively or elastically scattered electrons. By using Eqs. (2) and (3) with Eq. (4) and considering the fact that g must be independent of direction, but F^1 depends on direction, the solution to the Boltzmann equation becomes

$$F^1 = \frac{eE\tau}{m^*}\frac{\partial F^0}{\partial v_z}\{1 - (1 - p)\exp\{-[r\cos\varphi - (a^2 - r^2\sin^2\varphi)^{1/2}](\tau\sin\theta)^{-1}\}\},$$

(5)

where

$$\varphi = \sin^{-1}\left\{\frac{v_\theta}{\sqrt{v_r^2 + v_\theta^2}}\right\}.$$

(6)

Scattering from the fiber surface occurs only when $v_r > 0$ and when $0 < \varphi < \varphi_{max} = a/r$ [15, 16].

B. Composite Conductivity

Figure 1 shows the idealized cell used in the following calculations. The fibers are in a square array a distance D apart, and their radius is a. The mean free path Λ for scattering in the bulk metal is assumed to be $\Lambda \leq D/2$ so that there is no overlap between adjacent scattering regions. Also, A_b is the area of bulk metal not affected by scattering from nearby fiber surfaces, and A_{sc} is the area of the scattering regions centered on the fibers within the cell. Then, the cell area is $A_{cell} = A_b + A_{sc} + \pi a^2$, which assumes that $A_{sc} = \pi(r_{sc}^2 - a^2)$ and $A_{cell} = (D + 2a)^2$, $r_{sc} = \Lambda + a$. Denoting j as the current density in the bulk within the scattering region, the total current within the cell is

$$I = \int_a^{r_{sc}} [j_0 - \Delta j(r)]2\pi r\ dr + j_0 A_b,$$

(7)

and the conductivity of the cell is $\sigma = I/EA_{cells}$, where E is the applied longitudinal electric field. If σ_0 is the bulk conductivity, then $j_0 = \sigma_0 E$. It then follows that [12]

$$\frac{\sigma_c}{\sigma_0} = \frac{A_{sc} + A_b}{A_{cell}}\left\{1 - \frac{1}{A_{sc} + A_b}\int_a^{r_{sc}}\frac{\Delta j(r)}{j_0}2\pi r\ dr\right\},$$

(8)

and, using the values for A_{sc} and A_b results in

$$\frac{\sigma_c}{\sigma_0} = (1 - V_f)\left\{1 - \frac{1}{(1 - V_f)A_{cell}}\int_a^{r_{sc}}\frac{\Delta j(r)}{j_0}2\pi r\ dr\right\}.$$

(9)

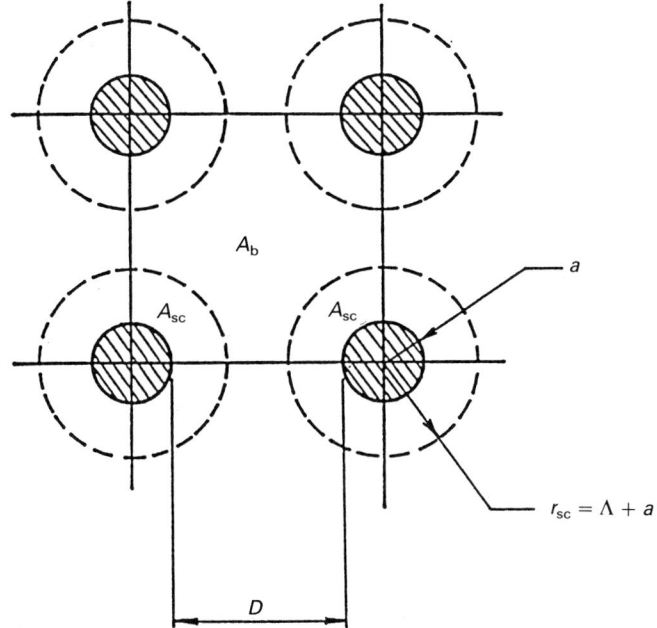

FIG. 1. Idealized composite cell used in deriving MMC conductivity model.

The expression for the change in the current is [12]

$$\Delta j(r) = 2e\left(\frac{m^*}{h}\right)\int_0^\infty v^2\,dv\int_0^\pi \sin\theta\,d\theta\int_{-\varphi_{max}}^{\varphi_{max}} \Delta F^1 v_z\,d\varphi \tag{10}$$

where $\varphi_{max} = \sin^{-1}(a/r)$. The current density due to the background scattering is [15]

$$j_0 = 2e\left(\frac{m^*}{h}\right)^2\int_0^\infty v^2\,dv\int_0^\pi \sin\theta\,d\theta\int_0^{2\pi} F^1 v_z\,d\varphi, \tag{11}$$

where h is Plank's constant. Substituting Eq. (10) and (11) into Eq. (9) gives

$$\frac{\sigma_c}{\sigma_0} = (1 - V_f)\Bigg\{1 - (1 - p)\frac{2\pi a^2}{(1 - V_f)A_{cell}}\int_1^{1+2/k} x\,dx\int_0^{\pi/2}\frac{3}{\pi}\cos^2\theta\sin\theta\,d\theta\Bigg\}$$

$$\times\Bigg\{\int_0^{\sin^{-1}(1/x)}\exp\{-k[x\cos\varphi - (1 - x^2\sin^2\varphi)^{1/2}](2\sin\theta)^{-1}\}d\varphi\Bigg\},$$

$$\tag{12}$$

where the following dimensionless variables have been used for convenience of numerical integration: $x = a/r$ and $k = 2a/\Lambda$. Three nested Simpson's rule integrations were used in the numerical integration. The calculations were performed on a Hewlett-Packard HP-41C hand-held calculator. The details of these computations are given elsewhere [15]. Equation (12) can be written as

$$\frac{\sigma_c}{\sigma_0} = (1 - V_f)\left\{1 - (1 - p)\frac{2V_f}{1 - V_f}\,I(k)\right\}, \tag{13}$$

where p is the probability for an elastic scattering at fiber surface.

It is now necessary to relate k to temperature. No simple model has been found that relates the electron mean free path in a real metal to its temperature. Such calculations are extremely complex [17, 18] and are not attempted here. But, a simple useful relation has been derived [15] that can be used to "calibrate" to composites. Figure 2 shows the value of the triple integral in Equation 12, represented by the function $I(k)$, as a function of k. A regression analysis shows that

$$I(k) = 0.27\,k^{-1.08}, \tag{14}$$

with a coefficient of determination of $r^2 = 1.00$. Now, a linear relationship between k and T can be written as $T = Ck$, where C is a constant. To obtain C, the argument is as follows: At room temperature, Λ is close to zero, making k large; this means that $\sigma_c/\sigma_0 = (1 - V_f)$, as it should. One can take $k = 100$ for room temperature (Fig. 2), which corresponds to small $I(k)$ ($\approx 2 \times 10^{-3}$) and, letting the room temperature $T_R = 300$ K, then

$$T = 3k. \tag{15}$$

Combining Eqs. (13) through (15) gives the composite conductivity in terms of temperature and fiber volume fraction as

$$\frac{\sigma_c}{\sigma_0} = (1 - V_f)\left\{1 - 1.77\left(\frac{V_f}{1 - V_f}\right)T^{-1.08}\right\}. \tag{16}$$

In Equation (13), it has been assumed that $p \equiv 0$ because microstructural examination of fiber surfaces shows them to be highly irregular compared with lattice dimensions; hence, electrons are scattered diffusively ($p \equiv 0$) [15]. Equation 16 has a lower limit for

$$T = T_{min} = \frac{6\sqrt{V_f/\pi}}{1 - \sqrt{V_f/\pi}},$$

which was derived elsewhere [15]. This means that $T < T_{min}$ has no physical meaning, because that would give $\rho_c/\rho_0 < 0$ ($\rho_0 = 1/\sigma_0$), or negative resistivity.

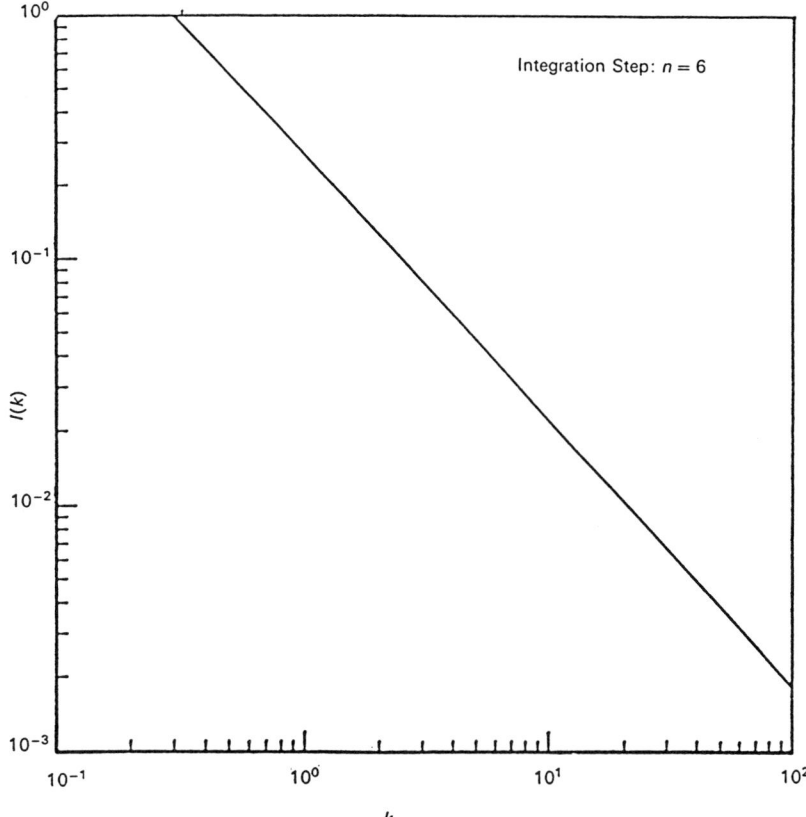

FIG. 2.   Values of the integral I(k) as a function of k ($k = 2a/\Lambda$).

Roig and Schoutens [12] solved the Boltzmann equation in the relaxation time approximation with parallel E and B fields parallel to the continuous fibers reinforcing a metal matrix. It is shown that this solution is identical to that described by them elsewhere, except for the addition of the cyclotron frequency. The addition of the cyclotron frequency term shows that the electrons follow helical paths as they drift down the composite. The boundary considered was either the external or the internal surface of a cylinder representing the fiber. To apply this solution to metal matrix composite materials, it was assumed that the cylindrical fibers were non conducting cylinders in a matrix of pure crystalline metal. The electron mean free path was never greater than half the fiber-separation distance.

In a subsequent paper, Roid and Schoutens [13] used the above mentioned solution to derive two integral expressions for the electrical conductivity of metal-matrix-composite materials when a magnetic field, B, is added to a small electric field also parallel to fibers. One expression applies to strong magnetic fields meaning that $R_0/a < 1$, where $R_0 = m^*V_f/eB$ when the Fermi velocity is perpendicular to the magnetic field. When $B \to \infty$, the integral expression reduces to the well-known conductivity value $\sigma = \sigma_0(1 - V_f)$, where σ_0 is the bulk matrix conductivity and V_f is the fiber volume fraction. For weak magnetic fields, $R_0/a > 1$, then the electrical conductivity is expressed by the sum of two integrals. When $B \to 0$, the electrical conductivity reduces to the integral expression obtained in the earlier results when there is only a longitudinal electric field. The paper by Roig and Schoutens corrects an incorrect derivation of the composite conductivity in the absence of a magnetic field published earlier [15].

C. Discussion

Figure 3 shows the normalized composite resistivity, ρ_c/ρ_0, as a function of temperature calculated with Eq. (16) for three values of the fiber volume fraction. This shows the steep increase in composite resistivity with increased fiber volume fraction and decreased temperature below about 10 K. This steep rise in resistivity is due entirely to the effects of electrons scattered from the fiber surfaces. At $T > 10$ K, the resistivity is essentially that at room temperature or $\rho_c/\rho_0 \to (1 - V_f)^{-1}$ as T increases, in agreement with experimental data [15].

The electrical conductivity of MMCs depends also on the postfabrication heat treatment, as discussed by Jenkins and Arajs [20]. They showed that matrix-annealing temperatures ranging from 298 to 873 K will result in a general reduction in the composite resistivity of about a factor of 2. These data are for 28 v/o C/Al where the C fiber was Thornel 50. They also showed that the decrease in resistivity with increasing annealing temperature for 28 v/o C/Al is linear. However, this appears to be incorrect. Despite the large error bars in their data, the distribution in the datum points would indicate that this decrease follows the recovery and recrystallization profiles for heat-treated alloys [21].

Aside from resistivity decreases due to annealing, Jenkins and Arajs [20] also observed a linear decrease in resistivity with temperature of about a factor of 2.5 for the same composite within a temperature range decreasing from 290 to 80 K. This linear decrease is due to thermal effects in the alloy and is not related to the electron scattering effects discussed in this chapter. This linear decrease was also observed for 60 vol-% B/Al in the axial and

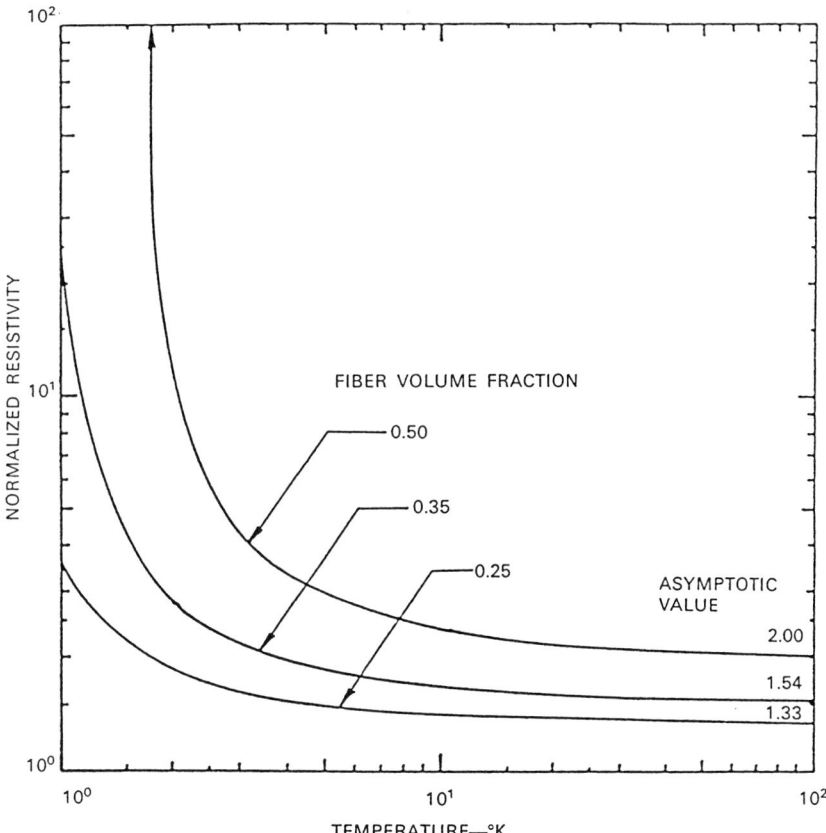

FIG. 3. Normalized composite resistivity as a function of temperature at cryogenic temperatures.

transverse directions [19]. The linear decrease can be well represented by a functional relationship of the form $\rho_0(T)/\rho_0(T_R) = (295/T)^C$, where $\rho_0(T_R)$ is the bulk matrix resistivity at room temperature, $\rho_0(T_R)$, taken as 3.3×10^{-6} $\Omega\,\mathrm{cm}$ [20], and the constant $C = -1.45$ for the same data [20]. This expression is not valid when $T/\Theta \leq 0.2$, or about 78 K for Al, where Θ is the Debye temperature [18].

The theory discussed in this chapter presents some difficulties. It does not account for the overlapping electron-scattering regions (Fig. 1), the effects of which have not yet been calculated. In real composites, the arrays of fibers are arranged in a somewhat random fashion [22] for very small-diameter fibers. This gives rise to overlapping regions. As the size of reinforcing fiber increases, order appears out of randomness, and the fiber distribution appears as quasisquare or hexagonal arrays [23]. This is due mostly to fabrication methods. For small-diameter closely-spaced fibers, many fibers are in physical contact. Consequently, the degree of overlap in the scattering regions varies a great deal. Within such overlapping regions, contributions to electron scattering can come from several fibers [15]. Consequently, the integral in Eq. (12) must be modified.

It is known from the theory of metals that electron transport is very sensitive to lattice defects and impurities. In a pure crystal, electron motion is unimpeded except by thermal effects at high temperatures. Thus, one would expect the mean free path in a pure crystal at low temperatures to be very large, i.e., on the order of 10^3 to 10^5 μm [24]. The value of the resistivity in pure metals is generally proportional to the absolute temperature for $T \approx \Theta$, where Θ is the Debye temperature. When $T \ll \Theta$, the resistivity is $\rho \approx T^5/M\Theta^6$, where M is the mass of the lattice atoms and for Al, $\Theta = 390$ K. On the other hand, the resistivity of a metal containing foreign atoms in solid solution is nearly always greater than that of the pure metal, and the increase is considerable in many cases. In general, this increase is independent of temperature (Matthiessen's rule) [18].

The greatest weakness of the theory presented in this chapter is the connection made between the integral $I(k)$ and T through an assumed linearity between k and T. Thus, some arbitrariness exists in selecting the lowest temperature, in other words, in "calibrating" the theory by means of a parameter in the absence of data. A physical approach is to relate Λ to T from a detailed theory supported by experimental data. As far as is known, this has not been done for MMCs at liquid-hydrogen or liquid-helium temperatures [25]. The mean free path Λ is a function of lattice defects and impurities and alloy structure. Typical grain dimensions in Al alloys are on the order of 1 to 10 μm. Consequently, at cryogenic temperature, one would expect the upper limit of the electron mean free path not to exceed 1 to 10 μm. Thus, for taking $\Lambda \approx 5$ μm, $k \approx 2a/\Lambda = 10/5 = 2$ when the fiber diameter is approximately 10 μm, and from the relation $T = 3k$, it is found that $T \approx 6$ K. According to the theory developed in this chapter, at this temperature, $\rho_c/\rho_0 \approx 1.17$ to 1.38 for $V_f = 0.25$ to 0.50, respectively. However, these estimates are based upon the constant $C = 3$, which was derived from heuristic arguments [15]. Therefore, T is not adequately defined. For obvious reasons, $T = 0$ cannot be used because then $\rho_c/\rho_0 < 0$ in this model. Then, the conclusion is that, in the absence of an adequate theory for Λ as a function of T and due to the apparent lack of experimental data [25], the present theory can only predict the trend expected in the composite resistivity at cryogenic temperatures.

III. Transverse Conductivity[3]

A number of transverse electrical conduction (resistivity) models for in situ and MMC-containing nonconducting continuous fibers have been proposed. All of these models are macroscopic. An excellent review of these models was recently given by Jenkins [27]. We will briefly discuss some of the more important models. Liebmann and Miller [28] proposed the following equation to calculate the electrical resistivity of InSb–Sb eutectic alloys,

$$\frac{1}{\rho_\perp} = \frac{1}{\rho_{InSb}}(1 - v^{1/2}) + \left[\rho_{InSb}\left(\frac{1 + v^{1/2}}{v^{1/2}}\right) + \rho_{Sb}\right], \qquad (17)$$

where $v = V_f/(1 + V_f)$, where V_f is the fiber volume fraction of the composite; $\rho_\perp$ is the transverse electrical resistivity; and ρ_{InSb} and ρ_{Sb} are the resistivities of indium antimonide and antimony, respectively. This theory explains the conductivity of InSb–Sb eutectic alloys by means of a simple electric analogue of the eutectic structure. Applying this equation to MMCs where the fiber resistivity is considered much greater than the resistivity of the bulk matrix, Eq. (17) reduces to

$$\rho_\perp \cong \rho_0(1 - v^{1/2})^{-1}, \qquad (18)$$

where ρ_0 is the matrix or bulk resistivity.

There are three theories based on the calculation of dielectric constants of a composite consisting of parallel fibers in a matrix with a different dielectric constant. These theories have been used to calculate the transverse electrical resistivity of composite materials, including MMCs. The reasoning behind this approach was that the dielectric constant can be replaced by the corresponding expressions for electrical conductivity. The theory of Rayleigh [29], which is based on a dilute suspension for a random dispersion of spheres, gives

$$\frac{\sigma_\perp - \sigma_0}{\sigma_\perp + \sigma_0} = \frac{\sigma_f - \sigma_0}{\sigma_f + \sigma_0}V_f, \qquad (19)$$

where $\sigma_\perp$ is the transverse conductivity of the composite; σ_f and σ_0 are the fiber and matrix conductivities, respectively; and V_f is the fiber volume fraction. For C/Al, B/Al, SiC/Al, and alumina/Al, $\sigma_f \ll \sigma_0$, so that Eq. (19) reduces to

$$\rho_\perp = \rho_0\left(\frac{1 + V_f}{1 - V_f}\right). \qquad (20)$$

[3] Adapted from the following source: J. E. Schoutens and F. S. Roig, *J. Mater. Sci.*, **22** (Chapman & Hall, 1987), 181–188.

Another theory due to Davies [30] gives the following implicit relation. This model is known as an effective medium theory, which is equivalent to the coherent potential approximation in the theory of disordered alloys or to the T-matrix approximation in scattering theory, and to a random array of parallel cylinders.

$$\sigma_\perp = \sigma_0 + 2V_f \sigma_\perp \left(\frac{\sigma_f - \sigma_0}{\sigma_\perp + \sigma_f} \right), \tag{21}$$

which leads to a quadratic equation in $\sigma_\perp$ with the following positive solution

$$\sigma_\perp = -\tfrac{1}{2}(1 - 2V_f)(\sigma_f - \sigma_0)\left(1 + \left\{ 1 - \frac{4\sigma_0\sigma_f}{[(1 - 2V_f)(\sigma_f - \sigma_0)]^2} \right\}^{1/2} \right), \tag{22}$$

and when $\sigma_f < \sigma_0$ is considered, Eq. 22 reduces to

$$\rho_\perp = \rho_0(1 - 2V_f)^{-1}. \tag{23}$$

The theory of Peterson and Hermans [31] is essentially the same as that of Davies and gives the same result. Keller [26] and Keller and Sachs [32] proposed another theory for an array of nonconducting cylinders in a conducting medium, given by

$$\rho_\perp = \rho_0 \frac{2}{(v - 2)^{1/2}} \tan^{-1}\left(\frac{v}{v - 2} \right)^{1/2}, \tag{24}$$

where $v = (\pi/V_f)^{1/2}$. This has the disadvantage of giving $\rho_\perp = 0$ at $V_f = 0$, instead of $\rho_\perp = \rho_0$, and becomes singular at $V_f = \pi/4$. This expression was shown by Crank [33] to be equivalent to an effective diffusion coefficient in cases of diffusion perpendicular to an array of parallel circular obstructions.

Calculations using these theories give the results shown in Table I for a B/Al composite. These results are compared with measurements made by Abukay et al. [34] and Yatsenko [35]. The numbers in parentheses are the factors by which the measured values are larger, equal, or smaller than those predicted by the models and equations indicated. Note that for volume fractions up to 35%, the models give fairly good predictions. This is expected because the models in question are known as dilute suspension models.

A model of transverse electrical conductivity of in situ composites and MMCs at temperatures above about that of liquid nitrogen is presented. Below that temperature threshold, and for very small-diameter fibers such as carbon fibers in an aluminum matrix and in situ composites, electron scattering begins to contribute additional resistivity to the composite that cannot be calculated by macroscopic models, as shown by Roig and Schoutens [15] for longitudinal conductivity.

TABLE I

COMPARISON BETWEEN MODEL PREDICTIONS AND MEASURED TRANSVERSE ELECTRICAL RESISTIVITY OF B/Al [40]

Model and equation	Volume fraction							
	0.15		0.223		0.35		0.60	
	Calculated[a] ($\mu\Omega$ cm)	Measured [35][b] ($\mu\Omega$ cm)	Calculated ($\mu\Omega$ cm)	Measured [35] ($\mu\Omega$ cm)	Calculated ($\mu\Omega$ cm)	Measured [35] ($\mu\Omega$ cm)	Calculated ($\mu\Omega$ cm)	Measured [34] ($\mu\Omega$ cm)
Liebmann and Miller [28] equation 18	5.20	4.6 (0.89)[c]	5.79	5.45 (0.94)	6.76	6.91 (1.02)	8.56	35.5 (4.14)
Rayleigh [29] equation 20	4.49	4.6 (1.02)	5.23	5.45 (1.04)	6.90	6.91 (1.00)	13.28	35.5 (2.67)
Davies [30] equation 23	4.74	4.6 (0.97)	5.99	5.45 (0.91)	11.07	6.91 (0.62)	−16.75[d]	35.5 (−)
Keller [26] equation 24	3.84	4.6 (1.20)	4.87	5.45 (1.12)	6.97	6.91 (0.99)	15.21	35.5 (2.33)

[a] In these calculation, we assumed $\rho_0 = 3.32$ $\mu\Omega$ cm for 6061 Al ribbons [34].

[b] The data from Yatsenko [35] were corrected to have a matrix resistivity of 3.32 $\mu\Omega$ cm.

[c] The numbers in parentheses are the factors by which the measured values are larger than, equal to, or smaller than the predicted values.

[d] When applied to MMC, Davies theory ($\sigma_f \ll \sigma_0$) gives negative values for $\rho_\perp$ when $V_f > 0.5$.

A. *Theory*

Table I shows that present theoretical treatments of the transverse electrical conductivity of MMCs predict values that are a factor of about 2 to 4 below measured values for high-volume-fraction B/Al, and other data to be discussed below. The measured resistivities for a volume fraction of 0.6 are the work of different authors [34] compared to those for volume fractions of 0.15, 0.223, and 0.35 [35]. This might account for the disagreement of the theoretical predictions for the high volume fraction, and also because all measurements were clearly not on the same samples. One hypothesis is that the transverse electrical resistivity in an MMC is the sum of two contributions: The first is due to the fact that the electrical resistivity varies periodically with the position of the fiber in the matrix; the second comes from disturbances in the electron drift as they proceed transversely across the array of fibers. In mathematical terms,

$$\rho_\perp = \rho_0 f(V_f)[1 + C(V_f, \alpha)], \qquad (25)$$

where $f(V_f)$ is a resistivity function dependent on the shape and volume fraction of fibers, and $C(V_f, \alpha)$ is a function that accounts for the fact that the transverse electric field in the bulk is nonuniform due to the presence of fibers. Moreover, from classical electrodynamics, such a nonuniformity has a range α, beyond which it is again uniform.

1.   DERIVATION OF THE RESISTIVITY FUNCTION

Figure 4 shows a square cell of composite material with a fiber of radius *a* at each corner and with each fiber separated by a distance D as shown. We conceptually divide the cell into four regions, marked I to IV, and proceed to calculate the resistivity of region I of unit depth in the direction of the fibers. It is assumed that an electric field is applied in a direction perpendicular to the fibers and along the horizontal axis joining two fibers, thereby creating a current as shown. In region I of the cell, the elemental resistance dR near the fiber is

$$dR = \rho_0 \frac{dx}{A(x)}, \qquad (26)$$

where dx is the infinitesimal matrix thickness normal to the current as shown in Fig. 4b, and $A(x)$ is the variable cross-section in that region when $0 \le x \le a$, or

$$A(x) = \left[a + \frac{D}{2} - Y(x) \right]L, \qquad (27)$$

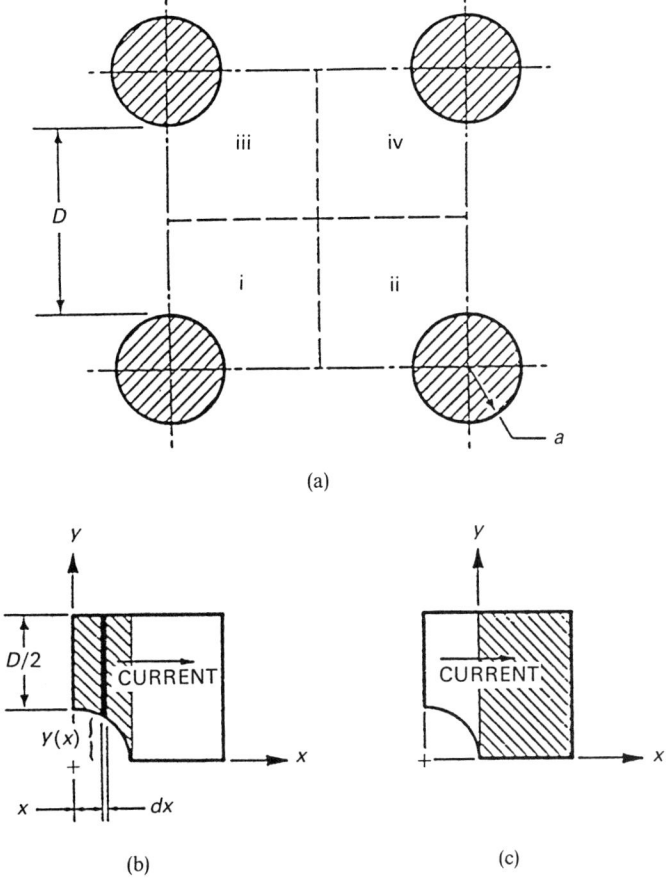

FIG. 4. Cell model used in calculating periodic variations in bulk conductivity: (a) cell; (b) variable cross-section; (c) constant cross-section [40].

where L is the distance along the fiber, and $Y(x) = (a^2 - x^2)^{1/2}$. Substituting Eq. 27 into Eq. 26 and integrating gives

$$R_1 = \frac{\rho_0}{L} \int_0^a \frac{dx}{a + (D/2) - (a^2 - x^2)^{1/2}}. \tag{28}$$

The integration of Eq. 28 is carried out by letting $1 + D/2a = \alpha_0$ and by successively using $w = x/a$ and $w = \sin\theta$. After integration, this gives

$$R_1 = \frac{\rho_0}{L} \left[-\frac{\pi}{2} + \frac{2\alpha_0}{(\alpha_0^2 - 1)^{1/2}} \tan^{-1}\left(\frac{\alpha_0 + 1}{\alpha_0 - 1}\right)^{1/2} \right]. \tag{29}$$

The resistance of the shaded region shown in Fig. 4c is

$$R_2 = \rho_0 \frac{D/2}{(a + D/2)L} = \frac{\rho_0}{L}\left(1 - \frac{1}{\alpha_0}\right), \tag{30}$$

and the total resistance of region I is $R_1 + R_2$, or

$$R_c = \frac{\rho_0}{L}\left[\left(1 - \frac{1}{\alpha_0} - \frac{\pi}{2}\right) + \frac{2\alpha_0}{(\alpha_0^2 - 1)^{1/2}} \tan^{-1}\left(\frac{\alpha_0 + 1}{\alpha_0 - 1}\right)^{1/2}\right]. \tag{31}$$

For convenience in computing R in the limit, we set $\beta = 1/\alpha_0$; then Eq. 31 becomes

$$R_c = R_0\left[\left(1 - \beta - \frac{\pi}{2}\right) + \frac{2}{(1 - \beta^2)^{1/2}} \tan^{-1}\left(\frac{1 + \beta}{1 - \beta}\right)^{1/2}\right] \tag{32}$$

where $R_0 = \rho_0/L$. When the fiber radius approaches zero $(a \to 0)$, $\beta \to 0$ and $R_c = R_0$, as it should. The second term in the bracket is greater than $\pi/2$ for $\beta < 1$, and consequently, the term in the bracket is always positive, giving R always positive. Moreover, as the space between fibers approaches zero $(D \to 0)$, $\beta \to 1$ and $R \to \infty$, meaning that when the fibers are in contact, the composite resistance is at least that of the fibers, which can be quite high for MMCs containing carbon or boron fibers.

Now, the resistance of regions I and II is $2R_c$, and the resistance of regions I to IV is then the total resistance given by

$$\frac{1}{R_T} = \frac{1}{2R_c} + \frac{1}{2R_c} = \frac{1}{R_c}, \tag{33}$$

Thus, the total resistance of the cell is given by Eq. (32).

Equation (32) can be converted to electrical resistivity in the following manner. The resistance of a sample of composite with respect to the resistance of a sample of bulk material without fibers $(V_f = 0)$ is

$$\frac{R(\beta)}{R_0(0)} = \frac{R_c}{R_0} = f(\beta), \tag{34}$$

where

$$f(\beta) = \left[\left(1 - \beta - \frac{\pi}{2}\right) + \frac{2}{(1 - \beta^2)^{1/2}} \tan^{-1}\left(\frac{1 + \beta}{1 - \beta}\right)^{1/2}\right] \tag{35}$$

and for the bulk, $R_0 = \rho_0\, l/A = \rho_0(D + 2a)/L(D + 2a) = \rho_0/L$, and for the composite $R_c = \rho_c\, l/A = \rho_c(D + 2a)/L(D + 2a) = \rho_c/L$, where ρ_c is the composite resistivity. Substituting these values in Eq. (34) yields

$$\frac{\rho_c}{\rho_0} = f(\beta). \tag{36}$$

Equation (36) is the transverse electrical resistivity of an MMC material due to periodic variations in the bulk cross-section in the direction of the current due to the presence of fiber boundaries, assuming the electric field remains uniform.

Now, a relationship between β and V_f must be found so that the above derivation has practical utility. From Fig. 4a, we note that the volume of fiber in the cell is

$$v_f = \pi a^2 L, \tag{37}$$

and the total volume of the cell is

$$v_T = (D + 2a)^2 L, \tag{38}$$

so that the fiber volume fraction is

$$V_f = \frac{v_f}{v_T} = \frac{\pi}{4}\left(\frac{1}{1 + D/2a}\right)^2, \tag{39}$$

where L is the length of the fiber. Recalling that $\alpha_0 = 1 + D/2a$ and substituting into Eq. (39) gives $V_f = \pi/(4\alpha_0^2)$ and using the definition of $\beta = 1/\alpha_0$, we obtain

$$V_f = \frac{\pi}{4}\beta^2. \tag{40}$$

Thus, Eq. (36) can be written as

$$\rho_c = \rho_0 f(V_f) \tag{41}$$

where V_f is substituted for β in Eq. (35) according to Eq. (40). Table II gives values of V_f as a function of β, and Table III gives the values of the

TABLE II
VALUES OF V_f AS A FUNCTION OF β FROM
EQUATION [40]

V_f	β
0.0	0.000
0.1	0.357
0.2	0.505
0.3	0.618
0.4	0.714
0.5	0.798
0.6	0.874
0.7	0.944
0.785	1.000

TABLE III
VALUES OF RESISTIVITY FUNCTION $f(V_f)$ AS A
FUNCTION OF V_f [40]

V_f	$f(V_f)$
0.0	1.000
0.1	1.144
0.2	1.357
0.3	1.657
0.4	2.092
0.5	2.770
0.6	3.977
0.7	6.993
0.785	∞

resistivity function $f(V_f)$ as a function of V_f. We see in Table III that $f(0) = 1$, as it should, and $f(V_f) \to \infty$ for $V_f \to \pi/4$. For 60 vol-% B/6061 Al composite, we note that according to Eq. (41), the transverse electrical conductivity is $\rho_c = \rho_0 f(V_f) = 3.32(3.977) = 13.204 \ \mu\Omega$ cm, or about the same value as that given by Rayleigh's Equation 20 [29] shown in Table I, which is a factor of about 2.7 below the value measured by Abukay et al. [34].

2. DERIVATION OF THE CORRECTION FUNCTION

An applied uniform electric field causes electrons to drift uniformly through a near-ideal metal in the absence of obstacles. In the presence of obstacles such as grain boundaries, dislocations, and defects, the electric field is locally nonuniform. When the metal is reinforced by nonconducting cylindrical fibers placed perpendicularly to the uniform electric field, that field becomes locally and periodically nonuniform. Thus, the electrons drift along periodically varying field lines. The field lines in the immediate neighborhood of the fiber terminate at the fiber to matrix interface—approximately normal if the fiber position is an empty cavity, or at an angle proportional to the dielectric constant if the cavity is filled with a dielectric fiber. In the following analysis, it is assumed that the matrix is a near-ideal metal so that the electric field in the absence of obstacles is uniform and that the cavity is empty. (Inside a perfect conductor, there cannot be electric fields.) Consequently, the dielectric properties of the fiber are not considered. In essence, then, this is the inverted problem of a conducting cylinder in a uniform electric field in empty space. From classical electrodynamics [36], the scalar potential at a point $P(r, \theta)$ in the neighborhood outside the fiber is given in cylindrical coordinates by

$$\phi = -E_0 r \cos \theta \left(1 - \frac{a^2}{r^2}\right), \tag{42}$$

where E_0 is the uniform electric field, a is the fiber or cavity radius, and r and θ are the polar coordinates of point P where the field components are determined. The radial and angular components of the electric field are obtained from the scalar potential by $E = -\nabla\phi$ or

$$E_r = -\frac{\partial\theta}{\partial r} = E_0 \cos \theta \left(1 + \frac{a^2}{r^2}\right), \tag{43}$$

$$E_\theta = -\frac{1}{r}\frac{\partial\phi}{\partial\theta} = E_0 \sin \theta \left(1 - \frac{a^2}{r^2}\right). \tag{44}$$

The total field at $P(r, \theta)$ is

$$E_T^2 = E_r^2 + E_\theta^2; \tag{45}$$

and in order to find the average value over the entire space from $a \le r \le \alpha a$ and for $0 \le \theta \le 2\pi$, where α is a dimensionless range parameter, we integrate both sides of Eq. (45) as follows:

$$\int_0^{2\pi}\int_a^{\alpha a} E_T^2 r \, dr \, d\theta = \int_0^{2\pi}\int_a^{\alpha a} (E_r^2 + E_\theta^2) r \, dt \, d\theta. \tag{46}$$

Substituting Eqs. (43) and (44) into Eq. (46) and integrating gives

$$\langle E_T^2 \rangle = E_0^2 \left(1 + \frac{1}{\alpha^2}\right). \tag{47}$$

Equation (47) shows that the total electric field calculated in this manner consists of two components, the uniform-field term plus a second-order term that depends on the range parameter α. It is this second-order term that, on average, accounts for the effects of nonuniformity upon electrons in the electric field in the neighborhood of a fiber. Therefore, we can write

$$\frac{\delta E_0}{E_0} \cong \frac{1}{\alpha}, \tag{48}$$

where E_0 is the uniform electric field, and δE_0 is the perturbation of that field. The electric field is nonuniform around the fiber out to a range $\alpha a \le 3a$ as can be seen in electric-field maps for this classical problem.

The range αa of this nonuniform field does not occupy the entire cell of interest but only a region of area A_F, as shown in Fig. 5. For high fiber volume fraction, the areas are expected to overlap. This overlap problem is

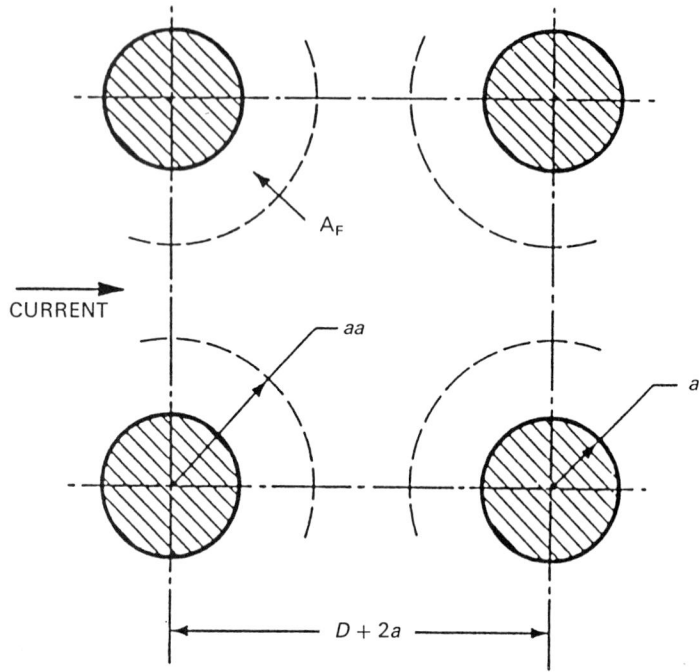

FIG. 5. Model for calculating the ratio A_F/A_c [40].

not treated in the simple model discussed in this chapter. Consequently, there is a region in the cell where the electric field is uniform, and other regions where it is not uniform. If the area of nonuniform field is A_F for $\alpha \leq 3$ and A_c is the area of the cell, as shown in Fig. 5, then the correction function is taken to be

$$C(V_f, \alpha) = \frac{\delta E_0}{E_0}\left(\frac{A_F}{A_c}\right) = \frac{1}{\alpha}\frac{A_F}{A_c}; \tag{49}$$

and we will see that this assumption can fit the data. Thus, from Fig. 5, $A_c = (D + 2a)^2$ and $A_F = \pi a^2(\alpha^2 - 1)$, so that Eq. (49) becomes

$$C(V_f, \alpha) = \left\{\frac{\alpha^2 - 1}{\alpha}\right\}V_f, \tag{50}$$

where Eq. (39) was used. This correction factor has the correct form, since for $V_f = 0$, $C(0, \alpha) = 0$. Therefore, in the absence of fibers, there are no nonuniformities in the electric field.

3. COMPLETE TRANSVERSE RESISTIVITY EQUATION

Returning to Eq. (25), we have

$$\rho_\perp = \rho_0 f(V_f)[1 + C(V_f, \alpha)],$$

and substituting Eq. (50) gives

$$\rho_\perp = \rho_0 f(V_f)\left[1 + \left(\frac{\alpha^2 - 1}{\alpha}\right)V_f\right], \tag{51}$$

where $f(V_f)$ is given by Eq. (32). Equation 51 gives $\rho_\perp = \rho_0$ for $V_f = 0$, since $f(0) = 1$, and the second term vanishes at $V_f = 0$. Moreover, $\rho_\perp \to \infty$ when $V_f \to \pi/4$, since $f(V_f) \to \pi/4 \to \infty$, which corresponds to the fibers being in actual contact with one another in the matrix. It should be emphasized that Eq. (50) applies only for the case where the fibers in the matrix form a regular array, either square or quasisquare. This is a consequence of the fabrication process employed in large-diameter fiber-reinforced MMC materials such as B/Al. For the case of very small-diameter fibers of the order of a few tens of micrometers, the distribution of fiber cross-sections in a plane normal to the fiber axis is nearly random. The electrical resistivity for such cases requires a different theory, which is discussed elsewhere [37].

B. Discussion

Returning to Eq. 51 and using the value $\alpha = 3$ results in

$$\rho_\perp = \rho_0 f(V_f)(1 + 2.667V_f). \tag{52}$$

The predicted values of the transverse electrical conductivity of MMCs are given in Table IV and plotted in Fig. 6 as a function of V_f. The predicted value of $\rho_\perp$ is seen to pass close to the datum point of 60 vol-% B/Al composite. Also shown are three data points, joined by a line of short dashes, from the work of Yatsenko [35]. The data published by Yatsenko were for a matrix resistivity of 3.21 $\mu\Omega$ cm. Consequently, these data were scaled in Fig. 6 by the factor $3.32/3.21 = 1.034$ to make comparison meaningful. As mentioned before and shown in Table I, the theories discussed in the introduction give predictions in good agreement with Yatsenko's data. The agreement with the datum point for 60 vol-% B/Al appears excellent: 34.33 $\mu\Omega$ cm predicted compared with 35.5 $\mu\Omega$ cm measured, or a difference of 3.3%. Such an agreement may mean the following, aside from being fortuitous: At high volume fraction, the electric field intensity around fibers

TABLE IV
VALUES OF $\rho_\perp$ PREDICTED FROM
EQUATION 51 [40]

V_f	$\rho_\perp$ ($\mu\Omega$ cm)
0.0	3.32[a]
0.1	4.81
0.2	6.91
0.3	9.90
0.4	14.35
0.5	21.46
0.6	34.33
0.7	66.56
0.785	∞

[a] $\rho_0 = 3.32$ $\mu\Omega$ cm for 6061 Al [34]

increases while the region of field uniformity between fibers decreases, resulting in a net increase in the composite resistance. As the fiber distance increases (lower values of V_f), this model seems to overpredict the field effect. We should recall that the model is rather simplistic, and more detailed description might be in order. It turns out that considering the field contributions from all fibers surrounding one fiber in a cell by assuming superposition of field does not improve predictions very much at low values of V_f [38].

Another group of data for C/Al and C/Mg [15] is shown in Fig. 3. An examination of photomicrographs of such composite materials reveals very irregular, almost random, fiber distributions in a plane perpendicular to the fiber axis. At low fiber volume fraction, these photomicrographs show numerous filaments in actual physical contact forming "strings" separated by irregular regions of matrix materials. Under such conditions, one would thus expect, on intuitive grounds, that the current follows tortuous paths with statistically varying resistivity. At high volume fraction, the number of filaments in physical contact is so large that it isolates islands of matrix material. Consequently, one would expect the transverse conductivity for these materials to be higher than that shown by the present model or the data for B/Al. Aside from matrix resistivity due to periodic variation in the matrix cross-section in between fibers, and aside from the added electric field effect discussed, it is likely that there is a capacitance effect in the immediate neighborhood of closely spaced fibers. To test this hypothesis, we have

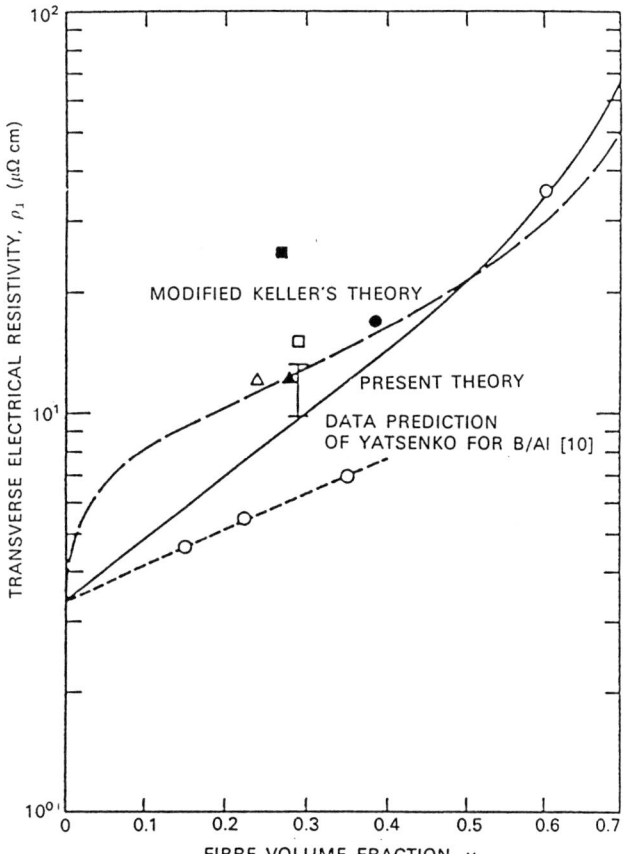

FIG. 6. Plot of the present theory and Keller's original and modified theories, and comparison with some experimental data. Room temperature data: (○) B/6061 Al; (□) 29 vol-% C/201 Al, with 0.004 in 2024 Al cladding, liquid-nitrogen quenched 3 times; (△) 24 vol-% C/Al, with 0.004 in 2024 Al cladding, liquid-nitrogen quenched 3 times; (●) 38.5 vol-% C/AZ31B Mg; (■) 27 vol-% C/AZ31B Mg; (▲) 27.75 vol-% C/201 Al with 0.002 in Al cladding; (I) 29 vol-% C/201 Al, with 0.002 in Al cladding [*40*].

modified Keller's theory [*26*], which is based on the capacitance that exists between two fibers along the direction of the current for conducting fibers embedded in a dielectric. Keller then inverted the result of his analysis to account for resistivity transverse to nonconducting cylinders embedded in a conducting matrix. This resulted in Equation 24 discussed earlier. If we neglect the electric-field nonuniformity at $0.1 \leq V_f \leq 0.4$ and consider only resistivity variations and capacitance effects, the modified Keller theory can be written as

$$\rho_\perp = \rho_0[f(V_f) + C(V_f)], \tag{53}$$

where $f(V_f)$ is given by Eq. (35) and

$$C(V_f) = \frac{2}{(v-2)^{1/2}} \tan^{-1}\left(\frac{v}{v-2}\right)^{1/2}, \tag{54}$$

where $v = (\pi/V_f)^{1/2}$. The predictions from Eq. (53) are shown in Fig. 6 as a line of long dashes that passes through the group of C/Al and C/Mg data but falls below the B/Al datum at $V_f = 0.6$. At higher values of V_f, one would expect larger values of $\rho_\perp$ for C/Al or C/Mg than for B/Al, as discussed earlier. At the present time, some theories are being considered to explain the phenomenon when the fiber arrays are highly irregular or random [37]. Equation (53) has the required behavior at the limits: when $V_f = 0$, $\rho_\perp = \rho_0$ and $\rho_\perp \to \infty$ when $V_f \to \pi/4$.

The model described here can be scaled to other temperatures (away from room temperature) above approximately liquid-nitrogen temperature and not too close to the matrix melting point. This can be done by using the following scaling rule.

$$\rho_\perp(T) = \rho_\perp\left(\frac{295}{T}\right)^a, \tag{55}$$

where $\rho_\perp$ is the room temperature resistivity, T is the temperature of interest in degrees K, and a is a noninteger. From the work of Abukay et al. [34], $a = -0.74$ for B/6061 Al.

IV. Summary

This chapter presented a discussion of the electrical conductivity in metal-matrix-composite materials for temperatures ranging from cryogenic to room temperature. It was divided into two parts: The first part discussed the electrical conductivity along the fiber direction in the absence of magnetic fields, and the second part discussed the electrical conductivity transverse to the fiber direction in the absence of magnetic fields. The effects of a magnetic field parallel to the fibers on the composite conductivity were mentioned briefly, but the reader is referred to the source literature. Wherever possible, the theories presented were compared with experimental data.

References

1. K. R. Karasek and J. Bevk, *J. Appl. Phys.*, **52**, 1370 (1981).

2. R. Simoneau and G. Begin, *J. Appl. Phys.*, **45**, 3828 (1974).

3. R. Simoneau and G. Begin, *J. Appl. Phys.*, **44**, 1461 (1973).

4. B. N. Aleksandrov and I. G. D'Yakov, *Soviet Physics JETP*, **16**, 603 (1963).

5. B. N. Aleksandrov, *Soviet Physics JETP*, **16**, 286 (1963).

6. B. N. Aleksandrov and M. I. Kaganov, *Soviet Physics, JETP*, **14**, 948 (1962).

7. R. B. Dingle, *Proc. Roy. Soc. Lond.*, **201A**, 545 (1949).

8. D. Abukay, et al., *Fibre Sci. Tech.*, **10**, 313 (1977).

9. E. H. Sondheimer, *Phys. Rev.*, **80**, 401 (1950).

10. R. G. Chambers, *Proc. Roy. Soc. Lond.*, **202A**, 378 (1950).

11. D. K. C. MacDonald and K. Sarginson, *Proc. Roy. Soc. Lond.*, **203A**, 223 (1950).

12. F. S. Roig and J. E. Schoutens, "The longitudinal electrical conductivity of metal matrix composites at cryogenic temperature in the presence of a longitudinal magnetic field. Part I: Solution of the linearized Boltzmann equation," *J. Mater. Sci.*, **22** (1987), pp. 3749–3754.

13. F. S. Roig and J. E. Schoutens, "The longitudinal electrical conductivity of metal matrix composites at cryogenic temperature in the presence of a longitudinal magnetic field. Part II: Application of the solution to metal matrix composites." *J. Mater. Sci.*, **22**, 4002–4010 (1987).

14. C. R. Chester, "Techniques in Partial Differential Equations," Chapter 8, McGraw-Hill Book, 1971.

15. F. S. Roig and J. E. Schoutens, *J. Mater. Sci.*, **21**, 2409–2417 (1986).

16. F. S. Roig and J. E. Schoutens, *J. Mater. Sci.*, **21**, 2767–2770 (1986).

17. J. M. Ziman, "Electrons and Phonons," Chapter 11. Oxford University Press, London, 1962.

18. N. F. Mott and H. Jones, "The Theory of the Properties of Metals and Alloys." Dover Publications, New York, 1958.

19. D. Abukay, et al., *Fibre Sci. Tech.*, **10**, 313 (1977).

20. T. A. Jenkins and S. Arajs, *Fibre Sci. Tech.*, 1982, **17**, 205.

21. J. D. Verhoeven, "Fundamentals of Physical Metallurgy," Figure 10.5, p. 331. Wiley, New York, 1975.

22. E. G. Kendall, "Development of metal-matrix composites reinforced with high-modulus graphite fibers," in *Composite Materials* (L. J. Broutman and R. H. Krock, eds.), Vol. 4, Figure 38, p. 368, Academic Press, New York, 1974.

23. K. G. Kreider and K. M. Prewo, Boron-Reinforced Aluminum, *ibid.*, Figure 9, p. 421.

24. J. M. Ziman, "Principles of the Theory of Solids," Chapter VII, Cambridge at the University Press, London, 1965.

25. S. Arajs, Department of Physics, Clarkson College of Technology, Potsdam, New York, Personal communication, May 1985.

26. J. B. Keller, *J. Appl. Phys.* **34**, 991 (1963).

27. T. A. Jenkins, PhD thesis, Clarkson College of Technology, Potsdam, New York, 1982.

28. W. K. Liebmann and E. A. Miller, *J. Appl. Phys.*, **34**, 2653 (1963).

29. J. W. Rayleigh, *Phil. Mag.*, **34**, 481 (1982).

30. W. E. A. Davies, *J. Phys.*, **D7**, 120 (1974).

31. J. M. Peterson and J. J. Hermans, *J. Comp. Mater.*, **3**, 338 (1969).

32. J. B. Keller and D. Sachs, *J. Appl. Phys.*, **35**, 537 (1984).

33. J. Crank, "The Mathematics of Diffusion," 2nd ed. Clarendon, Oxford, 1975.

34. D. Abukay, K. V. Rao, and S. Arajs, *Fibre Sci. Tech.*, **10**, 313 (1977).

35. M. I. Yatsenko, *Fizika i Khimiya Obrabotki Materialov*, **4**, 11 (1980).

36. L. D. Landau and E. M. "Lifshitz, "Electrodynamics of Continuous Media," Pergamon, New York, 1960.
37. J. E. Schoutens, "Model of transverse electrical conductivity of metal matrix composites above liquid nitrogen temperatures. Part 2: Irregular arrays of fibres," unpublished data, 1985.
38. J. E. Schoutens, unpublished data, 1985.
39. H. H. Armstrong and A. M. Ellison. Lockheed Missiles and Space Co., unpublished data, 1979.
40. J. E. Shoutens and F. S. Roig, *J. Mater. Sci.*, **22**, 181–188 (1987).

13

Corrosion

PATRICIA P. TRZASKOMA

Naval Research Laboratory
Washington, D.C.

I. Introduction

Corrosion is the chemical or electrochemical reaction of a metal with its environment that results in material deterioration and loss [1]. The effects of corrosion are costly, hazardous, and place a significant drain on our natural resources. In 1978, a study by the National Bureau of Standards [2] showed that the annual price of corrosion in the United States was about 4.2% of the gross national product; in 1985 the cost was 180 billion dollars [3]. This year, in one incident, a flight attendant was killed and several others injured when the upper fuselage tore off a commercial aircraft during operation over the Hawaiian Islands. Preliminary investigations indicate that failure of the fuselage was caused by corrosion fatigue experienced during

383

short-cycle takeoffs and landings in marine environments [4]. Thus, the problems associated with corrosion are not trivial, and the need to study and assess corrosion behavior along with the development of new materials is evident.

Even though there is extensive work being done on the development of metal matrix composites for their enhanced physical and mechanical properties, there are relatively few investigative studies addressing corrosion behavior. Because of their duplex nature, it is expected that metal matrix composites would be prone to accelerated corrosion reactions as compared with their monolithic counterparts. Rapid corrosion would limit service lifetimes, but of even more importance, minor corrosion of either constituent could adversely affect the mechanical or physical properties for which the composite is designed.

This chapter discusses the corrosion behavior of metal matrix composites in aqueous environments and is subdivided into eight sections. After an introduction, general and localized corrosion are discussed. Section III presents a review of selected studies concerning the corrosion behavior of metal matrix composites of current interest. Section IV discusses the effects of corrosion on the properties of metal matrix composites, and Section V deals with corrosion fatigue. Sections VI and VII are concerned with corrosion control and coatings, and in the final section, future directions for new research are identified and discussed. The intent of this chapter is to present a brief introduction to the subject of corrosion, provide information concerning the type and magnitude of corrosion problems that have been observed for metal matrix composites, suggest measures to minimize or prevent corrosion processes, and indicate important directions for future work.

II. Corrosion Processes

Corrosion processes fall into one of two categories depending on the extent of attack. General corrosion is uniform metal degradation and surface thinning, and localized corrosion is metal deterioration that is confined to specific regions. Localized corrosion often takes place at structurally (i.e., crevices) or compositionally (i.e., precipitates) inhomogeneous surface features.

A. General Corrosion

General corrosion reactions are ultimately controlled by the reactivity of a metal with some species in the environment. Two charge transfer reactions take place simultaneously. These are, the anodic reaction, which involves oxidation of metal atoms to soluble ions, and the cathodic reaction, which involves either reduction of hydrogen ion or water to hydrogen gas, or reduction of dissolved oxygen to hydroxide ion. These reactions can be represented by the following chemical equations.

(Anodic reaction)

$$M \rightarrow M^{+n} + ne \tag{1}$$

(Cathodic reactions)

$$2H^+ + 2e \rightarrow H_2 \tag{2}$$

or

$$2H_2O + 2e \rightarrow H_2 + 2OH^- \tag{3}$$

or

$$O_2 + 2H_2O + 4e \rightarrow 4OH^-. \tag{4}$$

Both the thermodynamics and kinetics of corrosion processes are determined by the metal and its environment. For example, in water, aluminum is passive and magnesium reacts vigorously, and while iron remains passive in solutions containing chromate (CrO_4^{-2}), nitrite (NO_2^{-2}), or molybdate (MoO_4^{-2}) ions, in the presence of chloride (Cl^{-1}), its dissolution rate increases sharply.

For metal matrix composites, there are three possible modes of general attack; selective dissolution of the matrix, selective dissolution of the reinforcement, or general dissolution of both matrix and reinforcement. For example, the copper matrix in Ta/Cu is attacked in 70% nitric acid, whereas, the copper remains passive while the tantalum is attacked in 45% hydrofluoric acid [5].

B. Localized Corrosion

In addition to general corrosion, metal matrix composites might be susceptible to three types of localized corrosion processes. These are

(1) Galvanic corrosion between the metal and reinforcement.
(2) Crevice corrosion at the matrix–reinforcement interface or within pores formed during fabrication.
(3) Pitting at secondary phases in the matrix or at active phases formed by reaction of the metal and reinforcement during consolidation.

FIG. 1. C/Mg after immersion in 0.001 N NaCl for five days. From Ref. [6].

Any of these processes would promote accelerated metal dissolution. Figure 1 shows a sample of C/Mg following immersion in dilute NaCl solution (0.001 N) for just five days [6]. Severe attack by general and galvanic corrosion is concentrated at edges where the fibers were exposed. Figure 2 shows the same sample at higher magnification. Here it is evident that carbon fibers (originally below a magnesium alloy face sheet) have been exposed as a result of magnesium dissolution.

A brief discussion of galvanic corrosion, crevice corrosion, and pitting follows. For a more extensive treatment of these topics, the reader is advised to consult various texts, for example, references [7], [8], and [9].

1. GALVANIC CORROSION

Galvanic corrosion is a result of current flow between two dissimilar conducting materials that are coupled by physical contact or through a conducting solution. The driving force of galvanic corrosion is the difference

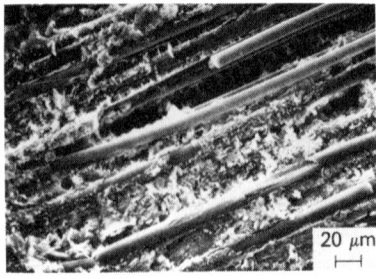

FIG. 2. C/Mg after immersion in 0.001 N NaCl for five days. This micrograph shows severe deterioration of the composite, which includes separation of the matrix and fibers. From Ref. [6].

in electrode potential between the two materials in the environment of interest. The tendency towards galvanic corrosion increases as the difference between electrode potentials increases. Tables for standard electrochemical potentials are often used to predict galvanic corrosion tendencies; however, these data can be misleading. An accurate assessment requires comparing electrode potentials that have been measured in the environment of interest. Lennox [10] has compiled electrode potentials for over 100 metals in sea water, and these data are used extensively in the design of equipment for marine applications. During galvanic corrosion, the anodic reaction (metal loss) takes place at the more active material (the more negative electrode potential), and the cathodic reaction takes place at the less active material. The dissolution rate of the active material accelerates relative to its rate without galvanic effects. Additionally, the galvanic corrosion rate increases as the ratio of the surface area of the cathode to anode increases.

2. CREVICE CORROSION

Crevice corrosion is an intense localized attack at crevices, pores, or voids. Metals that have a naturally formed protective film, such as aluminum or stainless steel, are particularly susceptible to crevice corrosion in chloride-containing solutions. Crevice attack is a result of restricted convection between the crevice and external solution. The lack of free movement of reaction species generates, in succession, depleted oxygen concentration and increased metal ion, hydrogen ion (by hydrolysis of metal ions), and chloride ion (by migration to neutralize metal ion charges) concentrations. Because these conditions catalyze metal dissolution, there is accelerated attack within the crevice.

3. PITTING

Pitting is localized corrosion that is focused at microscopically active features on the metal surface. These sites can be defects in protective films, impurities, reaction products, secondary phases, or inclusions. For example, it has been shown that manganese sulfide inclusions promote pit nucleation on iron, Fe–Cr–Ni alloys, and stainless steels [11]. Pitting reactions are affected by the environment. For example, chloride ion is known to promote pit initiation on aluminum and steel, and the local chemistry that develops within a pit can catalyze pit growth rates.

Localized corrosion processes can be particularly troublesome for metal matrix composites. In addition to increasing the rate of metal degradation, significant localized attack, particularly when concentrated at the metal–reinforcement interface, could adversely affect physical and mechanical properties.

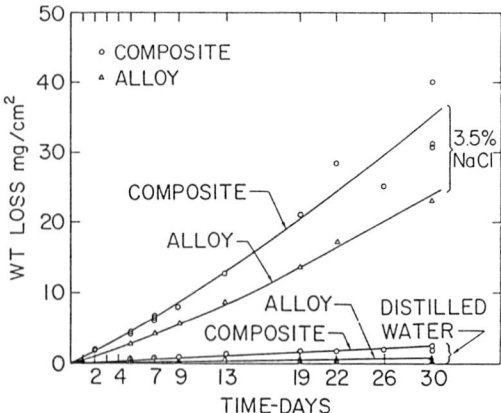

FIG. 3. Weight-loss-versus-time curves for DU-0.75 Ti and W/DU-0.75 in 3.5 wt% NaCl. From Ref. [12].

III.   Corrosion Studies of Selected Metal Matrix Composites

In this section, the results of corrosion studies on selected metal matrix composites are presented. Examples of metal-fiber-, carbon-fiber-, and ceramic-whisker-reinforced metals are discussed.

A.   *Tungsten/Depleted Uranium*

The corrosion behavior of W/DU[1] in 3.5% NaCl has been investigated by Trzaskoma [12]. The composite consisted of 50 vol-% tungsten fibers combined with alloy DU–0.75 Ti. This material was developed for applications requiring high density. Weight-loss measurements showed that the corrosion rate of the composite is greater than the alloy both in distilled water and 3.5 wt% NaCl. Figure 3 shows weight loss as a function of immersion time. After thirty days, the weight loss of the composite was about 50% greater than that of the depleted uranium alloy in NaCl. For this material, DU–0.75 Ti is the more active component and is therefore preferentially attacked.

In this study, the composite was modelled by coupling different lengths of tungsten wire to a sample of matrix alloy of fixed surface area. In this

---

[1] DU is depleted uranium.

way, the short-circuit (galvanic) currents for various surface area ratios of tungsten could be measured. Figure 4 shows these data. It can be seen that galvanic currents increase as the surface area ratio of tungsten (surface area tungsten/total surface area) to DU–0.775 Ti increases. This same effect would be observed for an actual composite material having different volume fractions of tungsten; that is to say, as the volume ratio of tungsten increases, the rate of metal dissolution increases. The dissolution rate of the composite studied (50 vol-% tungsten) was calculated from galvanic currents using Faraday's laws. This rate was the same as the depleted uranium corrosion rate calculated from immersion studies (Table I). It was therefore concluded that the principal corrosion process for W/DU–0.75 Ti is galvanic coupling of the matrix and reinforcement.

B. Carbon/Magnesium

The corrosion behavior of C/Mg was studied in 1000 ppm NaCl with techniques similar to those used for W/DU [6]. Samples were prepared by infiltrating carbon fiber tows with molten magnesium alloy AZ91C and then hot pressing between foils of alloy AZ31B to form a plate. Although the composite contained 26.5 vol-% carbon, only a small fraction of the fibers was exposed. When samples were immersed in dilute chloride solution, there was rapid dissolution of magnesium accompanied by hydrogen evolution (Fig. 1).

The potential difference between the magnesium alloys and carbon reinforcement was found to be 1.3 V in a borated buffer solution containing

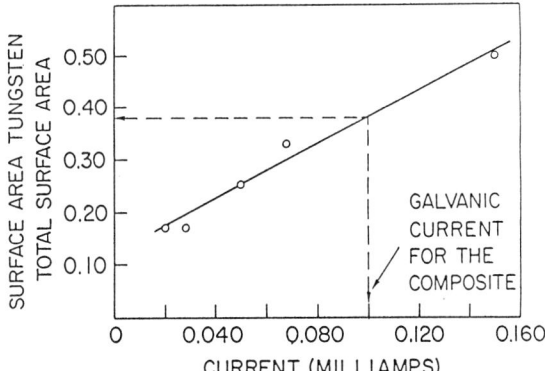

FIG. 4.   Galvanic currents for DU–0.75 coupled to W for various surface-area ratios of reinforcement (3.5 wt% NaCl solution). From Ref. [12].

TABLE I
CORROSION RATE OF W/DU–0.75Ti IN
3.5 wt% NaCl [*12*]

Total weight loss from (immersion test)	Total weight loss from galvanic current measurements
1.08 mg/cm²–day	1.01 mg/cm²–day

1000 ppm NaCl. Figure 5 shows galvanic currents between magnesium alloy AZ31B and carbon as a function of the surface-area ratio of carbon. The curves show that when only 10% of the surface is carbon, the galvanic current is 0.13 mA, which corresponds to a corrosion rate of 1.8 mg/cm²/day. This rate alone is ten times the acceptable rate for the corrosion of structural materials (0.18 mg/cm²/day) [*13*]. By comparing weight loss measurements with weight losses calculated from measured galvanic currents, it was later shown that galvanic corrosion represents about 10% of the total corrosion rate, whereas the surface area ratio of carbon is 0.04 (the ratio found on a coupon of about 1 by 3 centimeters, where only the edges have exposed carbon fibers) [*14*]. This study shows that despite the large potential difference between the metal matrix and reinforcement, only a small part of the total metal dissolution involves galvanic effects. Undoubtedly one reason for this behavior is that only a small portion of the carbon fibers are exposed. However, it is also possible that because magnesium reacts very efficiently with water, not all of the cathodic reaction is transferred to the carbon fibers. In any case, these studies show that unless they were coated, C/Mg composites would be unsuitable for service in aqueous environments.

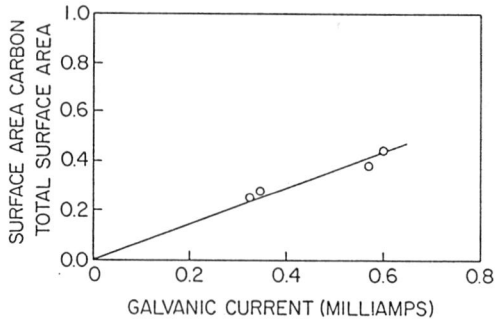

FIG. 5. Galvanic currents for AZ31B coupled to carbon fibers for various surface-area ratios of reinforcement (borated buffer solution containing 1000 ppm NaCl). From Ref. [6].

C. Carbon/Aluminum

C/Al is typically prepared by diffusion bonding of aluminum-coated carbon fibers. A layer of titanium boride is usually vapor-deposited on the fibers prior to coating to improve wettability. In many cases, the coated wires are sandwiched between aluminum alloy foils prior to diffusion bonding. The surfaces of these composites are therefore mostly aluminum, and carbon fibers are only exposed on the edges.

Various investigators have studied the corrosion behavior of C/Al by exposure tests in chloride environments (marine atmosphere, salt spray, seawater) [15–19]. General observations are that the deterioration of the composite is faster than that of aluminum alloys without the reinforcement and that preferential corrosion occurs at aluminum–carbon interfaces and diffusion bonds. Exposed samples show swelling and exfoliation at their edges as a result of accelerated corrosion and wedging of corrosion products in these regions. This behavior is attributed to crevice effects and galvanic coupling between the aluminum matrix and carbon fibers.

Dull *et al.* [16] compared the corrosion rates of C/6061 Al and 6061 Al in distilled water and 3.5 wt% NaCl at temperatures between 298 and 348 K by measuring weight losses after immersion. Figure 6 shows results of their work. It can be seen that in NaCl, at low temperature, the corrosion rate of the composite is about 15 times greater than that of the alloy. Above 325 K, the corrosion rates increase, and the rate of increase of the composite is greater than that of the alloy.

Payer and Sullivan [17] studied the corrosion behavior of C/Al by alternate immersion in seawater. Their samples consisted of carbon fibers coated with

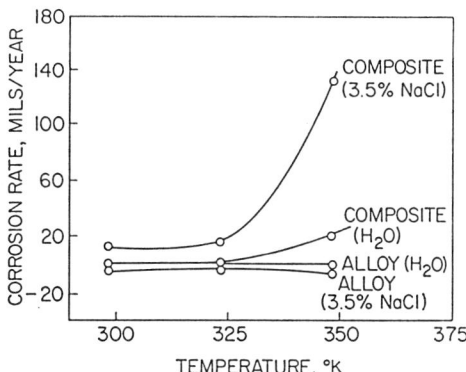

Fig. 6.   Corrosion rates after 150 hours of exposure for C/6061 Al and 6061 Al in distilled water and 3.5% NaCl at various temperatures. From Ref. [16].

201 Al and encased in 1100 Al interior and face foils. In this study, preferential corrosion was not observed at the interface after exposure for five weeks. The authors therefore concluded that galvanic effects were unimportant for this material. They also suggested that crevice effects were significantly diminished, because their samples had fewer voids and flaws than previous versions of the composite. The potential difference between aluminum and carbon in chloride solution is about 1 V [10]; therefore, it is difficult to understand why galvanic coupling does not occur on the composite. Perhaps the intermittent drying in Payer's tests tends to mitigate galvanic action.

Recently, by using a scanning vibrating electrode probe technique, Crowe [20] observed galvanic effects and increased anodic currents at interfacial flaws on a C/Al composite. Crowe's work indicates that galvanic corrosion dominates corrosion behavior in the absence of flaws, and crevice corrosion significantly increases the corrosion rate when interfacial flaws are present. In addition, Crowe suggests that selective dissolution of Al_4C_3 contributes to the high anodic current densities observed at the interfaces. Al_4C_3 is believed to form at the interface during consolidation of the composite. According to Crowe, corrosion control of C/Al is best achieved by using matrix alloys that are resistant to crevice corrosion and the formation of Al_4C_3 and additions of poisons to the carbon fibers to slow the cathodic reaction rate.

In summary, C/Al is unsuitable for use in moist air or saltwater environments unless it is coated. Corrosion resistance could be improved by eliminating crevices, interfacial reaction processes, and galvanic effects. These areas require further investigation.

D. Boron/Aluminum

B/Al composites usually consist of tungsten-cored boron fibers encased in an aluminum alloy matrix by powder compacting or hot pressing. The boron fibers are about 0.1 mm in diameter and are continuously aligned within the metal matrix. Hence, as is the case of C/Al, the reinforcement is only exposed at the edges of the material lying perpendicular to the fiber direction.

Porter and Wolff [21] determined the corrosion rate of B/Al in boiling, saturated chloride solutions by weight loss measurements. The composite was prepared by hot pressing 10 vol-% boron fibers in Type 5471 Al powder (99.95% + pure). Table II shows some of their data. It can be seen that the corrosion rate of the composite is about 1.2–1.9 times greater than that of aluminum. The authors also observed that the mode of attack was general corrosion of aluminum with no preferential degradation at the matrix–fiber interface.

TABLE II

AVERAGE CORROSION RATES FOR B/Al COMPOSITES
EXPOSED TO BOILING, SATURATED NaCl FOR
161 HOURS [*21*]

Sample material	Average corrosion rate (mpy)
Al	18.8
B/Al A[a]	22.2
B/Al B[a]	36.5

[a] A and B refer to samples from different batches.

Sedriks *et al.* [*22*] investigated the corrosion behavior of B/Al 2024. Their samples contained 40% boron and were fabricated by diffusion bonding of alloy foils wound with boron fibers. These investigators observed preferential corrosion both along the fiber–matrix interfaces and at diffusion bonds between the foils. Using electrochemical techniques, it was shown that anodic currents (metal dissolution rates) increase with increasing fiber content. The authors concluded that this effect is not a result of galvanic interactions of the boron and aluminum, but rather they suggested that crevice corrosion at the aluminum–boron interface breaks down the passive aluminum film and generates a large number of actively dissolving (anodic) sites.

In further work on B/2024 Al, Pohlman [*23*] measured corrosion rates and galvanic currents between the metal and boron fibers. Table III shows corrosion rates of B/Al in 3.5 wt% NaCl determined from electrochemical polarization data and the Tafel equation. It can be seen that addition of 18 vol-% boron doubles the corrosion rate of Al 2024, and the rate increases as the amount of boron increases. Galvanic currents were not observed between 2024 Al and unused fibers; however, they were observed between

TABLE III

CORROSION RATES OF 2024 Al AND B/2024 Al
IN 3.5 Wt% NaCl [*23*]

Material	Corrosion rate (mils/year)
2024 Al, no boron	0.9
B/2024 Al	
18 vol-% B	2.1
33 vol-% B	2.8
46 vol-% B	3.9

2024 Al and boron fibers that were extracted from the composite. X-ray diffraction analysis of extracted fibers showed they were covered with an aluminum boride intermetallic compound approximately 4 μm thick, presumably produced during consolidation of the composite. It was therefore concluded that galvanic coupling of the matrix and the aluminum boride intermetallic compound contribute to the preferential attack at the interface. Thus, galvanic reactions as well as crevice effects have been shown to contribute to the corrosion behavior of B/Al.

The major corrosion problem for B/Al is localized attack at the metal–fiber interface. This effect is a result of crevice corrosion and, depending upon the conditions of formation, galvanic interaction of the aluminum matrix and an interfacial aluminum boride reaction product.

E. Silicon Carbide/Aluminum

Silicon carbide is added to aluminum alloys in the form of whiskers, particles, or fibers to produce a material with greater strength and stiffness than the alloy alone. The corrosion behavior of this composite is by far the most widely investigated.

Aylor and Kain [18] tested the corrosion behavior of SiC/Al formed with three different alloy metal matrices. During long-term exposure in marine environments, pitting was found to be the principal form of corrosion for all materials. Some pits were observed at the silicon carbide–aluminum interface, and the authors suggest this could be caused by crevice effects.

Trzaskoma et al. [24] studied the effects of silicon carbide whiskers on pit initiation susceptibility by measuring pitting potentials in NaCl solution. The pitting potential is the electrode potential above which pits initiate and grow. The more positive the pitting potential, the lower the susceptibility to pit initiation. Table IV shows pitting potentials for three alloys and the corresponding composites. It can be seen that except for 2024 Al, the pitting potentials and therefore the pit-initiation susceptibility of the alloys and composites are the same in both aerated and deaerated solution. For 2024 Al, the pit-initiation susceptibility of the composite is greater than that of the alloy. In this work, it was concluded that silicon carbide does not affect the pit-initiation susceptibility of the composite. The results for 2024 Al can be explained in terms of the distribution of copper phases in the matrix. It is known that the pitting potentials of aluminum–copper alloys are dependent on copper content and heat treatment [25, 26]. As the concentration of Cu increases in the alloy, the pitting potential increases up to the limit of solubility of copper in aluminum. During heat treatment, copper precipitates,

TABLE IV
OPEN-CIRCUIT POTENTIALS (E_{corr}) AND PITTING POTENTIALS
(E_{pit} versus SCE) IN 0.1 N NaCl SOLUTIONS [24]

	Open to the air		Deaerated	
	E_{corr}	E_{pit}[a]	E_{corr}	E_{pit}
2024 Al	−0.546	−0.540	−1.065	−0.540
SiC/2024 Al	−0.652	−0.640	−0.899	−0.640
5456 Al	−0.704	−0.690	−1.091	−0.680
SiC/5456 Al	−0.705	−0.695	−1.097	−0.690
6061 Al	−0.683	−0.673	−1.272	−0.640
SiC/6061 Al	−0.669	−0.660	−1.108	−0.665

[a] As E_{pit} increases, pit-initiation susceptibility decreases.

and as the solid solution becomes depleted in copper, the pitting potential decreases. The fact that the pitting potential of powder-compacted 2024 Al, prepared in the same manner as the composite but without the silicon carbide whiskers, was the same as that of the wrought alloy [24] suggests the presence of silicon carbide whiskers increases copper precipitation in the matrix and thereby lowers the pitting potential relative to the unreinforced alloy.

Trzaskoma *et al.* [24] also studied pit morphology of SiC/Al. They observed a difference in the structure and distribution of pits on the composite and alloy. A greater number of pits formed on the composite, and compared with those on the alloy they were more uniform, shallow, and widespread. Furthermore, it was observed that pits on the composite were both adjacent to and separate from the silicon carbide whiskers. In another study, it was shown that the silicon carbide–aluminum interface is not a preferred site for pit initiation [27]. These results suggest that silicon carbide whiskers are not directly involved in pit-initiation processes. When they are present, however, more pits nucleate.

Paciej and Agarwala [28] studied the effects of processing on the corrosion behavior of SiC/7091 Al. They observed that preferential corrosion on the skins of extruded composite rods could be reduced by modifying tempering times. The authors suggest that active pitting on the composite is a result of elemental segregation and active phase precipitation during formation. When the composites are tempered, compositional inhomogeneities are reduced and therefore less pitting occurs.

Work of Reynolds *et al.* [29] supports Paciej's conclusions. Reynold's group employed AES (Auger electron spectroscopy), SEM (scanning electron microscopy), and XES (X-ray energy spectroscopy) techniques to relate pit-initiation sites to specific compositional or structural features. Their work shows that pits initiate at magnesium aluminum silicide precipitates in 6061 Al and the corresponding composite. Thus, it has been demonstrated that secondary phases play an important role in pitting of SiC/Al composites.

Pitting is the most important mode of corrosion for SiC/Al. Secondary phases on the metal surface seem to play a more important role in pitting events than the silicon carbide reinforcement. It is possible that the presence of silicon carbide affects the nature and distribution of these phases. Once the role of secondary phases in pitting is made clear and the effect of silicon carbide on microstructure is understood, improved pitting resistance can be achieved by alloy selection and careful control of fabrication and processing.

IV. Effects of Corrosion on the Properties of Metal Matrix Composites

Metal matrix composites are purposely designed for a broad range of specific properties, i.e., high strength, rigidity, and dimensional stability on thermal cycling. Because these properties evolve from a unique combination of the matrix and reinforcement, it is likely that accelerated degradation of either constituent could cause a deterioration in properties. For example, interfacial corrosion of a fiber/metal composite would be expected to result in a decrease in transverse load-bearing capability. Few studies have examined the effects of corrosion on the properties of metal matrix composites; however, Sedriks et al. [22] have investigated the effects of corrosion on tensile properties of B/Al for two-volume fractions of fibers. Table V shows ultimate tensile strengths of 2024 Al and B/2024 Al before and after six hours of exposure in a solution of 5.3 wt% NaCl with the addition of 30% H_2O_2. It

TABLE V

TENSILE PROPERTIES OF MATERIALS BEFORE AND
AFTER EXPOSURE TO NaCl SOLUTION FOR SIX HOURS [22]

Material	UTS (k.s.i) (before exposure)	UTS (k.s.i) (after exposure)
2024 Al	61.0	32.0
15/2024 Al		
longitudinal	72.0	59.0
long transverse	30.0	14.0
40B/2024 Al		
longitudinal	160.0	98.5
long transverse	16.5	6.8

can be seen that all samples suffered a degradation in tensile properties. The loss in strength in the transverse direction was about 50% and that of the composite was slightly higher than that of the alloy. The loss in transverse strength of the composite increased slightly as the volume fraction of fibers increased. However, in the longitudinal direction, the loss in strength of the composite, for the lower fiber fraction, was about half that of the alloy. This decrease in strength increased significantly as the volume fraction of fiber increased.

Sedriks's group also investigated the stress-assisted corrosion of B/Al in the same solution. Table VI shows the time to failure of various samples held at a stress level of 90% of the 0.1 YS in the long transverse direction. These data show that the composites are significantly less resistant to failure under stress than the alloy. For example, for the 15% reinforced composite, the time to failure was four times shorter than that of the unreinforced alloy. Additionally, as the volume fraction of boron reinforcment increases, the resistance to failure by stress corrosion decreases. Apparently, this behavior is a result of preferred dissolution at diffusion bonds and matrix–filament interfaces in the composite material. The results were different for longitudinally applied stress. Under these conditions, the alloy failed by a factor of 2 faster than the composite.

These studies show that a short exposure to chloride solution can significantly degrade the mechanical properties of metal matrix composites. The magnitude of this effect can be specific to the direction of bonding of the matrix and reinforcement. The severity of the problem, as illustrated for B/Al, clearly indicates the importance of testing and understanding the effects of

TABLE VI

TIMES TO FAILURE OF 2024 Al AND B/2024 Al
STRESSED AT 90% OF THE 0.1 YS IN THE LONG
TRANSVERSE DIRECTION [22]

Material	Times to failure
(A) Aerated solution containing 53 g/l NaCl	
2024 Al	Did not fail in 4000 hr
15B/2024 Al	1000 ± 300 hr
40B/2024 Al	40 ± 20 hr
(B) As above but containing 9 ml per liter of 30% H_2O_2	
2024 Al	2.7 ± 1.2 hr
15B/2024 Al	0.4 ± 0.1 hr
40B/2024 Al	0.2 ± 0.05 hr

corrosion reactions on the properties of metal matrix composites. Ideally, these investigations should accompany the development of composites.

V.   Corrosion Fatigue

Corrosion fatigue is the combined action of corrosion and cyclic stress. Resistance to corrosion fatigue is an important consideration in the selection of materials for aeronautical and aerospace applications. Hasson *et al.* [30] have studied corrosion fatigue of whisker and particulate-reinforced SiC/6061 Al in laboratory and salt-contaminated moist air. Figure 7 contains fatigue curves for the alloy and whisker composite. These data show that the fatigue behavior of the composite is better than the matrix alloy, in both lab air and salt-contaminated moist air. For example, the authors point out that at 10^7 cycles to failure, the composite has an advantage of at least 70 MPa in load-carrying capability in both environments. The fatigue behavior of the silicon carbide particulate-reinforced material was also significantly improved compared with the matrix alloy, but it was somewhat poorer than the whisker composite. The work further showed that for uncracked specimens, the corrosion fatigue behavior of the composite was substantially improved over the unreinforced matrix material. Conversely, for precracked specimens, the corrosion-fatigue performance of the composite was poorer than the alloy. It was concluded that the improvement in corrosion-fatigue properties of the composite results from a high resistance to crack initiation.

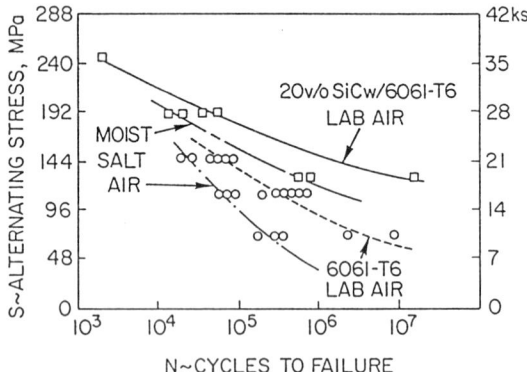

FIG. 7.   Flexure fatigue curves of 6061-T6 Al and 20 vol-% SiC$_w$/6061-T6 Al. From Ref. [30].

VI. Corrosion Control

Corrosion control can be achieved through material selection, careful design and fabrication, and attention to the compatibility of materials in the service environment. For example, service lifetimes of metallic components are significantly improved by avoiding crevices and the contact of dissimilar materials. For situations in which it is unfeasible to substitute materials for corrosion purposes, a multitude of corrosion-resistant coatings is available. The subject of coatings will be considered in Section VII.

The corrosion problems affecting metal matrix composites (i.e., galvanic, crevice, pitting) are amenable to control by materials selection and improvements in fabrication and processing. Once the problem is identified, corrective measures can be incorporated into the design of the composite. In the case of galvanic action, corrosion control is possible by avoiding contact of conducting components with large differences in electrochemical potential, maintaining a low surface area ratio of cathode to anode, or insulating conducting materials from each other and the environment. More advanced techniques might involve manipulation of cathodic and anodic reaction rates either through modifying the interface or by using inhibiting agents. Crevice corrosion can be controlled by improvements in fabrication to eliminate crevices, or by isolating surfaces from the environment with a coating. Finally, pitting can be controlled by selective alloying to mitigate galvanic coupling within the matrix, processing to disburse or eliminate deleterious second phases, and coating to isolate the material and environment. Corrosion control of metal matrix composites is best addressed at the time each system is designed and developed.

VII. Coatings

Portions of the following discussion are extracted from Uhlig's text *Corrosion and Corrosion Control* [*1*]. It is presented here to acquaint those involved in the development of metal matrix composites with the properties, effectiveness, and technicalities involved in the application of corrosion coatings.

Coating is one of the most commonly used methods to prevent corrosion. The purpose of coating is to provide a barrier between the metal and environment or, in some cases, to alter corrosion reactions to inhibit dissolution rates. The effectiveness of a coating depends on its continuity, porosity, degree of adhesion, and durability in the environment of interest. In addition to the technicalities of forming an effective barrier, such factors as cost, convenience of application, and ease of repair are important considerations in the selection of a coating for a particular application.

Numerous coatings and layering combinations are presently used for corrosion control. Essentially, each coating falls into one of three categories by composition: (1) metallic; (2) inorganic compound; and (3) organic compound. To provide an introduction to the broad range of options available for coating protection, the following discussion presents a brief description of each type.

Metallic coatings are relatively thick layers of metal (0.003 up to 6.25 mm) applied by one of several techniques such as hot dipping, electroplating, chemical vapor deposition and cladding. Some metals are used as coatings simply because they are less susceptible to chemical attack than the metal to be protected. For example, noble metals such as silver, chromium, and nickel might be used on steel for applications in acid- or chloride-containing environments. Should these coatings rupture, however, corrosion rates would tend to be very high due to galvanic interaction between the coating and metal substrate. Other metals are used for coatings because they are more reactive than the metal to be protected and sacrificially dissolve in the environment of interest. Zinc and cadmium coatings are applied to steel for use in industrial environments, and electrodeposited tin provides protection to iron used in food containers. The lifetime of materials coated with active metals depends upon the dissolution rate of the coating. Once the metal coating dissolves, the underlying material would be subject to attack.

Coatings of inorganic compounds are the porcelains, glasses, phosphate, and chromate conversion coatings, and oxides formed by anodizing. In general, their thickness ranges between 0.1–0.25 mm, and so dimensional changes are considerably less than with metallic coatings. Glass and porcelain coatings are unusually resistant to a wide range of chemicals, and because they are virtually impenetrable by water and oxygen, they are very long-lasting. Conversion coatings, on the other hand, provide only minimal improvement in corrosion resistance. Because they contain phosphates and chromates, which are known to inhibit corrosion reactions, conversion coatings are generally used to provide a base for paint applications. Thick metallic oxides formed by anodizing vary in depth and porosity. These properties depend upon the metal, anodizing conditions, and method of sealing. The protection afforded by anodizing is quite variable and once again the treatment is usually employed to provide a paint base.

Organic compounds are by far the most common form of coating for corrosion protection. An enormous variety of paints and synthetic resin compounds belong to this classification. Organic coatings are favored because they are conveniently applied, provide good protection, and minimally increase dimensions (0.13 mm paint is recommended for protection of steel that will be exposed to harsh environments [*31*]). In addition, inhibitors (chromates and molybdates) are easily combined with paints to increase the effectiveness of the coating. On the other hand, organic compounds do not always adhere well, break down, and are easily damaged.

Aylor and Kain [*18*] evaluated various coatings on C/Al and SiC/Al composites by marine exposure testing. In this study, it was found that organic coatings and thermal sprayed aluminum and alumina were the most effective coatings for both the composite and unreinforced alloy. Electro-deposited aluminum–manganese, and chromate–phosphate conversion coatings were also determined to be suitable if further improvements could be made in assuring complete surface coverage. Electroless nickel applied to C/Al showed degradation of the composite after exposure to filtered sea-water. This effect was attributed to severe galvanic action between the aluminum and nickel through pores in the coating. No effort was made to assess the effects of the carbon fibers on the formation and continuity of the coating. Payer and Sullivan [*17*] also tested various coatings on C/Al in seawater. Organic coatings, electroplated nickel, and chemical-vapor-deposited metallic coatings were applied using techniques similar to those used for unreinforced materials. Corrosion behavior of the coated composite was found to be about the same as that of the coated alloy.

Trzaskoma and McCafferty [*32*] studied the corrosion behavior of anodized SiC/Al for two-alloy matrices. Two anodizing solutions were considered: sulphuric acid (under conditions used to form a hardcoat) and ammonium tartrate solution. Table VII shows the pitting potentials of anodized specimens of 6061 Al, 2024 Al, and the corresponding composites containing 20 vol-% silicon carbide whiskers. For 6061 Al, the pitting potential of both the alloy and the composite is about 100 mV higher after anodizing in sulfuric acid; however, an increase in pitting potential was not observed for specimens anodized in ammonium tartrate. Thus, the pit-initiation resistance of both 6061 Al and SiC/6061 Al improves for samples anodized in sulfuric acid, but it does not change for samples anodized in ammonium tartrate. For 2024 Al, the pit-initiation resistance of the composite improves, but that of the alloy remains the same after anodizing in sulfuric acid. The authors suggest that the results for 2024 Al might be a result of the copper distribution in this alloy and its effect on the structure of the anodized coating.

Although coatings on SiC/Al and C/Al are as effective as they are on the unreinforced alloys, this may not be true for other metal matrix composites. Differences between surface properties of the metal and reinforcement could drastically affect the distribution and adherence of protective coatings. For electrochemically deposited coatings (anodized and electroplated coatings, for example), the reinforcement could conceivably affect composition, structure, and porosity. These changes would significantly alter the effectiveness

TABLE VII
OPEN-CIRCUIT AND PITTING POTENTIALS OF ANODIZED AND
UNANODIZED SiC$_w$/Al COMPOSITES IN 0.1N DEARATED NaCl [32]

	E_{corr} (Volts vs SCE)	E_{pit} (Volts vs SCE)
2024 Al	-1.065	-0.540
SiC$_w$/2024 Al	-0.899	-0.640
2024 Al HC[a]	(-0.728 to -0.822)	-0.530
SiC$_w$/2024 Al HC	(-0.585 to -0.638)	-0.450
6061 Al	-1.272	-0.640
SiC$_w$/6061 Al	-1.108	-0.665
6061 Al HC	(-0.568 to -0.945)	-0.550
SiC$_w$/6061 Al HC	(-0.686 to -0.899)	-0.530
6061 Al AT[b]	(-0.806)	-0.625
SiC$_w$/6061 Al AT	(-0.893 to -1.068)	-0.650

[a] HC Hardcoat (sulfuric acid)
[b] AT Barrier layer coating (ammonium tartrate)

of the coating. Additionally, the suitability of using coatings for corrosion protection of metal matrix composites depends on the effects of the coating on the mechanical, physical, and dimensional requirements of the application. Presently, we are a long way from making informed decisions concerning corrosion-protective coatings for metal matrix composites. The need for new information and testing is apparent and should be studied within the confines of the material requirements of an application.

VIII. Future Directions

Corrosion is an important consideration in the development of metal matrix composites for engineering applications. When both components are conductive, it has been demonstrated that galvanic effects increase corrosion rates. When the reinforcement is nonconductive, studies indicate that it could cause compositional and structural alterations in the metal matrix and thus affect pitting. Finally, investigations of the effects of corrosion on tensile properties show that degradation in strength can occur after short exposure times in harsh environments. The implications of these effects with respect to service lifetime have yet to be determined quantitatively. However, at this time, it is evident that in order to protect the mechanical and physical properties and prolong lifetimes of metal matrix composites, it is necessary to reduce corrosion rates.

From the material presented in this chapter, the following areas for further investigation can be identified.

(1) Methods to control galvanic interactions.
    (a) Separation of composite components by insulating barriers.
    (b) Reduction of galvanic reaction rates.
    (c) Reduction of volume fraction of reinforcement.
    (d) Reduction of metal-metal interfaces.
(2) Effects of microstructure on corrosion behavior.
    (a) Identification of deleterious microstructural features and reaction products.
    structural effects.
(3) Quantitative determination of the long-term effects of pitting.
(4) Testing for the effects of corrosion on mechanical and physical properties.
(5) Application and testing of coatings.
    (a) Adherence and continuity of coatings on metal matrix composites.
    (b) Effects of coatings on properties of composites.
    (c) Improvements in coating technology.
(6) Determination of corrosion behavior in nonaqueous environments.

Acknowledgments

The author gratefully acknowledges Mr. R. K. Everett for reading the manuscript and providing thoughtful suggestions for improvement.

References

1. H. H. Uhlig, "Corrosion and Corrosion Control," 2nd ed., p. 1. Wiley, New York, 1971.
2. "Economic Effects of Metallic Corrosion in the United States," NBS Special Publication 511. National Bureau of Standards, Washington, D.C., 1978.
3. R. M. Latanison, *Mater. Perform.*, **26**, 9 (1987).
4. J. Ott and R. G. O'Lone, *Aviat. Week and Space Technol.*, **129**, 29 (1988).
5. N. D. Greene, Jr., and N. Ahmed, *Mater. Protec.*, **9**, 16 (1970).
6. P. P. Trzaskoma *Corrosion*, **42**, 609 (1986).
7. M. G. Fontana and N. D. Greene, "Corrosion Engineering," 2nd ed., Chapter 3. McGraw-Hill, New York, 1978.
8. "Corrosion Basics, An Introduction," (L. S. Van DeLinder, ed.), Chapters 2 and 5. NACE, Houston, Texas, 1984.

9. G. Wranglen, "An Introduction to Corrosion and Protection of Metals," Chapters 6 and 7. Chapman and Hall, New York, 1985.

10. T. J. Lennox, Jr., "Marine Electrochemical Corrosion and Control Systems," NRL Report 3622. Naval Research Laboratory, Washington, D.C., 1977.

11. Z. Szklarska-Smialowska, "Pitting Corrosion of Metals," p. 77. NACE, Houston, Texas, 1986.

12. P. P. Trzaskoma, *J. Electrochem. Soc.*, **129**, 1398 (1982).*

13. H. H. Uhlig, "Corrosion and Corrosion Control," 2nd ed., p. 14. Wiley, New York, 1971.

14. P. P. Trzaskoma, unpublished data.

15. J. M. Evans and D. M. Braddick, *Corros. Sci.*, **11**, 611 (1971).

16. D. L. Dull, W. C. Harrigan Jr., and M. F. Amateau, "Proceedings of Triservice Corrosion of Military Equipment Conference" (F. H. Meyer, ed.), Vol. 1, p. 399. Air Force Materials Laboratory Report AMFL-TR-75-42, Dayton, Ohio, 1975.

17. J. H. Payer and P. G. Sullivan, "Bicentennial of Materials," National SAMPE Technical Conference Series, Vol. 8, p. 343. Society for the Advancement of Material and Process Engineering, Azusa, California, 1976.

18. D. M. Aylor and R. M. Kain, in "Recent Advances in Composites in the United States and Japan, ASTM Special Technical Publication 864," (J. Vinson and M. Taya, eds.), p. 632. ASTM, Philadelphia, Pennsylvania, 1983.

19. D. M. Aylor, DTNSRDC Report No. SME-86-71. David Taylor Naval Ship Research and Development Center, Bethesda, Maryland, 1986.

20. C. R. Crowe, "Localized Currents from Graphite/Aluminum and Welded SiC/Al Metal Matrix Composites," NRL Report 5415. Naval Research Laboratory, Washington, D.C., 1985.

21. M. C. Porter and E. G. Wolff, in "Advances in Structural Composites," Paper No. AC-14. Society of Aerospace Materials Process Engineering, 12th National Symposium Exhibit, Western Period Co., North Hollywood, California, 1967.

22. A. J. Sedriks, J. A. S. Green, and D. L. Novak, *Metall. Trans.*, **2**, 871 (1971).

23. S. L. Pohlman, *Corrosion*, **34**, 156 (1978).

24. P. P. Trzaskoma, E. McCafferty, and C. R. Crowe, *J. Electrochem. Soc.*, **130**, 1804 (1983).*

25. J. R. Galvele, S. M. de De Micheli, I. L. Muller, S. B. Wexler, and I. L. Alanis, in "Localized Corrosion" (R. W. Staehle, B. F. Brown, J. Kruger, and A. Agrawal, eds.), NACE-3, p. 580. NACE, Houston, Texas, 1974.

26. I. L. Mueller and J. R. Galvele, *Corros. Sci.*, **17**, 179 (1977).

27. P. P. Trzaskoma, Abstract 257. The Electrochemical Society Extended Abstracts, Vol. 86-2, p. 380, San Diego, California, meeting Oct. 19–24, 1986.

28. R. C. Paciej and V. S. Agarwala, *Corrosion*, **42**, 718 (1986).

29. G. H. Reynolds, L. Yang, and A. Joshi, "Fundamental Research on Corrosion Mechanisms in Discontinuous SiC/Al Metal Matrix Composites," SBIR Phase I Final Report, Contract No. N60921-86-C-0290. MSNW, Inc., San Marcos, California, 1987.

30. D. F. Hasson, C. R. Crowe, J. S. Ahearn, and D. C. Cooke, in "Failure Mechanisms in High-Performance Materials" (J. G. Early, T. R. Shives, and J. H. Smith, eds.), p. 147. Cambridge University Press, New York, 1985.

31. H. H. Uhlig, "Corrosion and Corrosion Control," 2nd ed., p. 251. Wiley, New York, 1971.

32. P. P. Trzaskoma and E. McCafferty in "Aluminum Surface Treatment Technology" (R. S. Alwitt and G. E. Thompson, eds.), p. 171. The Electrochemical Society, Pennington, New Jersey, 1986.**

* This paper was reprinted by permission of the publisher, The Electrochemical Society, Inc.
** This paper was originally presented at the Spring 1986 Meeting of the Electrochemical Society, Inc., which was held in Boston, Massachusetts.

Index

This index comprises entries from both volumes of *Metal Matrix Composites*. Page numbers in lightface indicate entries in the volume subtitled *Processing and Interfaces*; boldface page numbers denote entries in the volume subtitled *Mechanisms and Properties*.